Series:

Computational Product Design and Digital Manufacturing

Series editor: Prof. Panagiotis Kyratsis
University of Western Macedonia
Department of Product and Systems Design Engineering
Kozani, Greece

Series: Computational Product Design and Digital Manufacturing

Digital Product Design and Manufacturing

Editors:
Gheorghe Oancea
Department of Manufacturing Engineering
Transilvania University of Brasov, Romania

Panagiotis Kyratsis
University of Western Macedonia
Department of Product and Systems Design Engineering
Kozani, Greece

CRC Press
Taylor & Francis Group
Boca Raton London New York

CRC Press is an imprint of the
Taylor & Francis Group, an **informa** business
A SCIENCE PUBLISHERS BOOK

First edition published 2026
by CRC Press
2385 NW Executive Center Drive, Suite 320, Boca Raton FL 33431

and by CRC Press
4 Park Square, Milton Park, Abingdon, Oxon, OX14 4RN

CRC Press is an imprint of Taylor & Francis Group, LLC

Library of Congress Cataloging-in-Publication Data (applied for)

ISBN: 978-1-032-83559-4 (hbk)
ISBN: 978-1-032-85444-1 (pbk)
ISBN: 978-1-003-51819-8 (ebk)

DOI: 10.1201/9781003518198

Typeset in Times New Roman
by Prime Publishing Services

Preface

Worldwide, in the last few decades, digital manufacturing has been playing an important role in designing and manufacturing of products. It is defined as a computer-based process which integrates simulation, 3D visualization, analytics and collaboration tools to design entire manufacturing processes in a digital mode [Siemens n.d.]. In fact, digital manufacturing is the fastest and easiest way to design and manufacture products from different industries, in a collaborative manner among marketers, financers, designers, and engineers.

In the literature three key technologies for digital manufacturing have been presented [IndustryWeek 2017]: additive manufacturing is a process which consists of obtaining parts by adding materials layer after layer; laser cutting, which is like a subtractive manufacturing process used for cutting and engraving different materials; and Computer Numerical Control (CNC), which is a subtractive machining process where a computer controls the process of cutting of parts.

This book titled *Digital Product Design and Manufacturing* presents how different products can be designed and manufactured using different software tools and processes.

The first chapter discusses how digital tools from design and manufacturing areas can be used in automotive engineering, detailing the case of an instrument panel. Chapters 2 to 4 respectively offer a general overview on reverse engineering in digital manufacturing, additive manufacturing and multi-material VAT photopolymerization 3D printing.

The next three chapters are dedicated to additive manufacturing processes. Chapter 5 explores the role of additive manufacturing in the context of circular economy within an academic environment, presenting an optimized production planning framework for 3D printing processes. Chapter 6 is dedicated to the design and manufacturing framework of a 3D-printed, modularized and low-cost unmanned surface vessel. Chapter 7 discusses how selective laser melting technology can be implemented in different industrial environments based on research, case studies and recommendations.

Chapters eight and nine are dedicated to numerical modelling and simulation of cutting processes, respectively, of polytetrafluoroethylene orthogonal cutting in view of predicting and improving the machining performance, and a framework for designing and optimizing of solid end mill geometry using a CAD-based environment.

Chapter 10 is about artificial intelligence, from historical perspectives to its impact in contemporary practices, keeping in view discussions on the potentiality of these kind of tools from generative perspectives, and highlighting how human environment through technologies can be improved.

The last three chapters offer the readers an insight into how digital tools are used in textile and footwear industries. Chapter 11 offers a general overview of using different digital tools in the textile industry, reviewing the creative processes in design, related to design types, specific developments and digitizing CAD system integration. Chapter 12 discusses how digital tools are used to improve the aesthetics and functionality of footwear products, and Chapter 13 gives details on using computational design technology with an aim to automate sew pattern design for dress garment, based on an algorithm and a personalized software tool.

Professor Gheorghe Oancea
Professor Panagiotis Kyratsis

References

IndustryWeek. 2017. 'Digital manufacturing–The factory of the future is here today'. Available at http://www.industryweek.com/technology/digital-manufacturing-factory-future-here-today (accessed on 23.11.2024).

Siemens. n.d. 'Digital manufacturing'. Available at https://www.plm.automation.siemens.com/global/en/our-story/glossary/digital-manufacturing/13157 (accessed on 23.11.2024).

Contents

List of Contributors

Adelina Hrituc
Department of Technology of Machine Manufacturing, Gheorghe Asachi Technical University of Iasi.

Alexandru Sover
Department of Technology Laser Safety Officer, Ansbach University of Applied Sciences.

Anastasios Tzotzis
Department of Product and Systems Design Engineering, University of Western Macedonia.

Athanasios Manavis
Department of Product and Systems Design Engineering, University of Western Macedonia.

Catalin Fetecau
Center of Excellence Polymer Processing, Dunarea de Jos University of Galati.

Constantin Gheorghe Opran
Department of Manufacturing Engineering, The National University of Science and Technology Politehnica Bucharest.

Cosmin Cosma
Department of Manufacturing Engineering, Technical University of Cluj-Napoca

Cristian-Vasile Doicin
Department of Manufacturing Engineering, The National University of Science and Technology Politehnica Bucharest.

Dumitru Nedelcu
Department of Manufacturing Engineering, Gheorghe Asachi Technical University of Iasi.

Ermira Shehi
Polytechnic University of Tirana.

Felicia Stan
Center of Excellence Polymer Processing, Dunarea de Jos University of Galati.

Gheorghe Nagit
Department of Technology of Machine Manufacturing, Gheorghe Asachi Technical University of Iasi.

Gheorghe Oancea
Department of Manufacturing Engineering, Transilvania University of Brasov.

Giuseppe Lamanna
Department of Engineering, University of Campania Luigi Vanvitelli.

Gizem Karakan Günaydın
Department of Textile and Fashion Design, University of Pamukkale.

Harshit K. Dave
Department of Mechanical Engineering, Sardar Vallabhbhai National Institute of Technology, Surat.

Henrique Almeida
School of Technology and Management, Polytechnic Institute of Leiria.

Huseyin Kursat Celik
Department of Agricultural Machinery and Technology Engineering, Faculty of Agriculture, Akdeniz University.

İbrahim Üner
Department of Textile and Fashion Design, University of Pamukkale.

Keyur P. Desai
Department of Mechanical Engineering, Sardar Vallabhbhai National Institute of Technology, Surat.

Laurentiu Slatineanu
Department of Technology of Machine Manufacturing, Gheorghe Asachi Technical University of Iasi.

Liliia Chertenko
Kyiv National University of Technologies and Design.

Manoj Gupta
Department of Electrical Engineering, Guru Ghasidas Vishwavidyalaya.

Marius Andrei Mihalache
Department of Technology of Machine Manufacturing, Gheorghe Asachi Technical University of Iasi.

Marmik Dave
Yantra Innovations.

Mihaela-Elena Ulmeanu
Department of Manufacturing Engineering, The National University of Science and Technology Politehnica Bucharest.

Mohd. Khairol Anuar Bin Mohd Ariffin
Department of Mechanical and Manufacturing Engineering, Faculty of Engineering, Universiti Putra.

Nicolae Balc
Department of Manufacturing Engineering, Technical University of Cluj-Napoca.

Nikolaos Efkolidis
Department of Product and Systems Design Engineering, University of Western Macedonia.

Nikolaos Tapoglou
Department of Industrial Engineering and Management, International Hellenic University.

Oana Dodun-Des-Perrieres
Department of Technology of Machine Manufacturing, Gheorghe Asachi Technical University of Iasi.

Panagiotis Kyratsis
Department of Product and Systems Design Engineering, University of Western Macedonia.

Prodromos Minaoglou
Department of Product and Systems Design Engineering, University of Western Macedonia.

Ratnadeepsinh Jadeja
School of Engineering, RK University.

Sandip T. Patel
Department of Mechanical Engineering, Government Engineering College, Valsad.

Sever-Alexandru Haba
Department of Manufacturing Engineering, Transilvania University of Brasov.

Tatjana Spahiu
Polytechnic University of Tirana.

Valerio Perna
Department of Architecture, EPOKA University.

Vasile Merticaru
Department of Technology of Machine Manufacturing, Gheorghe Asachi Technical University of Iasi.

1 Digital Tools Used for an Instrument Panel Design and Manufacturing

Sever-Alexandru Haba[1], Panagiotis Kyratsis[2] and Gheorghe Oancea[3*]

1. Introduction

The automotive industry is constantly changing, especially because of the digital revolution. Industry 4.0 governs the development directions of the automotive industry. It is also important to highlight the changes in mass production parts. The automotive industry is moving toward the usage of a higher percentage of plastic parts, manufactured by injection molding for thermoplastic materials. Nowadays, injection molding development trends are leading to higher efficiency, greater profitability, and longer sustainability (Zhou et al., 2022).

A series of papers present the digital tools used in Instrument Panel engineering. This part is an iconic example of the digital techniques implemented in injection molding for mass production. Instrument Panel is included in the class of flat surface parts, with small wall thickness. Due to its mounting position inside the passenger compartment, the Instrument Panel is defined by very high-quality requirements for its surface. The final quality of Instrument Panel is strongly affected by the process parameters of injection molding (Yu et al., 2022). In order to decrease the cost incurred, high-quality simulation is used based on a new digital concept, the surrogate modeling technology (Saad et al., 2022).

Another challenge for visible injected parts is the high quality of color control. To increase the quality of color control and to eliminate the color flux lines, the static mixing nozzle is included in the injection machine nozzle (Artem Group, n.d.). In order to decrease the final warpage out of the mold tool for automotive glove box, a series of digital tools is used, i.e. Kriging models, micro genetic algorithms, and commercial software Moldflow™ (Kang et al., 2016).

In engineering the outer panel of the automotive plastic rear door, the Unigraphics NX™ software system can be used for designing the surfaces, ANSYS™ to perform mechanical analyses, and Hyper Works™ software to study the materials (Mao et al., 2022).

[1] *Lecturer*, Department of Manufacturing Engineering, Transilvania University of Brasov.
[2] *Professor*, Department of Product and Systems Design Engineering, University of Western, Macedonia.
[3] *Professor*, Department of Manufacturing Engineering, Transilvania University of Brasov.
E-mail: sevar.haba@unitbv.ro[1]; pkyratsis@uowm.gr[2]
* Corresponding author: gh.oancea@unitbv.ro

Automotive panels are improved using computer-aided analysis (CAE) software systems in case of simulating the mold details, i.e. gate locations, gate numbers, best gate location, filling analysis, weld lines displacement, injection speed, and filling pressure distribution (Zhao et al., 2010).

Although nowadays engineering is based on digital tools, the designer has a number of important tasks to perform when designing the 3D models for injection molding parts: general design (constant wall thickness, eliminating the mass accumulated area, rib design, sharp corners, draft angles, etc.) and design for assembly (mechanical fasteners, retention features, alignment features, tolerances, etc.) (Techcenter, 2007).

2. Materials and Methods

2.1 Digital Tools

In today's age, one of the most important resources is time. Recent studies show that during the last few years the range of products has doubled with a simultaneous reduction in the life cycle of products. One of the critical elements for successfully bringing new product designs to market is the time spent on design, prototyping, testing, and production. Current products have implemented a multitude of properties, so component development (design and manufacturing) is an increasingly difficult task for engineers. In order to ensure that all these tasks are incorporated in the final result, digital tools are used on a large scale in development and engineering of industrial products.

A short description of the digital tools categories is now presented.

Digital design systems:

- part design and design (CATIA™ V5, NX™, Autodesk Inventor™);
- manufacturing technology (CATIA™ V5);
- digital simulation systems (ANSYS™);
- databases (Product life cycle management - PLM systems);
- digital production rhythm assistance systems;
- industrial product quality control systems;
- logistic chain (PLM systems);
- interconnection systems (communications and data transfer).

Components of Industry 4.0:

- artificial intelligence to support and advise the human factor;
- digital ecosystems;
- product life cycle management.

In injection molding engineering, digital tools are widely used. Digital tools improve the quality of results for many product conditions: improved warpage, cavity pressure curves, dimensional prediction, pressure recorded by flush-mounted sensors, and simulation of injection mold quality for electronic industry.

Recent studies show the influence of cooling time in order to reduce the total product warpage, for example 500°C and 77% of the packing pressure are the optimal solutions (Wang et al., 2022). Usually, simulations are performed for both mold injections and test injections, but this solution increases the engineering time. In order to accelerate process simulation, researchers undertake strategies for adjusting process parameters in simulation with an aim to approach the real injection molding conditions (Li et al., 2022).

In order to enable dimensional prediction for injected parts, researchers successfully applied the Taguchi method and as a result achieved optimized results with significantly decreased cost (Kech and Maris-Haug, 2020). With an aim to compare the difference between the simulation and trial injection for a multi-pin connector part weight, engineers recorded the real pressure in cavity and compared it with simulation pressure. The results were successfully documented and presented (Cakmakci and Demirel-Ortabas, 2019). The Single-Minute Exchange of Die (SMED) tool is used in order to measure the performance of plastic injection parts for the electronics industry. This results in improvement of the final surface quality of the designed parts when manufactured in high volumes (Catoen and Rees, 2022).

2.2 *Virtual Environment*

Instrument Panel is the main part on top of the dashboard area. The following construction and design properties are important when designing the complete geometry of the dashboard and completed with engineering requests:

- panel whit visible surfaces;
- closing forces element for the entire upper dashboard area;
- closing panel for airducts area;
- the clock area, air flow to passenger and driver area, airbag in passenger area, etc., are positioned according to the Instrument Panel surfaces.

Due to its attributes, the Instrument Panel cannot be developed in isolation. It must be designed together with its surroundings, thereby demanding a complex virtual environment.

In this case study, the virtual environment for Instrument Panel is made of the following elements:

- CAD/CAE system (CATIA™ V5);
- environment input, original equipment manufacturer (OEM)—design and surfaces of the Instrument Panel's neighboring parts.

The CATIA™ system is used successfully because it can apply the "design from context" principles, based on the improved part quality and detailed 3D solid modeling, thus resulting in taking advantage of the space analysis capabilities offered.

The input design is imposed by the OEM. This is the surface visible to the car users. In general, the design is generated by using an advanced computer-aided design (CAD) software system (solid and surface modeling). For this reason, the input design usually encompasses many discontinuous surfaces, and surfaces not in tangent positions. The first stage includes redesigning the initial geometry introduced. Then the visible surfaces of Instrument Panel are readied for closing the main solid body (Figure 1a). If the final design surfaces are not in continuous position, or have sharp edges, the automatically added material option within CATIA™ does not work properly. In the current case study, the Thick Surface with First Offset was set to 3 mm, and the Second Offset was set to 0 mm, which resulted in achieving an acceptable CAD model quality.

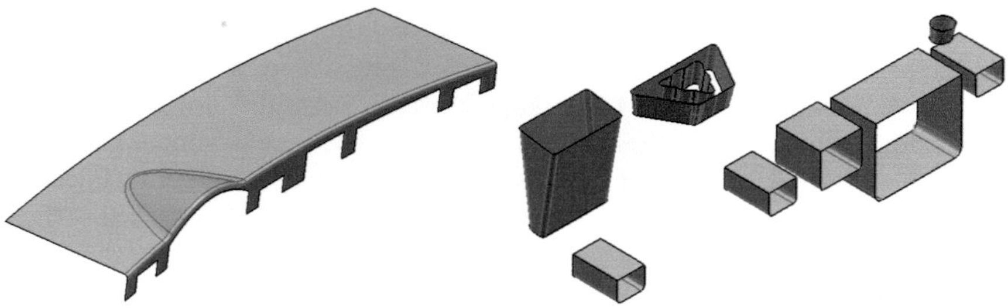

(a) (b)

Fig. 1 (a) Input Design-after design rework stage; (b) Input surfaces from context

The best solution to easily check the surface position to tangent solution, and without sharp edges in unwanted places, is by taking advantage of the View Mode Customization tool. In this case, all the surface edges with no tangency position are colored in dark green and the designer can effectively control the geometry produced.

When virtual environments are used, the best accessory for designers is the 3D mouse. This is because the designers can use the Render Style in Perspective mode and, as a result, they can actually navigate through the solid part of the design and separate the normal hidden surfaces and details.

In order to ensure the predefined position for all the neighboring parts (head up display, center speaker, air defrosters, clock, etc.) and the centering surfaces (3-2-1 centering system) according to the OEM list tolerances, the input surface from context must be imported with active links (Figure 1b). Obviously, all these surface sets must be of high geometric quality, i.e. continuous, without any gaps.

2.3 Material and Technologies

Due to the mass production strategy in the automotive industry, injection molding technology is very commonly used in the manufacturing process of Instrument Panels. A complete set of technical parameters should be defined with accuracy, i.e.:

- single cavity for active area;
- mechanical ejectors;
- mechanical actioned sliders;
- multi gate injections (not in visible area);
- hot runner injection;
- vacuum robot end effectors solution (for extracting the part from its mold);
- digital feedback from mold to injection machine (for quality surveillance).

Instrument Panels are made from polycarbonate/acrylonitrile butadiene styrene (PC/ABS) material, with standard material parameters, i.e.:

- high heat resistance;
- high impact resistance;
- good filling quality;
- dark color;
- no reinforcement;
- storage condition—pellets.

From the point of view of logistics, it is important to maintain the PC/ABS pellets containers in an area with low humidity. This procedure results in avoiding degrading of the material itself.

2.4 Part Design and Part Optimization

Acquiring a 3D model of the part through injection mold process involves providing numerous construction details to the mold tool, including the assembly system between the part and its neighboring parts and the geometric deviations and tolerances (GD&T) involved (Keranen et al., 2005):

- plastic parts' specific design for drafts, corners, fillets, ribs, sidewall pockets, undercuts;
- parting lines;
- hooks design, holes for passing screw, holes for in-screw.

The part design stage allows engineers to create a 3D model for the Instrument Panel. Due to the use of digital tools for 3D modeling, the Instrument Panel is divided into multiple primary 3D bodies, and finallly it is used the boolean operations types "Add" or "Assembly" to create the complete body. All these bodies are combined into one using Boolean operation: add, trim, and remove.

Figure 2a depicts the trimming operation between the top surface and the front wall, both having an identical wall thickness of 3 mm. In addition, the center radius design is implemented when 3D modeling the necessary CAD geometries. In order to implement the "ECE—R21, Regulation No. 21 of the Economic Commission for Europe of the United Nations (UN/ECE)—Uniform provisions concerning the approval of vehicles with regard to their interior fittings requirement", it is important to check the final design, with an aim to ensure that all possible head impact zones have the minimum deceleration (Cheng et al., 2012).

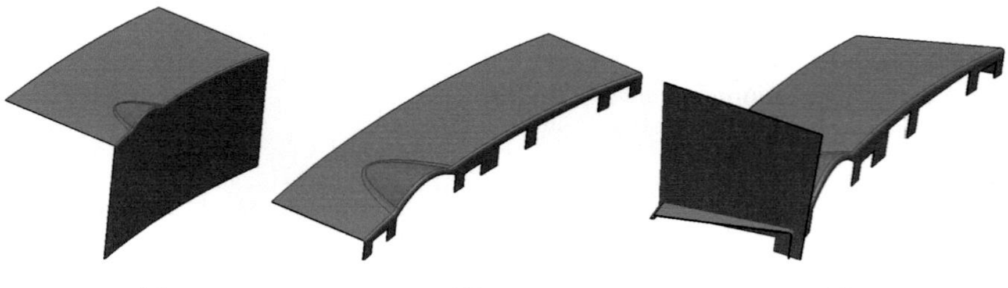

(a) (b) (c)

Fig. 2 (a) Main body; (b) Main body split with environment; (c) Left panel before trim

The Instrument Panel front wall was split with different environmental surfaces: watch area, left and right air ducts, glove box hole, and control panel pocket. Figure 2b shows the main trim part after the split stage.

In order to reach the final shape for the Instrument Panel, the central trim was combined (Boolean operation) with the left and right panels (Figure 2c). Due to this modular construction, design changes could be easily implemented, if needed.

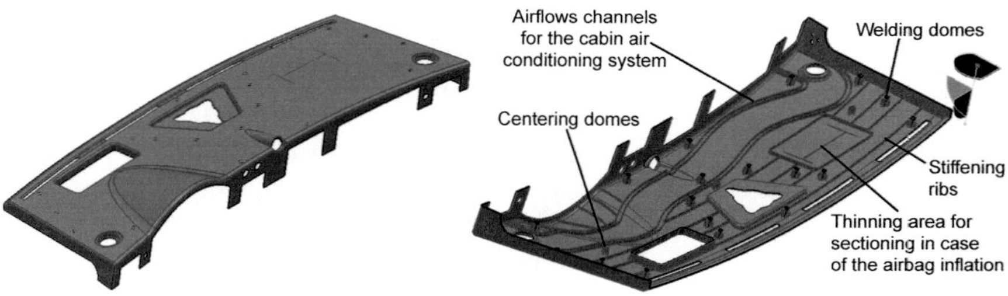

(a) (b)

Fig. 3 Instrument panel part top view; (b) Instrument panel final part inner detail

After obtaining the shape of the Instrument Panel (Figure 3a), including all the radii, pockets and draft angles, the next step was the optimization of the Instrument Panel's stiffness and inner functionality (centering items, neighboring assembly items, ribs, air ducts channels, etc.). The Instrument Panel is a complex and important plastic part because it should provide high visibility to the user, and it must also fulfill multiple technical functions.

Figure 3b shows the inner geometry of the Instrument Panel, with all the additional parts relevant to its functionality:

- left and right inner air channels (with double wall the seat for the gasket);
- center speaker pocket;
- Head Up Display (HUD) pocket;
- center clock pocket;
- ribs for increased mechanical stiffness;
- left and right pockets for air defroster;

- windshield defroster long pockets;
- passenger air bag frame (with discrete cutting area);
- centering domes (3-2-1 system for Degree of Freedom);
- welding domes (with seat ribs for neighbor part);
- left and right screw holes.

2.5 Part Analysis

A number of digital tools used were able to ensure that 3D analysis of the parts with complex geometry was optimized and the designers could make use of tools such as the compare tools, thickness analysis tools, and geometry curvature analysis.

In order to compare the Instrument Panel shape and the mechanical items added, the Digital Mockup (DMU) space analysis tool highlighted the new added construction. This tool must be used when a design change takes place. This is the appropriate way to ensure that designers optimize the final geometry, and they are certain about the results of every modification (Figure 4).

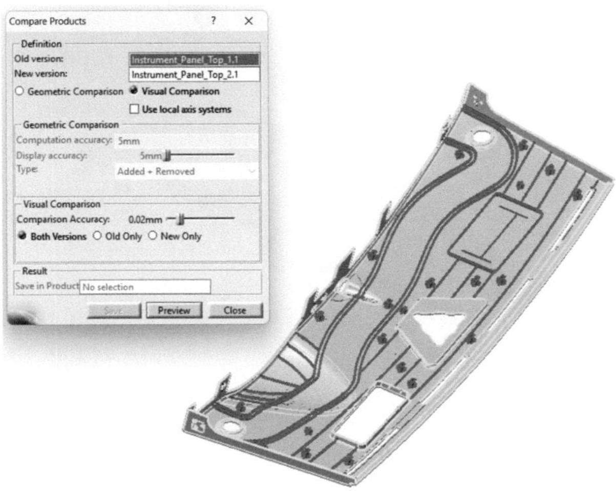

Fig. 4 DMU Space Analysis: Inner geometry compare

For plastic parts made by using the injection mold tool, it is very important to use the following digital tools to check part accommodation with the complete mold process:

- *Draft analysis tool:* This tool checks the all-entities angle along the chosen direction, called draft direction in technical terms. Draft direction is the direction for opening the mold tool, the minimum required angle being 0.5 degree. In Figure 5a, the green area has the correct draft angle along the main draft direction. In the area where the draft direction changes, the mold closing area can be depicted (technically, this zone is named parting line).

- *Wall thickness analysis tool:* This tool helps verify the all-part wall minimum and maximum thickness. If the wall thickness is increased over the general thickness, the mold part will be deformed; and if the thickness is decreased less than 1 mm, the injection will be incomplete in that area. In Figure 5b, the blue area means that the part is in good condition, all the radii are performed using the constant wall thickness rules (inner radius = outer radius – wall thickness).
- *Surface curvature analysis tool:* This digital tool scans all the radii from the complex body. In this mode, it is easy to check the minimum radius in all the head impact boundaries (Figure 5c).

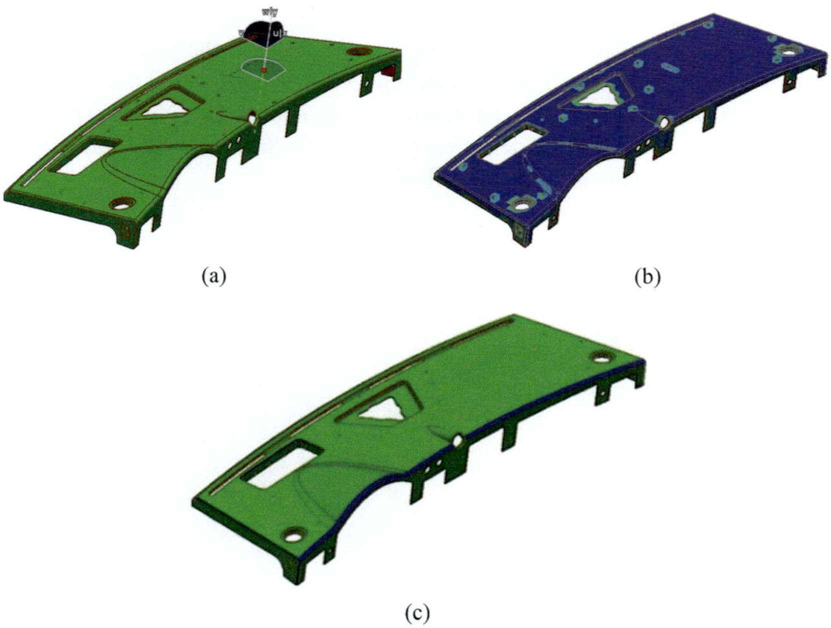

Fig. 5 (a) Draft analysis; (b) Wall thickness analysis; (c) Surface curvature analysis

2.6 *Transfer of Information of Part Design to Mold Design*

In Instrument Panel development, transfer of information among the project teams is a core activity. The transfer of information is ensured by the Project Lifestyle Management (PLM) digital platforms. These platforms ensure direct data transfer, data management, and management of all feedback among the project teams and the OEM office.

Generally, the information regarding tool design is sent together with the 3D part model, structured in standard groups or layers for output. The information for the mold tool design includes a great deal of data, i.e.:

- main draft direction (1): direction for mold open/close main movement (Figure 6a);
- sliders direction (2): front sliders, and (3) left and right sliders (Figure 6a);

- parting lines: the sealed metal-to-metal closing surface of the molds (Figure 6b);
- visible surfaces: marked onto the 3D model surfaces;
- grained surfaces: marked onto the 3D model surfaces;
- manufacturer's information label: closed contour positioned in non-visible area;
- 3-2-1 centering items: domes and rips with tolerances;
- GD&T for part visible surfaces;
- projected area: this information is needed for clamping force dimensioning;
- part material and manufacturer classes.

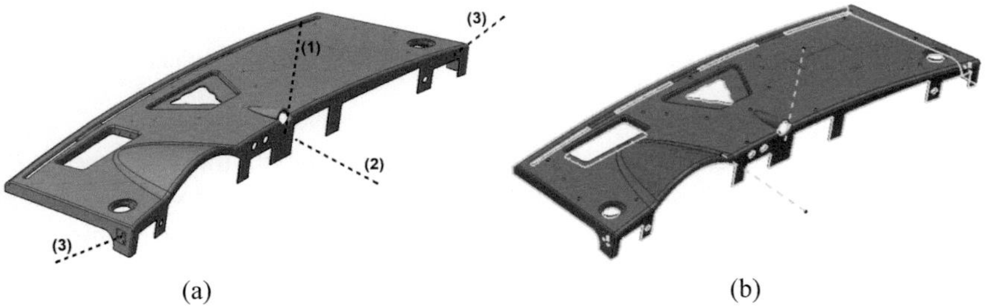

Fig. 6 (a) Main draft direction-1 and sliders direction-2; (b) Parting Lines

Figure 7 shows the particular traces of the parting lines around the central clock area. The main parting line in this area goes from the outside of the part toward the inside of the part. Attention should be given to the changed direction points, and the necessary corner area, without any radius.

Fig. 7 Parting line clock area detail

2.7 *Mold Injection for Instrument Panel*

As it was mentioned earlier, for engineering of the Instrument Panel, the mold injection multi-gate technology with hot runner system is used; the model of inner channel of hot runner is presented in Figure 8.

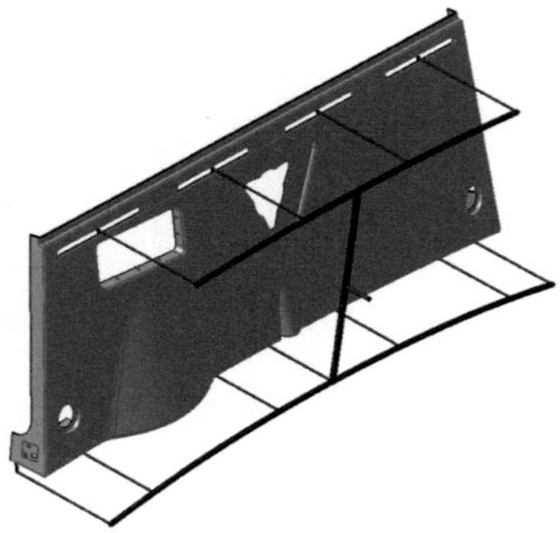

Fig. 8 Hot runner detail—inner flow channels between the injection machine gate and part gates

Because the Instrument Panel has a long flat surface with thin wall, the multi-gate injection is the best solution; otherwise the melt front does not have enough time to fill the mold cavity. Obviously, the injection points (or gate injection), parting lines, sliders, and ejectors are positioned on the non-visible surfaces.

This manufacturing process can generate many issues regarding the part dimensional and geometry deformation (warpage, shrinkage, sign mark) or the inner structure of part (weld line, holes, incomplete injection). In literature, this category of the long flat surface with thin wall thickness is named thin-walled parts. In order to reduce the warpage deformation for this category of parts, use of the finite element analysis is necessary, and researchers focus on the relationship between the gate number and warpage deformation (Cheng et al., 2012).

One of the most common issues is dimensional shrinkage, and the final shape of the injected part could be affected. This phenomenon is influenced by the melt temperature, the construction tool, the gate position, and the injection velocity (Nasić et al., 2022).

The inner non-uniform structure is generated by the weld line phenomenon. This issue decreases the quality of visible surfaces and mechanical stiffness of the injected part. In order to eliminate this issue, researchers investigate the influence of the polymer properties, injection pressure, principal stress, and molecular orientation on the weld phenomenon (Liu et al., 2020). For increasing the cooling mold performance, conformal cooling channels (CCCs) are used instead of the classic straight drilled channels (Kanbur et al., 2020).

3. Results

3.1 Parameters for Mold Tool Design

The thermoplastic material used for Instrument Panel is the PC+ABS. This material has excellent moldable properties, and the final part color is dark (Table 1). In the final stage, depending on the OEM request, the IP will either be black or it will be covered by synthetic leather.

Table 1 Properties of polycarbonate+acrylonitrile butadiene styrene (PC+ABS) (ESTER, n.d.)

Description	*Typical Value*	*Unit*
Filler/Reinforcement	Nil	
Forms	Pellets	
Additive	Mold Release	
RoHS Compliance	RoHS compliant	
Features	Good flow	
Appearance/Color	Black	
Processing Method	Injection molding	
Specific Gravity	1.14	
Mold Shrinkage—Flow	0.7–0.8	%
Tensile Strength	50	MPa
Tensile Modulus	2100	MPa
Flammability Classification 1.50 mm	HB	Rating
Drying Time	3–4	Hour
Injection Mold Temperature	60–90	^{0}C
Injection Melt Temperature	260–270	^{0}C

Figure 9a depicts the Instrument Panel and the hot runner core. This assembly was used for mold flow simulations. The all-gates and connection channels were positioned on the non-visible area, and the injection gates rest pins should not disturb the neighboring element. For example, in the central speaker pocket, the injection gates rest pins must not come in contact with the central speaker (taking into account the necessary tolerances with respect to the gaps needed).

Figure 9b provides a clear view of the result in the projected area digital tool. This area must be generated perpendicular to the main draft direction, and its measure units are in m^2.

The equation of mold tool opening force for injection machine, Fd (Neagoe, 2016; Seres, 2002) is:

$$Fd = \text{Pi}*\text{Apr} \qquad ...(1)$$

where:

Fd = mold tool opening force (force that tends to open the mold plate A and mold plate B along the injection process);

Apr = projected area for part and injection channels from common planes.

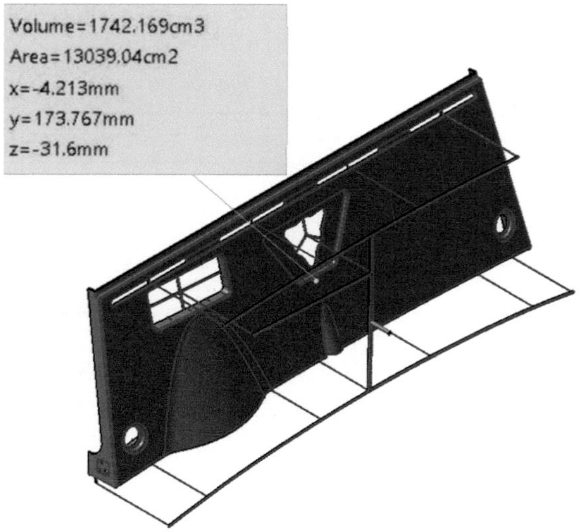

(a)

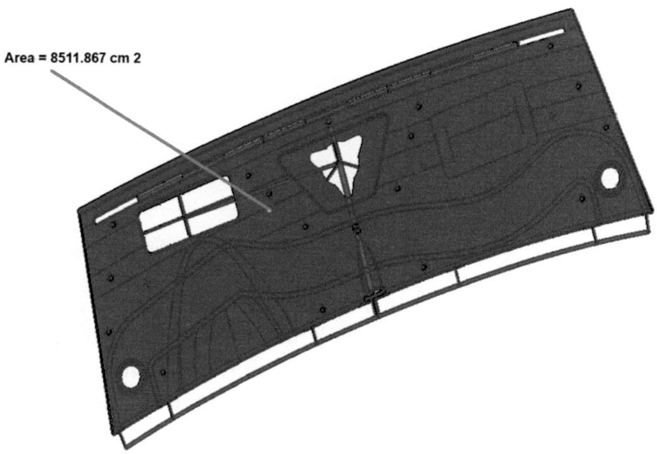

(b)

Fig. 9 (a) Instrument Panel properties with hot runner core – final version; (b) Projected area – final version

Indicative values for such parameters are Apr = 0.851187/2 [m^2] = 0.4255935 [m^2]. CATIA^{TM} Affinity tool returns the Apr = Apr to Mold plate A + Apr to Mold plate B + Lateral area. The lateral area is insignificant due to OZ component = 0.001 [mm]. Pi = inner mold core pressure 148 [MPa], assigned with Drop pressure from Mold flow simulation for the worst-case scenario.

For clamping systems actuation mechanism with levers, Fm is:

$$Fm = (1.2 \ldots 1.3)*Fd \qquad ...(2)$$

where:

Fm = maintaining force (the force needed to maintain the mold tool close along the injection process), the chosen value for safety coefficient = 1 (in order to not use worst case combined with worst case), and the final value needed for clamping system is $Fm = 62.31$ [MN]. For fine tuning of Fm, this value is compared with a set of real test injections.

The Maximum shot size needed from injection machine (Neagoe, 2016; Seres, 2002) is:

$$N = 0.8*[Vm/(Vp + Vr)] \quad ...(3)$$

where:

Fd = mold tool opening force (force that tends to open the mold plate A) and
N = number of mold tool cavities;
Vm = maximum shot size needed from injection machine;
Vp, *Vr* = part and hot runner core volume.

Having in view that: $N = 1$, $Vp + Vr = 1742.17$ cm^3, and the fact that 80% is used from the injection screw volume capacity, the final value needed for maximum shot size is $Vp = 1394$ [cm^3].

Plasticizing capacity needed from injection machine *Q*[Kg/h] (Neagoe, 2016; Seres, 2002) can be calculated using the relation:

$$N = (3.6*t*Q)/[(Vp + Vr) *\rho] \quad ...(4)$$

where:

N = number of mold tool cavities;
Vp, *Vr* = part and hot runner core volume;
t = injection total cycle time;
Q = plasticizing capacity;
ρ = density of material.

Considering the following values: $N = 1$, $Vp + Vr = 1742.17$ cm^3, $t = 6$ [*s*], $\rho = 1.1$ [g/cm^3], the value for plasticizing capacity needed is $Q = 320$ [Kg/h].

3.2 Simulation in Moldflow™

In order to ensure a significant mold filling capacity, the most important stage is the mold flow simulation. This simulation gives information to the project teams for optimizing the mold tool or feedback to the design teams to change the 3D model details.

Standard simulation has as output information the following parameters:

- fill time;
- injection pressure;
- confidence of fill;
- weld lines;
- pressure drop.

For mold flow simulation, the Instrument Panel was joined with the hot runner core. In the 3D modeling of the hot runner core, attention must be paid to leave enough space in the mold tool for hot runner integral construction and enough metal wall in mold. Figure 10a presents the Instrument Panel and the hot runner inner core for the first injection solution. It has been used for simulating only the hot runner core because the fluid movement was sufficient for the final part properties simulations.

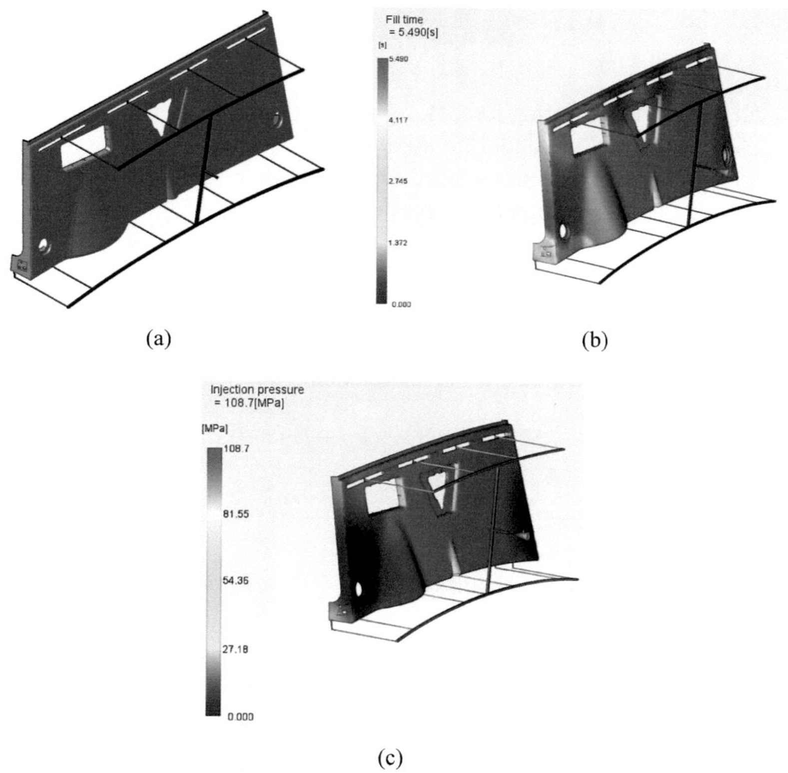

Fig. 10 (a) Initial version 1; (b) Part and hot runner, fill time; (c) Injection pressure

The simulation results provided analytically and pictorially the information about the fill time (Figure 10b) and the injection pressure (Figure 10c). In similar regions, the properties of the final part could be different. This situation results in obtaining an injected part with geometric deformation after the part cooling.

3.3 Injection Strategy Optimization

The next step toward injection optimization was to add the island injection strategy (the injection channels in the boundary of the central speaker pocket). This added geometry is marked in Figure 11a. The fill time (Figure 11b) for the central area

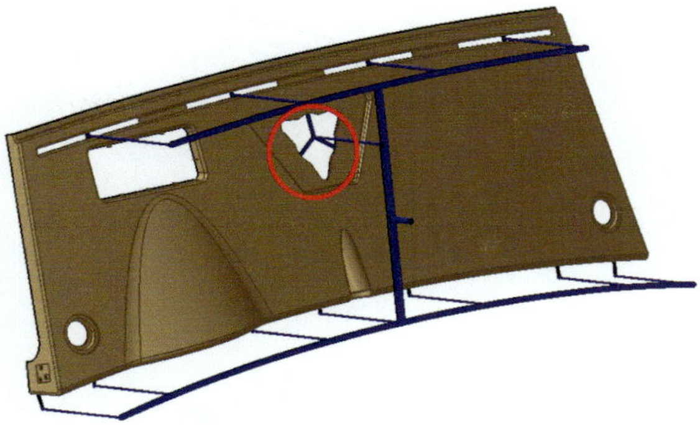

(a)

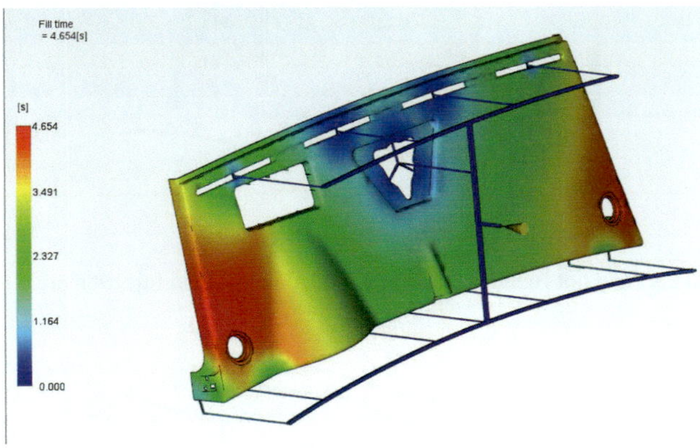

(b)

Fig. 11 (a) Part and hot runner – version 2 (added island injection area – marked in red); (b) Fill time

of part was decreased and consequently the dimensional result in this area was improved. The center of Instrument Panel is a critical area because it is in front of the user, without any design masks.

In version 3 of injection optimization, the island injection for Head Up Display (HUD) pocket was added. This added geometry is marked in Figure 12a. A channel was created at the same level, in order not to increase the value of hot runners. The filling time (Figure 12b) for this area was decreased because the melted material wall no longer had to surround the HUD area. Although a number of optimizations took place, red areas in simulations continue to exist.

In the next design version, optimization was continued for the left and right Instrument Panel walls, and a singular injection gate was added (marked in Figure 13a). These long left and right injection channels increased the hot runner value, and finally, after the fill time simulation (Figure 13b), red areas were still present in the simulation. Another possible issue was the welding lines that were moved to the center area of the Instrument Panel.

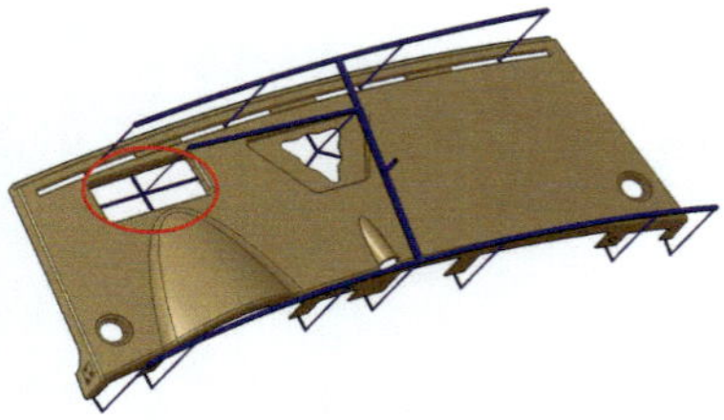

(a)

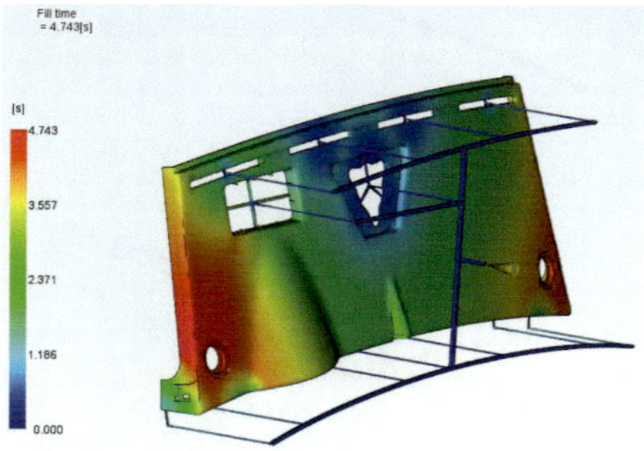

(b)

Fig. 12 (a) Part and hot runner—version 3 (added new island injection area—marked in red); (b) Fill time

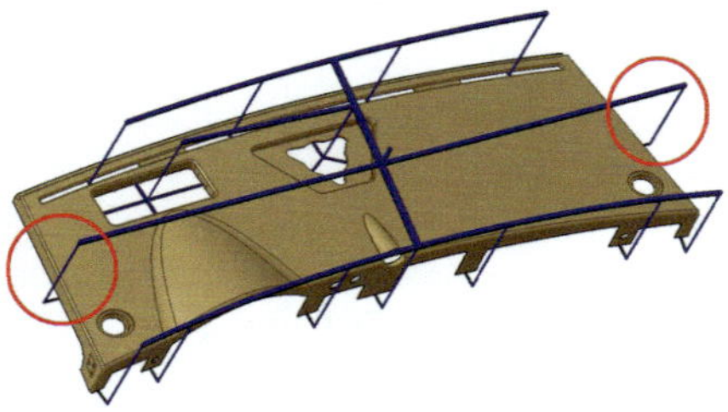

(a)

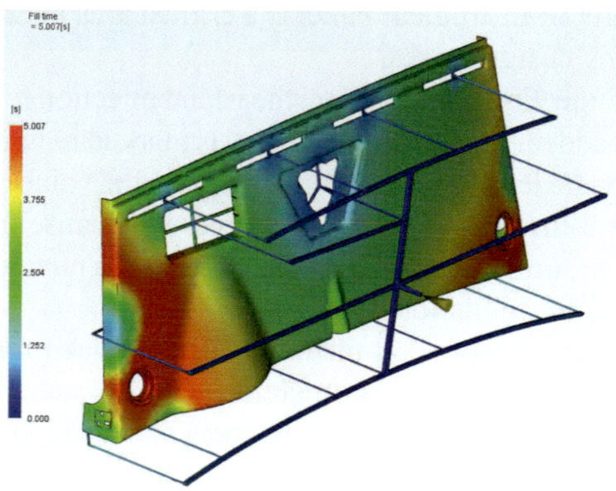

(b)

Fig. 13 (a) Part and hot runner – version 4 (added left and right injection gates – marked in red) – increase the hot runner construction; (b) Fill Time

3.4 Solution for Production

In this final version, both left and right long channels and injection gates from left and right Instrument Panel walls were removed (Figure 14a). An optimization of the flow rates was carried out in this version of the injection molding solution. The first optimization of the flow rates was performed only on the injection gates at the part level, without obtaining the desired results in full form.

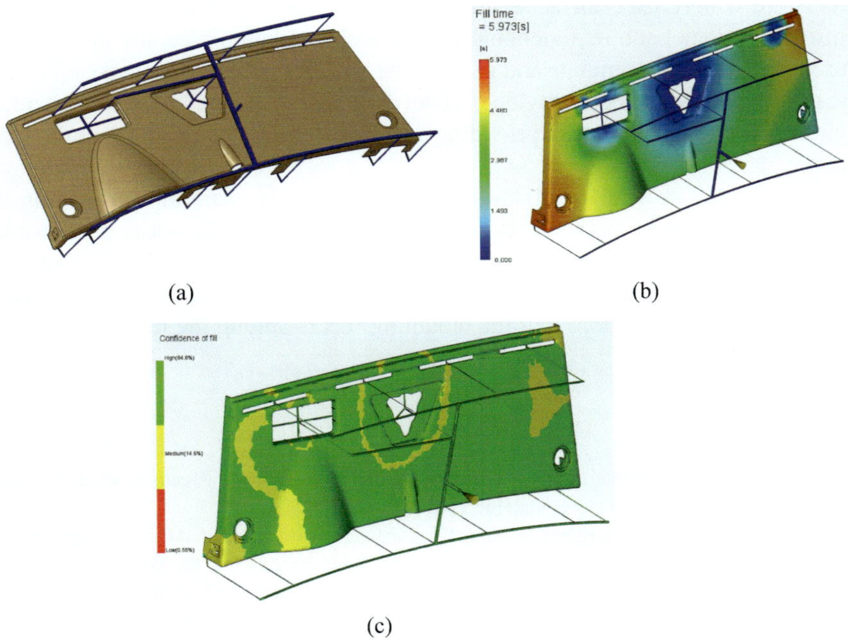

Fig. 14 (a) Part and hot runner – version 5 (final solution); (b) Fill time; (c) Confidence of fill

The desired results were obtained only after a flow optimization that was carried out at the level of the hot runner channels. Figure 14b depicts the fill time in a balanced form, and as a result no red areas were present, while at the same time the weld lines did not affect the Instrument Panel mechanical structure. The confidence of fill (Figure 14c) provides a solid basis that there will be no areas of incomplete injection in the Instrument Panel production. This digital tool is extremely important in checking the centering domes and stiffness ribs onto part because these constructions have elements with thin wall at the top end zone (due to the draft angle presence).

Conclusions

In this study, the authors showed how digital tools are applied in design and engineering of the Instrument Panel in the automotive industry. Digital manufacturing, standard solution for the automotive industry, requires the extensive use of digital tools in order to work within the limited time for obtaining the 3D parts, tools design and parts engineering releases.

Digital manufacturing is the best solution in the current engineering environment because the large-scale work-from-home and remotely located project teams need real-time links for project performance and for data management.

When designing the functional parts related to the automotive industry, digital tools are widely used for accommodating the OEM's design requests with the Instrument Panel and its functionalities. In tool simulation area, digital tools are used with an aim to simulate and validate the injection molding processes. Thanks to the digital environment provided, all development stages are linked to each other through feedback loops, thus providing a high performing platform for design and optimization.

The ultimate goal of all these digital tools is to start the production of automotive parts in the shortest possible time, with as little waste as possible. Future research regarding the application of different digital tools in mechanical/manufacturing engineering is expected to focus on studying the links between the geometrical deviations and the mounting forces among the Instrument Panel mechanical parts.

References

Artem Group. n.d. Available at https://www.artemgroup.eu/images/presa/Technomarket/ARTEM_octombrie-noiembrie%202015.pdf (accessed on 15.08.2023).

Cakmakci, M. and Demirel-Ortabas, N. 2019. Performance Measurement of SMED Improved Plastic Injection Molding Production by Using Process Capability Analysis for Attribute Data. In A. Hamrol, A. Kujawińska and M. Barraza (eds), *Advances in Manufacturing II. MANUFACTURING 2019. Lecture Notes in Mechanical Engineering*, 188–205. Springer, Cham.

Catoen, B. and Rees, H. 2022. *Injection Mold Design Handbook*. Carl Hanser.

Cheng, X., Liu, S., Zheng, X. and Seng, N. 2012. Effect of Gate Number on the Warpage in Injection Molding. In Y. Zhang (ed.), *Future Communication, Computing, Control and Management. Lecture Notes in Electrical Engineering*, vol 142, 241–246.Berlin, Heidelberg: Springer.

ESTER, n.d. Product Data Sheet ESTOPLAST ZK8508BB01; ESTER Industries LTD. Available at https://www.esterindustries.com/sites/default/files/ep-products/ZK8508BB01.pdf (accessed on 10.07.2023).

Kanbur, B.B., Suping S. and Duan, F. 2020. Design and Optimization of Conformal Cooling Channels for Injection Molding: A Review. *The International Journal of Advanced Manufacturing Technology*, 106: 3253–3271.

Kang, G.J., Park, C.H. and Choi, D.H. 2016. Metamodel-based Design Optimization of Injection Molding Process Variables and Gates of an Automotive Glove Box for Enhancing Its Quality. *Journal of Mechanical Science and Technology* 30: 1723–1732.

Kech, A. and Maris-Haug, P. 2020. A Study on the Determination of Virtual Process and Quality Data in Injection Molding Simulation. In C. Hopmann and R. Dahlmann (eds), *Advances in Polymer Processing 2020*. Berlin, Heidelberg: Springer Vieweg.

Keranen, M., Krishnaraj, S., Kulkarni, K., Lu, L. and Thyagarajan, R., 2005. Automating Instrument Panel Head Impact Simulation. SAE Technical Paper 2005-01-1221, 2005. Available at https://doi.org/10.4271/2005-01-1221 (accessed on 23.12.2024).

Li, Y., Chen, J.C. and Ali, W.M. Process optimization and in-mold sensing enabled dimensional prediction for high precision injection molding. *Int J Interact Des Manuf* 16, 997–1013 (2022). https://doi.org/10.1007/s12008-021-00800-1

Liu, Q., Liu, Y., Jiang, C. and Zheng, S. 2020. Modeling and Simulation of Weld Line Location and Properties during Injection Molding based on Viscoelastic Constitutive Equation. *Rheologica Acta* 59: 109–121.

Mao, H., Wang, Y. and Yang, D. 2022. Study of Injection Molding Process Simulation and Mold Design of Automotive Back Door Panel. *Journal of Mechanical Science and Technology* 36: 2331–2344.

Nasić, E., Sprečić, D., Halilović, J. and Kovačević, D. 2022. Dependence of Basic Geometric Characteristics of Polymer Components on Parameters of Injection Molding and Gate Position. In I. Karabegović, A. Kovačević and S. Mandžuka (eds), *New Technologies, Development and Application V. NT 2022. Lecture Notes in Networks and Systems*, vol 472, 202–208. Springer, Cham.

Neagoe, I., 2016. Injection of Plastic Materials. University Course, Transilvania University of Brasov.

Saad, S., Sinha, A., Cruz, C., Regnier, G. and Ammar, A. 2022. Towards an Accurate Pressure Estimation in Injection Molding Simulation Using Surrogate Modeling. *International Journal of Material Forming* 15: 72.

Seres, I. 2002. *Thermoplastic Materials for Injection.* Oradea: Imprimerie de West Publishing House.

Techcenter, 2007. Engineering Plastics—Part and Mold Design: A Design Guide. Available at https://techcenter.lanxess.com/scp/americas/en/docguard/Part_and_Mold_Design_Guide.pdf?docId=77015 (accessed on 15.08.2023).

Wang, B.W., Nian, S.C. and Huang, M.S. 2022. Strategies for Adjusting Process Parameters in CAE Simulation to Meet Real Injection Molding Condition of Screw Positions and Cavity Pressure Curves. *The International Journal of Advanced Manufacturing Technology* 122: 1339–1351.

Yu, S., Zhang, T., Zhang, Y., Huang, Z., Gao, H., Han, W., Turng, L.S. and Zhou, H. 2022. Intelligent Setting of Process Parameters for Injection Molding based on Case-based Reasoning of Molding Features. *Journal of Intelligent Manufacturing* 33: 77–89.

Zhao, Z., Liu, M., He, X. and Liu, B. 2010. Injection Mold Design and Optimization of Automotive Panel. *2010 Third International Conference on Information and Computing*, Wuxi, China, pp. 119–122.

Zhou, Y., Cao, Z., Lu, J., Zhao, C., Li, D. and Gao, F. 2022. Objectives, Challenges, and Prospects of Batch Processes: Arising from Injection Molding Applications. *Korean Journal of Chemical Engineering* 39: 3179–3189.

2 Modern Trends of Reverse Engineering in Digital Manufacturing

Marius Andrei Mihalache[1*], Alexandru Sover[2],
Oana Dodun-Des-Perrieres[3], Gheorghe Nagit[4], Vasile Merticaru[5],
Adelina Hrituc[6] and Laurentiu Slatineanu[7]

1. Introduction

Reverse engineering encompasses an array of technologies and methods that produce results in terms of design, simulation, optimization, and manufacture. Options were pushed a little further by a new set of digital tools and software that can virtually mimic the shape of a scanned object in a precise way and even fight back the effect of spring back in designing a mold or when 3D printing. Finite element methods can validate the model and propose an optimized geometry with improved efficiency. The entire additive manufacturing process can be simulated, thus making it possible to pinpoint problematic areas before the actual process takes place. Effects such as warping, skewing, or tolerance deviation can now be overcome by choosing the right additive printing strategy ranging from plastic filaments to metal ones or others.

The applicability of reverse engineering varies from medicine to mechanical engineering, biology, and many other fields of study. For product design and digital manufacturing, reverse engineering uses some techniques, tools, methods, and means of optimization that ease the process of remanufacturing. Of course, there are legal and ethical concerns that must be met within legal boundaries. Some limitations pose several provocations but the actual trends are looking more

[1] *Lecturer*, Department of Technology of Machine Manufacturing, Gheorghe Asachi Technical University of Iasi.

[2] *Professor*, Department of Technology Laser Safety Officer, Ansbach University of Applied Sciences.

[3,4,7] *Professor*, Department of Technology of Machine Manufacturing, Gheorghe Asachi Technical University of Iasi.

[5] *Associate Professor*, Department of Technology of Machine Manufacturing, Gheorghe Asachi Technical University of Iasi.

[6] *PhD student*, Department of Technology of Machine Manufacturing, Gheorghe Asachi Technical University of Iasi.

E-mail: a.sover@hs-ansbach.de[2]; oana.dodun-des-perrieres@academic.tuiasi.ro[3]; gheorghe.nagit@academic.tuiasi.ro[4]; vasile.merticaru@academic.tuiasi.ro[5]; adelina.hrituc@student.tuiasi.ro[6]; laurentiu.slatineanu@academic.tuiasi.ro[7]

* Corresponding author: marius-andrei.mihalache@academic.tuiasi.ro

promising than ever, especially by using modern approaches such as artificial intelligence (AI).

Reverse engineering can be corroborated with other techniques of data interpretation and manipulation in such a way that results are both time- and cost-effective. Easy integration with the Industry 5.0 concept is an ongoing process that proposes sustainable optimization of energy consumption, materials processing, and product life cycles but centers on the individuals and their well-being.

2. Methods, Equipment, and Their Applicability in Reverse Engineering

There are two main types of methods when referring to reverse engineering: non-contact and contact-based. Each presents a series of advantages and disadvantages. A map of the workflow that reverse engineering implies is presented in Figure 1. It is regarded as an essential part in transition between Industry 4.0 and 5.0.

2.1 Non-contact Methods

Non-contact methods are addressed to objects that are sensitive and intricate and are based on technologies that use imaging, scanning, and sensing. In this category, it is worth mentioning key methods such as:

- *3D Laser Scanning:* It relies on a beam directed to an object, its reflected light being captured by a sensor. The 3D model can be obtained by measuring the time-of-flight needed by the light to touch one surface and come back as well as the angle of reflection. Equipment includes laser scanners are not portable as well as systems that are stationary: FARO Laser Scanner (Faro, 2024a), Creaform HandySCAN (Creaform/Ametek, 2024) as well as optical sensors that are able to capture the light being reflected and help in building point clouds.
- *Structured Light Scanning:* To build a 3D model, it uses a light pattern that could take the form of stripes or grids that are being projected onto an object, and with cameras it measures distortion in the pattern provoked by irregularities of the object's surfaces. Equipment is based on projectors that emit patterned blue light as well as cameras with sensors able to capture deformations: Zeiss ATOS system (Zeiss, 2024a).
- *Photogrammetry:* It uses photos taken from different angles to build a 3D model by triangulating the positions of geometric features. Equipment involves high-resolution cameras for capturing multiple perspectives and relies on software solutions for image processing: Agisoft Metashape (Agisoft, 2024).
- *Computed Tomography (CT) Scanning:* The technology has gone beyond the medical domain, and it is widely used for internal and external scans. X-rays are sent to an object in slices and then combined into a 3D model that includes internal structures as well. Equipment is based on industrial CT scanners designed for metrology purposes: Nikon Metrology CT scanners (Nikon, 2024) and X-ray detectors for capturing data.

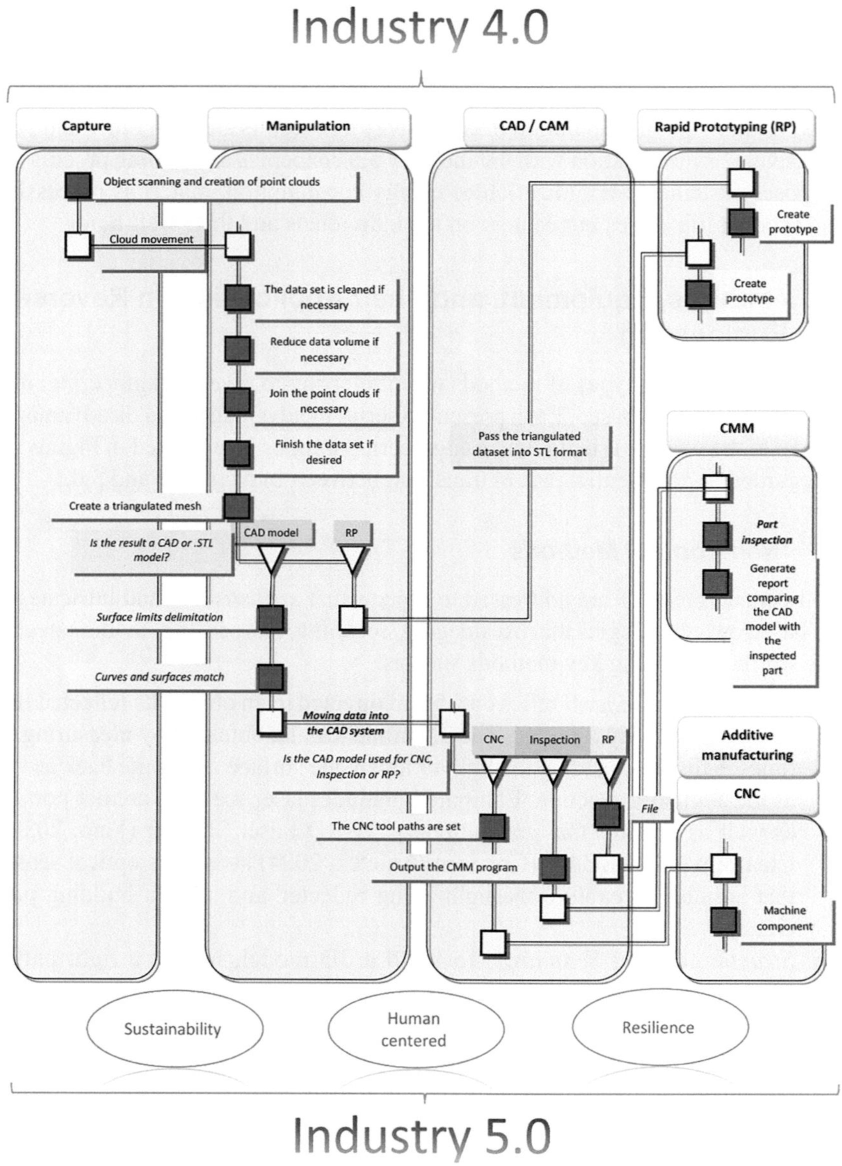

Fig. 1 Workflow and integration of reverse engineering

- *Optical Coherence Tomography (OCT):* It captures very detailed images at the microns level by using light waves that penetrate the object's material and then are reflected back. OCT systems are equipped for subsurface features capturing: Thorlabs OCT systems (Thorlabs, 2024).
- *White Light Scanning:* It uses white light instead of wavelengths to measure an object by projecting light toward it and analyze distortions as it builds up a

3D map of surfaces being scanned. Equipment includes scanners that emit and capture white light patterns: Steinbichler COMET (Inspection Engineering, 2024) as well as optical sensors and cameras used for capturing details.

- *Terrestrial LiDAR (Light Detection and Ranging):* The method senses remotely and uses laser to measure distances to objects and compute them into a 3D model. It is used mainly in the case of large-scale objects such as buildings or even terrains. LiDAR equipment may be stationary or mobile and it is often used for surveying: Trimble LiDAR systems (Trimble Geospatial, 2024) uses Global Positioning System (GPS) and inertial measurement unit (IMU) to track scanner position.

The areas of application of non-contact methods are aerospace and automotive industry for analyzing geometries of complex parts without having them disassembled, or preservation of cultural heritage, for example, for artifacts that require digitalization. Other appealing domains are medical and biomedical because of the constant need to replicate implants or even anatomical structures as well as various medical devices. The applicability of non-contact methods has been well served in product design and quality control also where they have to reproduce intricate objects and inspect them for flaws afterwards.

The advantages of non-contact methods include speed and efficiency as many can capture large amounts of data quickly. It is worth mentioning the fact that these techniques can scan complex surfaces in a fraction of the time required for contact methods. Another advantage refers to their ability to measure delicate or soft objects because they prevent potential damage, very useful, for example, when working with biological samples. Another advantage is their ability to capture complex geometries and freeform surfaces as objects may have irregular shapes, including surfaces with deep cavities, undercuts, or detailed textures such as organic surfaces (bionic), which otherwise would be difficult or impossible to access with a physical probe. Their suitability for large and inaccessible objects is one major advantage as there is no need for repositioning. Also, the equipment often has no wear and tear signs of extensive use as it does not touch objects.

The disadvantages involve a lower accuracy in some cases because of factors like lighting, surface reflectivity, or environmental conditions such as dust or vibrations. Also, there are challenges with certain materials because objects made from transparent, shiny, or reflective materials have surfaces that absorb or refract light, such as glass or polished metal. That can result in incomplete or erroneous data, which means there is a need for prior surface preparation (colored layer). Another major disadvantage is the high cost of the equipment in comparison with the one used for contact methods. Let us not omit the complexity of data that needs processing, which requires advanced software solutions and more computing power to reconstruct and analyze 3D models.

2.2 Contact Methods

Contact methods involve physical touching of objects and, often enough interaction with the objects. They too provide highly accurate measurements by recording point coordinates of a touching tool as it surveys the object's surface. Some of the most important contact methods worth mentioning are:

- *Coordinate Measuring Machine (CMM):* It uses probes that physically touch points on the object as it records their precise coordinates to build up a 3D model. There are three types of CMM machines: bridge-type CMMs (Zeiss, 2024b) that are stationary and primarily used for small to medium-sized objects; Cantilever CMMs (Creaform/Ametek, 2021) that are used for very small, precise parts; and Gantry CMMs (Hexagon, 2024a) used for larger, heavier objects. The equipment also includes touch-sensitive probes that can be either mechanical or using styluses with ruby or ceramic tips. Some advanced CMMs use a multi-sensor system with both tactile and optical probes.
- *Portable CMM (Articulated Arm CMM):* They use articulated arms equipped with touch-sensitive probes either handheld or portable versions of CMMs. They gain flexibility when objects are too big to move or when scanning is needed on a client's premises. Equipment involves an articulated arm that typically has several joints for a full range of motion: FARO Arm (Faro, 2024b), Romer Arm (Hexagon, 2024b) as well as contact probes for point-by-point measurements.
- *Profilometers (Contact Profilometry):* It measures surface roughness and builds profiles by moving a stylus on an object's surface, and afterward it creates texture surface-related maps. The equipment has a built-in mechanical stylus that is dragged across the entire surface in such a way that it detects even minor surface deviations: Taylor Hobson Talysurf (Taylor-Hobson/Ametek, 2024) as well as stylus tips that are made from diamond or tungsten carbide for durability.
- *Manual Gauging Tools:* The traditional tools have not lost their basic ability to measure and dimension objects and still are very useful for basic reverse engineering. The equipment includes calipers used for the measurement of internal and external dimensions: vernier, dial, or digital calipers, micrometers used for measuring small dimensions with high precision, height gauges that are used for vertical measurements, and dial indicators used to measure small distances and displacements with high accuracy.
- *Touch Probe Scanning:* It uses a touch probe that scans the surface of more complex objects as it records multiple points along its way. The equipment includes probing systems that can be part of Computer Numerical Control, CNCs or CMMs as well as probes which are stylus tip-shaped that make physical contact with the part and thus capture point data during the scan.
- *Fixed and Portable Gauge Systems:* They are customized measuring tools designed to inspect specific part dimensions as part of the quality control processes as well as reverse engineering. From an equipment standpoint, they include fixed gauges that are specifically designed for a part's features, go/no-go gauges that are simple mechanical devices used to check tolerances, and thread gauges used to check threaded components.

The applications of contact methods too are in the automotive and aerospace industries as they are the means for analyses and reverse-engineering mechanical parts, engines, and components that require tight tolerances, manufacturing quality control procedures that ensure that parts and components meet precise specifications by direct measurement; mold and die industries that use reverse engineering for the replication of complex tooling surfaces; and medical devices because of their ability to create accurate models of prosthetics, implants, and surgical tools.

The advantages of contact methods are high precision and accuracy at the micron level, particularly for measuring small, detailed features, or geometrically complex parts. Another advantage is minimal impact from environmental variables such as lighting, surface reflectivity, or dust as they are suitable for various surface types. Their ability to measure directly means they can provide direct and reliable measurements of surface points, ensuring consistent accuracy without the need to account for factors like light scattering.

The disadvantage of contact scanning methods is a slower measurement process as each point must be touched individually. They are also limited to accessible surfaces because of their inability to access deep internal cavities, complex undercuts, or highly irregular geometries. There is a potential for objects to be damaged since they physically touch surfaces. Another disadvantage is their confinement inside stationary systems which means objects have to be moved to the system and not the other way around.

3. Software Used in Reverse Engineering

Most producers offer their software for acquisition, interpretation, and refinement when selling scanning equipment. However, there are software that better address the growing special needs for handling large datasets, process point clouds or mesh data, converting them into Computer Aided Design, CAD models and supporting complex geometries in an intuitive framework.

- *Geomagic Design X (Oqton, 2024):* It is one of the industry-leading reverse engineering software as it provides advanced surfacing, solid modeling, and parametric design capabilities, and is also easily integrated with all major CAD platforms. It works seamlessly with most 3D scanning technologies, benefits from highly automated workflows, and offers robust tools for surface and feature recognition.
- *SolidWorks with ScanTo3D (Dassault Systemes, 2024):* It is a widely used 3D CAD software that includes tools for reverse engineering through its ScanTo3D module, thus allowing users to import and manipulate scan data directly in the SolidWorks environment. It also benefits from great integration with parametric CAD tools and is found to be suitable for users familiar with SolidWorks in need of reverse engineering tools. It offers a large ecosystem of plugins and support for various industries.

- *Autodesk ReCap (Autodesk, 2024a):* It is specifically designed when working with laser scans, photogrammetry, and point clouds by allowing easy manipulation of point clouds and seamless integration with Autodesk platforms like AutoCAD and Inventor. It was found to be powerful for large-scale projects, easy to use for users already in the Autodesk ecosystem, and effective for creating models from drone or LiDAR scans.
- *Siemens NX with Reverse Engineering Module (Siemens, 2024):* It also supports direct import and manipulation of scan data and point clouds, and offers advanced tools for surfacing, mesh creation, and converting data into parametric CAD models. The software is fully integrated with CAD, Computer Aided Manufacturing, CAM" and "Computer Aided Engineering, CAE workflows as well as Siemens' comprehensive engineering suite. It was found to be suitable for complex and high-precision reverse engineering tasks but also sustains robust modeling, assembly design, and simulation.
- *CATIA with Digitized Shape Editor (DSE) (3DS, 2024):* It is widely used in aerospace and automotive industries and is capable of importing point clouds and meshes from 3D scanners and providing sophisticated surface fitting, mesh editing, and parametric tools. It is considered one of the industry's highest grades in terms of precision and control over complex surfaces and features. By providing seamless integration with other CATIA modules, its performance was found to be excellent when dealing with complex geometries and assemblies.
- *PolyWorks (Innovmetric, 2024):* The solution is a powerful metrology and reverse engineering software suite that focuses on 3D scanning, inspection, and CAD modeling as it features inspection, measurement, and model verification capabilities, making it suitable for quality control. That makes it great for both reverse engineering and dimensional inspection because of its ability for highly detailed mesh processing and integration of point cloud management tools.
- *3D Systems Geomagic Wrap (3D Systems, 2024):* It converts point clouds and polygon meshes into 3D surface models for CAD processing by including tools for mesh editing, cleaning, and surface generation. It presents an efficient and easy-to-use interface, exhibits strong capabilities for mesh-to-surface conversions, and benefits from integration with 3D printing workflows.
- *MeshLab (MeshLab, 2024):* It is an open-source, free software for mesh editing and processing, primarily used for cleaning, editing, and converting 3D scan data into usable formats that offer basic tools for mesh simplification, point cloud alignment, surface reconstruction, and texture mapping. It is renowned as a lightweight and free solution for basic mesh processing which provides a wide range of tools for mesh refinement and optimization.
- *Artec Studio (Artec 3D, 2024):* Even though it was designed to work in tandem with Artec 3D scanners, it is also compatible with other formats. It provides high-accuracy 3D scan data processing, alignment, and conversion into 3D models, and includes automatic features for noise reduction, alignment, and

meshing. One of its key features is that it is highly automated, thus reducing manual intervention for scan data processing and making it ideal for product design and industrial applications.

- *Rhino with RhinoReverse (RhinoReverse, 2024) or Mesh2Surface (Mesh2Surface, 2024):* Rhinoceros (Rhino) is used for creating, editing, analyzing, and translating 3D geometry that, when combined with plugins like RhinoReverse or Mesh2Surface, becomes a powerful reverse engineering tool. It was found to be great when handling complex, organic geometries as well as in the case of freeform modeling, being addressed for industries such as architecture, jewelry design, and product development. Compared to competition one key advantage is represented by its affordability among other CAD/CAM or standalone reverse engineering modules.
- *Autodesk Fusion 360 (Autodesk, 2024b):* It seems to integrate CAD/CAM/CAE-related tools in just one solution that streamlines the processes of reverse engineering. It features an automatic mesh-to-BRep (boundary representation) conversion, thus helping engineers to design precise solid models from scanned data. Parametric modeling allows the ability to impose dimensional constraints, which can be easily modified, but at the same time it possesses freeform modeling tools (T-Splines) that enable users to reshape and modify surfaces organically, all that being shared among users employing the cloud. It has built-in tools for simulation. It integrates 3D printing and can generate toolpaths for CNC.

These are just a few of the software solutions available that comply with the reverse engineering concept.

4. Present Trends in Reverse Engineering

Companies, researchers, and scholars have proposed different methods that are suitable for reverse engineering products, processes, or procedures that offer valuable insights into what can be done at the present moment. This study spans multiple directions ranging from basic reconstruction techniques to 3D scanning possibilities for optimization by pairing with others, minimization of scanning errors, use of scientific methods as well as finite element methods (FEMs) for failure analyses and prediction purposes, leverage of additive manufacturing for circular economy, and quality control alongside multi-criteria decision-making approaches. The content presents findings and conclusions from different research papers for a better understanding of the applied methods and strategies in context.

We see that reverse engineering usually pairs with 3D scanning for optimal results. Some researchers have added data from measurement as a means for prediction. Coordinate Measuring Machine (CMM) equipment has been used for years in production since it can provide valuable data regarding roughness, tolerances, and even post-machining surface conditions. Kim has used such a hybrid method for the prediction of performance for an industrial gas turbine. His study relies on operational data without considering performance in use. He uses an initial design based not on a physical actual part but on existing information in literature about design and performance of gas turbine components. The gas turbine's physical model was obtained through measurement data using limited design information at a reference point using five measured parameters. Calibration was performed on the same datasets retrieved using measurements and the final model was subjected to a series of comparisons for verification and validation. Different components were evaluated in terms of efficiency by analyzing detailed component performance information and the calibration aimed to achieve the same response as the one given by the monitoring sensor. The author has managed to adjust the performance curve and, as a result, develop performance maps. The accuracy of the obtained model was verified against three cases that consisted of measurements under different conditions, which validated the proposed procedure. Error-wise, most values were below 1%. By reverse engineering general data regarding design and performance-related parameters retrieved from literature and not inside data owned by companies, the author managed to obtain a functional virtual model. Parameters other than the measured operating data are not contained within the study, which limits its effectiveness (Kim, 2023).

Fan et al. showed that reverse engineering does suffer from poor versatility when dealing with deformed parts with many irregularities. Their methods add to point cloud segmentation by growing specific regions and thus maintaining accuracy throughout the entire cloud. A grid offset method is used to refine the massive data and filter the point cloud. By smoothing data across u and v directions of the grid, the authors achieved little data fluctuation but the problem of scattered points or even disordered features remains. By splitting and reconstructing features, accurate Non-Uniform Rational B-Splines Modeling, NURBS surfaces were produced where mesh size plays a significant role. Greater values lead to pronounced errors whereas average still produce wrinkles at surface joints. The best results are obtained under 0.4 mm of grid resolution. The method does address parts with little deformation that can be managed when dealing with batches of products (Fan et al., 2022).

In some cases, a key benefit of reverse engineering might be the absence of tooling because the product can be obtained through additive manufacturing. This greatly balances a multi-modal business model in favor of reverse engineering because it relies on the possibility of manufacturing decentralized through automation concerning cost, time, and quality. Kyaw et al. have highlighted the use of a Multi-Criteria-Decision-Making (MCDM) process when assessing the three aforementioned approaches for remanufacturing. Quality-wise, there is a need for obtaining a similar functional part as the original, if not better. This means that thresholds must be imposed as several benchmarks may be considered. When assessing costs, one must consider purchasing costs, installation, and training, as well as effective machine manufacturing time. That depends on the type of technology that is intended for remanufacturing purposes. If you consider Fused Filament Fabrication, FFF, for example, expenses are going to be amortized faster than metal printing or even CNC. Time assessment is divided by data acquisition, post-processing, refinement, and actual manufacturing. Those add up to non-manufacturing times, trial and error, or even additional scanning. These three parameters are at the base level. The next one summarizes results in three categories: manufacturing time, manufacturing cost, and manufacturing quality. The third one will offer a well-documented choice for remanufacturing. By employing an Analytical Hierarchy Process (AHP), the authors have valued the results that are weighted in favor of laser scanning as the most precise, cost-effective, and on-time result provider (Kyaw et al., 2023).

Reverse engineering usually refers to the possibility to obtain an as-much-as-accurate 3D replica of a physical model. But there is more to it than tackling point clouds and meshes. Qie et al. propose a functionality-oriented approach for surface reconstruction, thus making it possible to reassign the purpose of a part. They iterate several functions such as save, transfer, change, and possibility to connect stating that models may be reassigned or gain multiple functionalities by choosing from among different reconstruction techniques such as pre-defined rules, machine learning, or similarity search. This produces labeled features which are at the base of a labeled part that goes in the assembly. The benefits refer to the possibility that the database thus created can be used for the generation of automatic tolerances with real-time digital update of the model because components are assigned to pre-defined classes using a convolutional neural network (CNN). The machine learning filters even further as it can be taught to distinguish a bearing cover as a sealing part. Assembly information can add up to the function-based reconstruction as joints may be recognized and treated as such, imposing all subsequent constraints on the other components that participate in the movement produced by it. Similarity search will find models with the same descriptors or even labels that may be taken into account and transferred after multiple checkpoints check out. The authors, however, mention the "lack of explicitly defined rules for knowledge reuse" for proper training of neural networks, which marks their approach as a work in progress (Qie et al., 2021).

After scanning, a significant number of points are gathered in dependence with the number of scans and the complexity of the object. Almost every time in this type of situation, the engineer needs to refine and downsize the point cloud to achieve convergence. It does so by overlapping point clouds but that only increases furthermore the number of points. Other methods take only part of point clouds to better describe a certain feature but this too adds to the large number of points. Then segmentation can be imposed but that only works for points that meet the same criteria. Good meshes are obtained by using curvatures because even primitives use curvature approximation. Through refinement, one can delete duplicates and detect noise or bad points. This stage of reverse engineering requires skilled workers, good knowledge of CAD, expensive software, and powerful computers. Water-tight geometry can be obtained when using descriptors such as curvatures. Gauthier et al. proposed a method that analyses the histograms of 3D mesh curvatures for peaks and valleys. Then a statistical method is applied for prediction. When trying to approximate geometric features, a good way is to apply, whenever possible, primitives, which can be extracted by fitting profiles or from surfaces that are free-formed. Many software solutions for reverse engineering rely on region growth to compensate for inadequate data or even missing data. Noise, under the form of Gaussian or isotropic, on the other hand, is strictly related to the equipment used for scanning. Many times a visual comparison between the object and its virtual replica can help in identifying real noise apart from the one produced by the scanner's technology. But another issue arises next. When in need to retain surface roughness, one cannot just disregard all noise because some of it could potentially quantify roughness. This makes some procedures prone to failure. One way to achieve that is by normalizing curvatures so histograms exhibit the same range by multiplying them with the mean edge length of the mesh, as Gauthier et al. did. Lengths of edges are not so different as to distort a histogram, but wherever the case there is only a small number of those which can be disregarded without affecting the overall quality. The authors have identified three cases as applications for their methodology: segmentation, primitive detection, and mesh quality. They are improving results by recursively considering homogenous regions of a mesh. Second, they use thresholds to eliminate non-conforming values, being able to achieve homogeneity. Third, when growing regions, they used triangles instead of points and grouped similar elements basing their procedure on seeds. When dealing with primitives, one has to take into account tight tolerance values when best-fitting, and one way to do this is by determining the dominant surface type inside of a mesh. By using mean standard deviation, one can quantify the amount of noise in a dataset, and eliminate noise by doing so. The results address primitive adjustment to account for deviation or deformation, which paves the way for model reconstruction from little datasets (Gauthier et al., 2017).

Suzuki et al. offer another angle in curvature-based estimation by relying on gradients. They analyze CT sinograms and propose a methodology for segmentation by applying B-spline patches on desired regions. Modern equipment, CT included, can capture gradients and textures of scanned objects, thus providing additional data to process when trying to reverse engineer certain objects. If automation is in place, any CAD software will interpret the mesh data as it is and will provide the same distribution on a virtual solid, making it very difficult to reconstruct its basic shape. That does not happen when segmentation is applied to the same object. Entire surfaces or regions are mapped as a single face of a cube, for example, and that face will be recreated on the virtual solid model. So, instead of having an array of nodes with a random distribution that cannot separate the cube's face limits, the 3D solid may have a distinguishable facet as if it was created using CAD tools. In general, continuous curvature is being represented by just one patch and thus treated as discontinuities. Fillets in general pose a lot of problems because their depiction changes in terms of vectors normal to surfaces that change orientation much more slowly than the ones related to sharp edges. Several researchers proposed methods for the generation of a smooth principal vector field, some of them using integers and other iterations for local smoothing. The authors have proposed a three-step procedure that generates two-fold surface mesh surface mesh, computes a high-order derivative for each mesh vertex, and then computes the curvature gradient on the derivative's basis. Their study shows that sinogram-based computation outperforms the mesh-based one. They performed segmentation based on the Morse theory, resulting in multiple Morse cell-like regions. Then all cells were merged as a single region. The algorithms that Suzuki et al. are proposing offer satisfying results in terms of visual cues as opposed to mesh-based segmentation which seems a bit rougher. On rounded intermediate surfaces and fillets, results are showing a significant improvement as the models are easy to be transformed into solids that retain such a level of accuracy. The versatility of the proposed method makes it suitable to transform an entire array of intricate parts using reverse engineering, something that is very difficult by conventional means (Suzuki et al., 2022).

Yanjun Zhang and Junhua Sun tackled a 3D reconstruction technique named fringe projection profilometry (FPP). Because fringes often appear in overexposed images, they proposed a method based on π phase-shifting fringes. By using an adaptive projection method, they observed an overall reduction of saturated images but some areas remained overexposed due to a custom projection grey level based on empirical values rather than the highest reflection. That produced errors below 0.1 mm for standard deviation as the reconstructed model had little overexposed areas. The method does not suppress phase error in overexposed areas but adds additional π phase-shifting fringes that overcome this issue. One benefit is represented by the possibility of doubling the dynamic range which would translate into an improved capability (Zhang et al., 2023).

Reverse engineering often pairs with optimization using statistical methods such as Grey Wolf Optimization (GWO), Whale Optimization (WOA), Particle Swarm Optimization (PSO), and Genetic Algorithm (GA) using MATLAB software. Pourmostaghimi et al. applied them to determine the geometry of a pure involute of a cylindrical helical gear. The 3D scanning provides data required for the definition of the objective function, and an optimal result is achieved by processing. This technique requires high accuracy and the authors have considered the cost function's minimum value as a measure of resolution in searching for the optimum within the all-possible answers. By analyzing performance in achieving convergence and providing solutions, it was found that GWO was faster. Error-wise, PSO and WOA exceeded the threshold value whereas GA showed a pronounced weakness. Even when algorithms were expected to respect predefined error criteria, GWO still shadowed the others. A Shapiro–Wilk test was performed which revealed that the zero hypothesis was rejected. In all scenarios, swarm-based optimization methods claimed efficiency as opposed to others such as GA (Pourmostaghimi et al., 2023).

Che et al. proposed a method that improves the efficiency of milling tools using reverse engineering when they analyzed the entire working process. After obtaining a 3D model of the MX97 blade milling tool by laser scanning, they introduced a full-size virtual model for simulation and compared results. The Johnson–Cook damage criteria were used to evaluate failure strain as well as compressive stress by considering several ideal conditions such as homogeneity of the material, the toll was set as rigid and only two Degrees of Freedom, DOFs were assumed, vertical and rotational motion of the tool as it cuts into the material by milling. The simulation was replicated in an experimental setup as the data acquisition module recorded the rotating speed, torque, and pressure. Due to the abrasive nature of the cutter, different circular stress plots were obtained in both cases. The authors verified results and came up with a 9.9% error within five groups of repeated experiments, making their method reliable enough for general purpose assessments. By recreating the geometry of the milling tool that underwent a series of machining periods and accumulated wear and abrasive grains on its surface, the authors gained access to stress and strain distribution plots in various time steps that otherwise would be very difficult to visualize. Moreover, they were able to predict failure under the form of cracks which in real life are produced by easy skid of the milling tool during machining and the accumulation of chips around the cutting edges. By using systematic analysis, the authors assessed tool efficiency depending on the index of working safety, thus making it possible to avoid high torques by adjusting the other working parameters and improving the overall values. This would not have been possible if the milling tool had not been engineered and instead a 3D ideal model would have been used (Che et al., 2021).

Researchers are striving to predict deformation that appears during extensive usage in harsh environments, deformation caused by sudden changes in temperature, or deformation caused by additional procedures such as welding. Liu et al. compared results obtained using finite element analysis (FEA) of a deformed continuous spot-welded panel after extensive usage and the original, being able to assess residual stress and deformation at the nugget level. Being able to properly represent electro-thermal contact conditions is a very tricky thing to do in FEM. Some have previously tried to introduce a surface between the electrode and the sheet of metal to emulate the electrical contact resistance. Others disregarded material-dependent temperature variation and allowed mismatches of interface elements, thus obtaining a nugget morphology that allowed them to address areas affected by heat. The authors used reverse engineering to find predominant factors for deformation. When they retrieved results, it was possible to determine causes and effects, and by that they were able to further study the development of deformation. The results were compared against real-life measurements. Simulations considered clamping pressure, electrode force, and spot welding sequence that influence the deformation process. Reverse engineering had to provide dense meshes around the nuggets that represent the spot welding process. Furthermore, the representation of electrodes and clamps is very difficult to introduce in a finite element-destined model because it severely affects the computational effort. From a designated software solution a suitable material was proposed that would encompass electrical, mechanical, and thermal parameters needed for coupled simulation. Simulation achieves warping as well as leftward and downward deformation as is the case in real life. Without having an accurate representation of nuggets, the simulation would have been purely theoretical based on a CAD design rather than the real thing. Thus, we can see how using reverse engineering, scientists are now ready to determine residual deformation within welded structures and compare results with the ones from experimental testing. In such a manner, software can integrate their methods to accurately predict the life cycles of products and offer solutions to most known defects that would appear in usage (Liu et al., 2020).

Lazzarini et al. also studied the importance of the FEM when assessing a reversed-engineered 3D model. The aim was to optimize a motocross knee brace to prevent injuries. Reverse engineering was employed based on scans of the part and limb. The limb underwent additional reconstruction; then both were subjected to impact. Bone and tissue were left aside but the 3D model was further refined and simplified in Space Claim. Results show a decrease in displacement when using the bracket but the connection element on the femoral side takes the impact a bit harder in terms of pressure distribution in a 1 millisecond time frame. This is another example of how reverse engineering can successfully be paired with FEM to assess or predict the failure of product design. Analyzing insertion points where the hyperextension most likely will occur, authors have presumed a 37% decrease in distance among them which demonstrates the effectiveness of the protective bracket. Stress distribution revealed the most affected areas and could help in future re-design of the product to overcome it (Lazzarini et al., 2023).

Hwang et al. proposed an enhanced method for the prediction of lamina properties using reverse engineering and FEM by assessing their elastic modulus. When running simulations, it was found that tensile environments alongside modal will predict more accurately the designated values than tensile on its own. Findings showed an abrupt decrease in maximum prediction error. The problem is that composites exhibit highly anisotropic behavior which is very difficult to assess and simulate, not to mention the way they are stacked or the variation of the lamination angle in a sequence. Usually, further tests such as burnout for post-resin combustion and optical inspections are performed. FEM accounted for the anisotropic nature of fiber and performed modal analysis for vibration instances in multiple directions. However, an ideal study environment was employed. Thus, the authors proved that reverse engineering can be applied at a theoretical level also by inverse calculation (Hwang et al., 2023).

When they considered reverse engineering a mechanical rotating shaft, Engel B. and Hassan Al-Maeeni S.S had to take into account the load fluctuation, vibrations, and combined solicitations such as bending or torsion which can affect its structural integrity and functionality. By extracting valuable original data design information, one can establish inspection procedures that prolong the life cycle of that product by use of FEM and experimental methods. In this manner, reverse engineering opens up a new set of opportunities for manufacturers that now can look at more manufacturing technologies such as CAM by CNC or additive printing with metal filaments or powders. Strictly related to the case study of Engel and Hassan Al-Maeeni, they proved that manufacturing of a part was possible by retrofitting. It was first measured the surface roughness and afterward, through

FEA was set a safety factor. Analyses focused on yield values for stress when loadings were the highest. Microscopy revealed the type of original machining that the part was originally subjected to. Most surfaces retained their roughness values but there were areas where changes were observed. After 3D non-contact scanning was performed, the digitized model was aligned by bets-fit method. After this step, colored maps were possible to be projected which included deviation data from the original design. FEM results were plotted for bending, von Mises stresses, and shear stresses. By experimentally testing the original shaft under loading and retrieving stresses before reuse, the authors were able to assess deformation in that particular time frame that the shaft worked on the machine. Their approach highlighted that while the shaft still behaves in the required parameters, it does suffer from wear and pitting, making it a valuable candidate for re-carburization and surface re-harden (Engel and Hassan Al-Maeeni, 2019).

Reverse engineering encapsulates a wide range of methods and technologies that aim to recreate or predict failure, one of them being a combination of non-destructive testing and FEMs. Katunin et al. studied the possibility of evaluating residual stress in composite parts that were impacted at low velocities. Damage and even failure of such parts are difficult to assess. Pre-damaged structures were tested ultrasonically and using non-destructive techniques. By understanding the degree of damage inside a part and its distribution pattern, one may propose correction which would be at the base of an accurate prediction of life cycles that are left until failure. Numerical results may be plotted as maps for failure stages and show delamination areas and cracks. Thresholds may be imposed to find yield limits and excess stress. By overlapping maps retrieved from tests, the authors were able to flatten all 3D images and thus highlight the most affected areas after impact. When compared to the standard part design in a CAD environment, the true scale of damage could be observed by converting delamination into extrusions. The procedure may be then parameterized (Katunin et al., 2022).

Wei Zhou and Chao Ke proposed a method for the process of remanufacturing scheme design (RSD). High demand requires shorter design cycles in a market where there are a lot of faulty products that lead to scrapes and even disposable products. The restorations are reverse engineering based and mostly involve remanufacturing. Products obtained by this method undergo a series of tests but, more importantly, tools like case-based reasoning (CBR) showcase best matches while grey relational analysis (GRA) algorithm helps in finding similarities for the proposed cases. Using laser cladding a broken part (i.e., gear) was quickly restored and put back in service. The process involved the used model of the gear to be introduced in a software solution that could accurately predict the processing path based on node coordinates and volume. MATLAB was able to predict above 0.9 similarity for one of the analyzed cases, thus proofing that the products almost match perfectly. Further operations involved sandblasting and high-frequency hardening which has prolonged the service life of the remanufactured part—a critical aspect in inherent remanufacturing demand (IRD). By combining RSD and

IRD, the authors demonstrated the key benefits of reverse engineering. They aim to achieve a mass customized-based RSD for classes of products (Zhou et al., 2022).

Raza et al. investigated the way white light–based scanners reduce scanning error. This is a new generation of scanners apart from those that use blue light and suffer from an array of optical faults that affect the accuracy of acquired point clouds. Various methods are proposed to deal with such issues, one of them being the use of spray which coats the assembly and/or component to be scanned, thus reducing reflections. In many industries, laser-based scanners are considered above structured light ones in terms of efficiency and accuracy. However, the authors set to prove that LEDs are as accurate as any. Most equipment comes with its software solution which also requires high-performance computers (HPC). Of course, there are a few dedicated software that address reverse engineering challenges in post-processing, which can be summed up as follows: water tightness, simplification, smoothness, and sharpness. Not surprisingly, the authors found that the usual 3D scanning allows an unnecessary number of errors when set for smoothness or sharpness with or without texture. That suggests that for straightforward simple parts, it would be much faster to re-create the model in CAD than handling a large number of points from a cloud. This is not the case for intricate parts or assemblies which benefit the most from reverse engineering. The main issue though remains the volumetric error which can cause dimensional misleads. If options like sharpness or water tightness are on, they will influence the accuracy of scanning to a greater degree than in the case of the non-texture modes. ANOVA revealed that in some cases samples gained water tightness combined with a simplification of up to 70%. The authors concluded that combined effects of 3D scanning factors lead to pronounced errors (Raza et al., 2023).

Helle and Lemu offered a new perspective by showcasing a case study on the use of 3D scanning in quality control. 3D scans can capture intricate enough details that may be specific to manufacturing errors or at the end of life use–related errors. This means that engineers can now tackle these dimensional outcomes by better predicting life cycles or by improving design. The most problematic areas are going to be those inadequately captured in the form of holes or noise in the point clouds which have to be properly reconstructed to deformation or service life. The straight-forward design usually imposes tolerance limits which are translated into the final product. By reverse engineering that product, one acquires dimensions with those manufacturing-related tolerance values without knowing what the initial nominal dimensions were. Moreover, 3D scans may include measurement errors depending on the accuracy of the equipment used, not to mention the errors that are propagated by remanufacturing the product. Depending on the technology used, accuracy of 3D scans spans from 0.00005 mm for modern contact equipment up to 1.5 mm for non-contact ones. So we see that the degree of uncertainty regarding dimensional accuracy has a great influence on the outcome of reverse engineering using 3D scanning. Surface reconstruction may be achieved either by using primitives where they fit or by approximating using NURBS in the area in question. Both are expected to incorporate errors but a hybrid method between the two tends to minimize the values recorded for standard deviations. Another significant aspect of error management is related to the post-processing of manufactured parts like surface treatment (Helle et al., 2021).

Aviation and automotive industry represent domains where reverse engineering is most effective. Continuous innovation drives the main manufacturers to find a better life span for ways to reuse multiple intricate parts made of expensive materials. Rozesara et al. studied multiple case studies that show an immutable bond between the ways production systems react to changes driven by innovation. It stands at the very basis of research and development (R&D) and can even be applied to forward engineering. Cases presented in the paper show that innovation occurs even at the latest stages of design or at the end of a product's life cycle because new technologies make it possible for it to have an extended time in service or by re-design can become a whole new part. By moving back and forth between new technologies included under the umbrella of reverse engineering, R&D now stands out as a mandatory procedure (Rozesara et al., 2023).

Circular economy relies on reverse engineering and additive manufacturing (AM) to fulfill its purpose. Chirita et al. have highlighted the importance of the aforementioned technologies and techniques in prolonged sustainability and the use of resources efficiently and in environmentally friendly manner. Waste management plays a huge role in the circular economy concept and it may be improved by remanufacturing or repurposing used items. Another key benefit is represented by the ability to extend the service life of a product by remanufacture. Scanning methods are mostly solid feature–based and surface feature–based. The choice of one of them is important because it directly affects acquired data size and management. Of course, whenever needed, there is always the possibility to add another set to better fit a specific feature or to approximate a faulty one. 3D printing plays its part in green manufacturing because it uses recyclable or eco-friendly materials. That can translate into reduced energy consumption. By optimizing printing parameters, various strategies may be considered that can contribute to emissions diminishment while producing with little waste. By scaling reverse engineering paired with additive manufacturing, we might address big environmental issues as the spectrum of applications widens with technology advances (Chirita et al., 2023).

Sustainability has been the front-runner for many years now, being mandatory in most modern production processes. A study carried out by Saiga et al. proposes a procedure for reverse engineering in a sustainable manner in five steps. For relatively simple objects it is possible to use standardized features from software libraries which would speed up the process of reconstruction and matching. The last step produces a valid 3D-printed prototype. Waste reduction can be achieved through multiple methods some of which are related to prevention and recycling. The reused parts obtained using reverse engineering are more and more favored as they are not only similar and sometimes better in terms of functionality but they also greatly contribute to the reduction of waste impact and CO_2 footprint in the atmosphere. Prediction of product life cycles implies the fact that many times there is going to be a waste of valuable resources such as time and energy, not to mention the amount of material that goes into trial and error. Digitalization has managed to overcome most of these drawbacks by simulation. However, there are still flaws in achieving a valid 3D replica of the scanned object which can produce inaccurate results and directly affect the system's efficiency. In particular, accurate representation of an intricate feature represents the primary goal of any scanning procedure. Many times, depending on the equipment used as well as the software, feature extraction from a large number of point clouds is tedious and very specialized work as it often involves advanced skills that have to do with remodeling and patching missing areas. Some software does it better than others and can even predict effects such as spring back effects in the case of car doors or even correct skewed ones by approximating the original shape of scanned models. Some researchers have proposed methods that can remove as-cast and machined surfaces from the point cloud, thus easing the computational effort of transformation to a solid (Saiga et al., 2021). Others have dealt with welding quality control by means of reverse engineering. By considering weighted neighbors, it created an algorithm that deletes points that are specific to a welded part from the point clouds but that just add up to an already significant computational effort. Nowadays AI plays a huge role in feature recognition by means of machine learning. The iterations are carried out in a cloud and take into consideration multiple sources. In this way, solids are now possible to obtain even from plain sketches by using a rule for topological recognition applied to edges and vertices. By understanding the relationships within the drawing, AI systems can generate full models at the same dimensions and tolerance values as the ones contained in the sketch. This technology is now available for certain CAD systems as installation packages. Other researchers have gone a bit further and created their own set of point clouds by analytical means free of scanning procedures, and by doing that they have reduced redundancy and noise, two of the biggest problems with 3D scanning up-to-date. But they still had to deal with misalignments and hidden points to produce a clean, refined point cloud ready for accurate meshing. This shows the potential of new technologies that lie within reverse engineering (Saiga et al., 2021).

Zhang et al. addressed the effect of disruptive innovation led by reverse engineering. In the case of emerging markets, it represents a valuable tool for innovating alongside giants in the fields. The new entries in a domain must catch up by aligning their strategies as much as possible at the firm level with the bigger ones in terms of their technological capabilities. Frameworks were proposed as most studies suggest that knowledge could be retrieved from steady and reliable informational sources worldwide. But a mesh of license agreements, paid services, and literature that limit access, does make it hard for some entities to compete. Private companies or research laboratories have the advantage of big finances, strategic partnerships, alliances, and logistics that are just unattainable to newcomers. But the balance has shifted with the introduction of reverse engineering which could provide models that proved their value in time to the ones with no access to initial design or technical data whatsoever. "*Reverse engineering is often defined as extracting declarative and procedural product knowledge from the process of disassembling the final product into observable technological units and conducting detailed analysis, usually to construct similar but different or improved products*" (Zhang et al., 2023). Thus, the learning curve from design to prototype is greatly reduced to the understanding of various components' functionality and assessment of their current status. But trial and error are just skipped as it is only assesses the final product which is the pinnacle of success after numerous fails and re-design situations. Using reverse engineering, firms are now able to scale their business and get a fair share from industries that otherwise would be very difficult to operate within. By investing in reverse engineering tools and by applying their methods, newcomers may gain knowledge and propose competitive products in a short amount of time by innovation or re-interpretation of other proofed designs from the market. The next step is to individualize the innovation process on its own toward independence. Thus, disruptive innovation may be implemented which can facilitate access to mature technologies at low costs. But that works only for a few short periods because competitiveness cannot be sustained just by acquiring knowledge using reverse engineering. There is a certain degree of impact over knowledge accumulation but that has to be enhanced from different sources and especially through R&D activities carried out inside the firm. But the benefits are undeniable as now firms can accelerate their learning curves by having access to valuable data (Zhang et al., 2023). There are governmental programs that encourage firms to use external technologies and incorporate them into their products to gain more significant market shares and even tax reductions. Universities play a huge role in filtering the knowledge gained by reverse engineering. They must offer feedback by proofing designs and methods. And there is, of course, the dysfunctional competition that allows certain entities to gain important knowledge without playing by the rules or facing legal penalties. There are copyright infringements, patents overlooked, legal concerns, and others, which some competitors tend to fight back by classifying under business secrecy. However, the short-term gains cannot be fought where there are no government penalty predictions or a strong law enforcement environment. It was found that mistakes are often introduced by misinterpretation or design overlooks that are translated into faulty products offered at very low prices. 3D printing has done nothing but deepen this phenomenon by the use of different chemicals in percentages way beyond safety limits for certain filaments. But the real concern lies in mass production where hazardous substances are used for low-quality products destined for public use, that resemble more expensive ones. Health issues are the frontrunners of the endeavors to have a sustainable framework that would standardize activities contained within reverse engineering. Long-term development should not rely only on external knowledge using reverse engineering but should also focus on digestion, absorption, and integration on their way to significant and innovative products.

5. Future Trends in Reverse Engineering

Researchers and generally people involved in various industries praise the benefits of Industry 4.0. They consider it to be a game changer if we are referring to AI, robots and their role in automation, machine learning models which now are part of so-called smart systems, and let us not forget about the internet of things (IoT). All these were merged and are intricate parts of the concept as they evolve into autonomous systems that can interact with each other. But now there are already talks about Industry 5.0. According to the European Union, Industry 5.0 “provides a vision of an industry that aims beyond efficiency and productivity as the sole goals, and reinforces the role and the contribution of industry to society” (European Commission, 2024). That means that a step further is being taken with a focus on the human operator whose well-being should be considered throughout the manufacturing stages and product life cycles as another desiderate besides sustainable production technologies, waste management, and integration of more green energy consumption. This type of emphasis aligns with global efforts for preserving planetary resources and reuse whenever possible but contravenes with profit-driven businesses whose sole purpose is to maximize their presence in certain markets and achieve global supremacy. The European Union identified three pillars of Industry 5.0: resilience, sustainability, and human-centered, where the third one promotes “talents, diversity and empowerment;” the resilient strategy addresses the influence of companies that are driven to profit optimization by organic growth over flexibility, making them more likely to succumb in crisis rather than anticipate it; and last but not least, sustainability that has to be kept in mind in all actions because it is not sufficient to reduce environmental footprints but there also must be a positive impact. It is yet too early to quantify if the concept will be or not embraced. Companies are expected to rely more and more on self-learning systems as robotics are widely used for maintenance and monitoring tasks. The use of augmented reality will make operations and even processes interactive and intuitive. This builds up a new ecosystem that will grow with every new interaction as automation, AI, and intelligent manufacturing are more present than ever (Barataa et al., 2023). The transition between industries would bring numerous benefits to humans and will have a significant impact on the use of resources toward a healthier planet. Of course, reverse engineering will become a self-sustainable means of knowledge accumulation and innovation.

References

3D SYSTEMS. 2024. Software. Available at https://www.3dsystems.com/software (accessed on 24.08.2024).

3DS. 2024. CATIA V5, Driving Design Excellence, Dassault Systèmes. Available at https://www.3ds.com/products/catia/catia-v5 (accessed on 24.08.2024).

Agisoft. 2024. Discover Intelligent Photogrammetry with Metashape. Available at https://www.agisoft.com/ (accessed on 20.08.2024).

Artec 3D. 2024. 3D Scanning Solutions, Artec Studio 18. Available at https://www.artec3d.com/3d-software/artec-studio (accessed on 24.08.2024).

AUTODESK, 2024a. Autodesk Fusion 360 for Personal Use, Autodesk Inc. Available at https://www.autodesk.com/products/fusion-360/personal (accessed on 25.08.2024).

AUTODESK, 2024b. Autodesk ReCap Pro: Turn the Physical into Digital, Autodesk Inc. Available at https://www.autodesk.com/eu/products/recap/overview?term=1-YEAR&tab=subscription (accessed on 23.08.2024).

Barataa, J. and Kayser, I. 2023. Industry 5.0—Past, Present, and Near Future. Procedia Computer Science 219: 778–788.

Che, J., Zhang, Y., Wang, H., Liu, Y., Du, M., Ma, S., Zhao, Y. and Suo, C. 2021. A Novel Method for Analyzing Working Performance of Milling Tools based on Reverse Engineering. *Journal of Petroleum Science and Engineering*, 197: 107987.

Chirita, A.P., Bors, A.M., Radoi, R.I., Dumitrescu, I.C. and Popescu, A.M.C. 2023. Leveraging Additive Manufacturing and Reverse Engineering for Circular Economy-driven Remanufacturing of Hydraulic Drive System Components. *Applied Sciences* 13(22): 12200.

CREAFORM/AMETEK. 2024. Portable 3D Measurement Solutions, HandySCAN 3D, BLACK Series. Available at https://www.creaform3d.com/en/portable-3d-scanner-handyscan-3d (accessed on 17.08.2024).

Creaform/Ametek, 2021. What is a CMM? Everything You Need to Know about Coordinate Measuring Machines and Their Types, Written by Creaform. Available at https://www.creaform3d.com/blog/what-is-cmm-and-their-types/ (accessed on 22.08.2024).

DASSAULT SYSTEMES. 2024. ScanTo3D Overview, Dassault Systèmes. Available at https://help.solidworks.com/2021/english/SolidWorks/scanto3d/c_Scanto3d_overview.htm (accessed on 23.08.2024).

Engel, B. and Hassan Al-Maeeni, S.S. 2019. An Integrated Reverse Engineering and Failure Analysis Approach for Recovery of Mechanical Shafts. *Procedia CIRP*, 81: 1083–1088.

European Commission. 2024. Research and Innovation, Industry 5.0, Directorate-General for Research and Innovation. Available at: https://research-and-innovation.ec.europa.eu/research-area/industrial-research-and-innovation/industry-50_en (accessed on 07.09.2024)

Fan, L., Wang, J., Xu, Z. and Yang, X. 2022. A Reverse Modeling Method Based on CAD Model Prior and Surface Modeling. *Machines* 10(10): 905.

FARO, 2024a. FARO Focus Laser Scanners. Available at https://www.faro.com/en/Products/Hardware/Focus-Laser-Scanners (accessed on 16.08.2024).

FARO, 2024b. Introducing FARO Sphere® XG. Your Digital Reality Platform, Quantum Max FaroArm® Series. Available at https://www.faro.com/en/Products/Hardware/Quantum-FaroArms (accessed on 22.08.2024).

Gauthier, S., Puech, W., Bénière, R. and Subsola, G. 2017. Analysis of Digitized 3D Mesh Curvature Histograms for Reverse Engineering. *Computers in Industry*, 92–93: 67–83.

Helle, R.H. and Lemu, H.G. 2021. A Case Study on Use of 3D Scanning for Reverse Engineering and Quality Control. *Materials Today: Proceedings* 45: 5255–5262.

HEXAGON, 2024a. Gantry CMMs, Hexagon AB and/or Its Subsidiaries. Available at https://hexagon.com/products/product-groups/measurement-inspection-hardware/coordinate-measuring-machines/gantry-cmms (accessed on 22.08.2024).

HEXAGON, 2024b. ROMER is Now Hexagon, Hexagon AB and/or Its Subsidiaries. Available at https://hexagon.com/company/divisions/manufacturing-intelligence/romer (accessed on 22.08.2024).

Hwang, M.-Y., Park, J.H., Song, J., Hwang, S.W., Kang, H.H. and Lee, H.C. 2023. Enhanced Reverse-engineering Method for Accurately Predicting Lamina Properties in Laminated Composites via Combined Static and Dynamic Finite Element Simulations. *Journal of Composites Science* 7(12): 518.

INNOVMETRIC. 2024. PolyWorks Inspector, InnovMetric Software Inc. Available at https://www.innovmetric.com/products/polyworks-inspector (accessed on 24.08.2024).

Inspection Engineering. 2024. Steinbichler. Available at https://inspectionengineering.com/steinbichler/ (accessed on 21.08.2024).

Katunin, A., Wronkowicz-Katunin, A., Danek, W., Dragan, K. and Wyleżoł, M. 2022. A Reverse Engineering Approach for Modeling of Barely Visible Impact Damage by Combining Results of Non-destructive Testing and Numerical Simulations. *Procedia Structural Integrity* 37: 195–202.

Kim, S. 2023. Generating a Virtual Physical Model through Measurement Data and Reverse Engineering: Applying a Performance Prediction Model for an Industrial Gas Turbine during Start-up. *Applied Thermal Engineering*, 232: 120927.

Kyaw, A.C., Nagengast, N., Usma-Mansfield, C. and Fuss, F.K. 2023. A Combined Reverse Engineering and Multi-criteria Decision-making Approach for Remanufacturing a Classic Car Part. *Procedia CIRP*, 119: 222–228.

Lazzarini, L., Civera, M., Burgio, V., Rodriguez Reinoso, M., Antonaci, P. and Surace, C. 2023. Analysis of a Motocross Knee Brace: From the Real Model to the Numerical Finite Element Model via 3D Scanning and Reverse Engineering. *Applied Sciences* 13(8): 5186.

Liu, X., Wei, Y., Wu, H. and Zhang, T. 2020. Factor Analysis of Deformation in Resistance Spot Welding of Complex Steel Sheets based on Reverse Engineering Technology and Direct Finite Element Analysis. *Journal of Manufacturing Processes* 57: 72–90.

Mesh2surface. 2024. Convert Your 3D Scan to CAD Models with Mesh2Surface. Available at https://www.mesh2surface.com/ (accessed on 25.08.2024).

MeshLab. 2024. Available at https://www.meshlab.net/ (accessed on 24.08.2024).

Nikon. 2024. X-Ray and CT Metrology Solutions, Nikon Metrology NV. Available at https://industry.nikon.com/en-gb/products/x-ray-ct/ (accessed on 18.08.2024).

OQTON. 2024. Geomagic Design X. Available at https://oqton.com/geomagic-designx/ (accessed on 23.08.2024).

Pourmostaghimi, V., Heidari, F., Khalilpourazary, S. and Qazani, M.R.C. 2023. Application of Evolutionary Optimization Techniques in Reverse Engineering of Helical Gears: An Applied Study. *Axioms* 12(3): 252.

Qie, Y., Bickel, S., Wartzack, S., Schleich, B. and Anwer, N. 2021. A Function-oriented Surface Reconstruction Framework for Reverse Engineering. *CIRP Annals* 70: 135–138.

Raza, S.F., Amjad, M., Ishfaq, K., Ahmad, S. and Abdollahian, M. 2023. Effect of Three-dimensional (3D) Scanning Factors on Minimizing the Scanning Errors Using a White LED Light 3D Scanner. *Applied Sciences* 13(5): 3303.

RhinoReverse. 2024. Rhinoceros©. Available at https://rhinoreverse.icapp.ch/english/ (accessed on 25.08.2024).

Rozesara, M., Ghazinoori, S., Manteghi, M. and Tabatabaeian, S.H. 2023. 'A Reverse Engineering-based Model for Innovation Process in Complex Product Systems: Multiple Case Studies in the Aviation Industry. *Journal of Engineering and Technology Management* 69: 101765.

Saiga, K., Ullah, AMM S., Kubo A. and Tashi. 2021. A Sustainable Reverse Engineering Process. *Procedia CIRP* 98: 517–522.

Siemens. 2024. NX CAD Industrial Design Software. Available at https://plm.sw.siemens.com/en-US/nx/cad-online/mcad-software/industrial-product-design/ (accessed on 24.08.2024).

Suzuki, S., Ohtake, Y. and Suzuki, H. 2022. Curvature Gradient-estimation Using CT Sinogram and Its Application to Reverse Engineering. *Computer-Aided Design* 148: 103240.

TAYLOR-HOBSON/AMETEK. 2024. Form Talysurf® i-Series PRO Range, AMETEK.Inc. Available at https://www.taylor-hobson.com/products/surface-profilers/i-series-pro-range/talysurf-pro (accessed on 22.08.2024).

THORLABS. 2024. OCT Imaging Systems & Components, Thorlabs, Inc. Available at https://www.thorlabs.com/navigation.cfm?guide_id=2039 (accessed on 21.08.2024).

Trimble Geospatial. 2024. Laser Scanning, Terrestrial 3D Laser Scanning and Imaging Systems, Trimble Inc. Available at https://geospatial.trimble.com/en/products/hardware/laser-scanning (accessed on 21.08.2024).

ZEISS, 2024a. Industrial Quality Solutions, 3D Scanning with ATOS. Available at https://www.zeiss.com/metrology/en/systems/optical-3d/3d-scanning.html (accessed on 17.08.2024).

ZEISS, 2024b. Industrial Quality Solutions, Bridge-type Coordinate Measuring Machines. Available at https://www.zeiss.com/metrology/en/systems/cmms/bridge-type-cmms.html (accessed on 22.08.2024).

Zhang, F., Zhu, L., Xu, Z. and Wu, Y. 2023. Moving from Reverse Engineering to Disruptive Innovation in Emerging Markets: The Importance of Knowledge Creation. *Technovation* 125: 102791.

Zhang, Y. and Sun, J. 2023. A Phase Retrieval Method for 3D Shape Measurement of High-reflectivity Surface based on Π Phase-Shifting Fringes. *Sensors* 23(21): 8848.

Zhou, W. and Ke, C. 2022. A Mass-customization-based Remanufacturing Scheme Design Method for Used Products. *Sustainability* 14(16): 10059.

3 A Brief Review of Additive Manufacturing Technologies

Ratnadeepsinh Jadeja[1] and Marmik Dave[2]

1. Introduction

A collection of additive manufacturing (AM) technologies, sometimes called 3D printing, constructs 3D items using digital models or computer-aided design (CAD) data (Gibson et al., 2010). In contrast to conventional subtractive manufacturing techniques, which entail shaping or cutting material to form a product, AM builds the finished object by deposition of a thin film of the material. The computerized slicing of the CAD model into tiny cross-sectional layers and the successive deposition or solidification of material to produce each layer are the fundamental ideas of AM. In AM procedures, a wide range of technologies and materials are utilized, such as composites, metals, polymers, and ceramics. The capacity to construct complicated geometries, quick prototyping, personalization, and less material waste are just a few benefits of AM. This technology is revolutionizing the design, prototyping, and manufacturing of items across a variety of sectors, including consumer goods, healthcare, automotive, and aerospace.

The revolutionary technology of 3D printing is clearly going to have a major impact on contemporary manufacturing. Some of the significant outlines of its contributions to sustainability, efficiency, and innovation across a wide range of industries are:

- *Innovation and Design Flexibility:* Incredibly intricate forms, which would be impossible or extremely difficult to manufacture using traditional methods, are within the realm of possibility using AM. This leeway in design promotes innovation in product creation and allows for the creation of items and components with optimal performance (Gibson et al., 2010).
- *Rapid Prototyping and Time-to-Market:* AM facilitates fast prototyping, which shortens the time it takes to build a product. Because prototypes can be made rapidly and iteratively, designers and engineers may test and improve concepts at a lower cost (Chua et al., 2010).
- *Customization and Personalization:* Customized and tailored products that meet the demands of individuals can be produced thanks to AM. This is

[1] *PhD Scholar*, School of Engineering, RK University.
[2] *Project Manager*, Yantra Innovations.
E-mail: marmik.dave44@gmail.com
* Corresponding author: ratnadeepsinh.jadeja@rku.ac.in

especially important for consumer goods, and automotive and healthcare industries (customized implants, customized accessories, and customized components) (Ligon et al., 2017).

- *Complex Assemblies and Lightweight Structures:* AM makes it easier to create intricate assemblies and lightweight structures by combining several parts into a single, cohesive design. By doing this, the product performs better overall and uses less material (Kruth et al., 1998).
- *Sustainability and Waste Reduction:* Compared to subtractive manufacturing, AM technologies minimize material waste due to their ability to add material layer by layer. This is in line with the increasing focus on environmentally friendly production methods (Murr et al., 2012).

The purpose of this chapter is to showcase different AM technologies, materials, and applications.

1.1 Historical Development

When numerical control (NC) machines were introduced in the 1950s and 1960s, the seeds of AM were sown. Initial computer-based systems established the foundation of digital industrial process control. In the 1970s, stereolithography (SLA), a crucial AM process, became conceptualized. A method that used ultraviolet (UV) light to harden a photopolymer, a forerunner to SLA, was patented in 1981 by Dr Kodama of Japan (Kodama, 1981). A significant turning point in AM progress was the 1986 patent of the SLA equipment by C.W. Hull. The commercialization of 3D printing technologies occurred in the 1990s. Chuck Hull founded 3D Systems, and Stratasys was among the companies that brought out the first commercial 3D printers for product development and prototyping. The 2000s saw advancements including the development of new AM technologies, such as selective laser sintering (SLS), and the expansion of materials beyond polymers to include ceramics and metals (Gibson et al., 2010). With the popularity of desktop 3D printers, 3D printing became more accessible in the 2010s. A community-driven approach was fostered by the RepRap project, which was started in 2005 and resulted in the production of inexpensive open-source 3D printers (Amanda and Al'Aref, 2018). The future will see the continued development of AM as things are incorporated into conventional manufacturing procedures. It helps usher in a fresh period of enhanced production through its participation in Industry 4.0 and digital manufacturing methods.

2. Additive Manufacturing Processes

A variety of 3D printing methods have been created, each with its own unique function. The following methods are classified as 3D printing technologies according to American Society for Testing and Materials, ASTM Standard F2792: material extrusion, sheet lamination, powder bed fusion, binder jetting, and vat photopolymerization. Each piece of technology has a certain purpose, so there is no use in arguing over which is better. The use of 3D printing technology to create a wide range of products, not only prototypes, is on the rise at the moment (Yuanbin et al., 2017).

2.1 Sheet Lamination

According to ASTM, sheet lamination is the process of fusing together several sheets of material to form a single component (Silbernagel, 2018). Laminated object manufacturing (LOM) and ultrasonic additive manufacturing (UAM) are two processes that use sheet lamination (Syed et al., 2018). This approach has several advantages, such as being able to print in full color, being inexpensive, being easy to work with, and having the ability to recycle surplus material by sheet lamination. Complex geometric parts may be mass produced more quickly and at lower fabrication costs using LOM (Vijayavenkataraman et al., 2017). An innovative method known as ultrasound-assisted metal fusion (UAMF) combines metal layers of thin film derived from featureless foil stock using sound.

2.2 Vat Photopolymerization

The main 3D printing process that is often used is photopolymerization, which basically involves curing photo-reactive polymers using a laser, light, or UV radiation (Low et al., 2017). Techs that use photopolymerization for 3D printing include digital light processing (DLP) and SLA. The photoinitiator, the conditions of irradiation, and the presence of any extra UV absorbers, including pigments or dyes, all have a role in determining the SLA (Stansbury and Idacavage, 2016). Meanwhile, DLP is similar to SLA in that it uses photopolymers. The main difference is the illumination. DLP makes use of a more conventional light source, such as liquid crystal display screen atop an arc lamp. Since it just needs a single pass to cover the whole surface of the photopolymer resin vat, it is often quicker than SLA (Chaudhari et al., 2022). Critical variables in Vat photopolymerization are exposure time, wavelength, and power source. Materials in a liquid state will solidify when exposed to UV radiation. Producing high-end goods with precise geometric information and a smooth surface is a good fit for photopolymerization (Muller and Karevska, 2016).

2.3 Powder Bed Fusion

The powder bed fusion technique makes use of two procedures: electron beam melting printing (EBM) and SLS. Using a laser or electron beam, this method melts or fuses the powdered substance. Various materials are used in this process, including metals, ceramics, polymers, composites, and hybrids. Among AM systems that use powders, SLS is by far the most prevalent. SLS is a fast, precise, and customizable 3D printing method that allows for a variety of surface treatments (Tiwari et al., 2015). Objects made of ceramic, plastic, or metal may be created with SLS (Ventola, 2014). Powdered polymers are sintered with a powerful laser in SLS to produce an item or part. The energy source used to heat the material is improved by EBM (Tumbleston, 2015).

2.4 Binder Jetting

Binder jetting is a rapid prototyping and 3D printing method that uses the selective deposition of a liquid binding agent to combine powder particles. A chemical binder is sprayed over the scattered powder using a binder jetting technique to produce the layer. Casting patterns, incomplete sintered objects, and similar large-scale sand products are produced via binder jetting. Binder jet printing is possible with a wide variety of materials, including metals, sands, polymers, hybrids, and ceramics. Some materials, such as sand, do not need any further processing. Additionally, binder jetting is a simple, fast, and cheap process due to the bound powder particles. Binder jetting has finally made it feasible to print very large objects (Ze-Xian et al., 2017).

2.5 Material Extrusion

A variety of colors and materials may be printed utilizing extrusion-based AM techniques on plastics, food, and even living cells (Muller and Karevska, 2016). This method has seen a lot of usage, and it is not costly either. A further benefit of this process is that it may make fully operational product parts (Syed et al., 2018). Extrusion of materials may be seen in processes like fused deposition modeling (FDM). The process of FDM involves the layer-by-layer extrusion of thermoplastic filaments; however, it is not a stereolithographic method (Yee et al., 2017). Nonetheless, the idea of material deposition is in line with the concepts of AM. Fast prototyping and the manufacturing of functioning parts are two areas where FDM is widely used and acknowledged for its versatility (Stansbury and Idacavage, 2016).

2.6 Directed Energy Deposition

Directed energy deposition is a more complex printing technique that is often used to repair or enhance an existing object with new material (Syed et al., 2018). In terms of product quality and controllability over grain structure, directed energy deposition may be a game-changer. While directed energy deposition allows the nozzle to move in several directions rather than being fixed to just one, the process is conceptually similar to material extrusion. Although the process is most often used with metals in wire or powder form, it may also be used with ceramics and polymers. One instance of this technique is laser deposition (Syed et al., 2018). A new technique called laser deposition can produce or fix items having measurements that range from millimeters to meters. The laser deposition technique is gaining popularity due to its ability to provide scalability and several capabilities in one system (Lang, 2017).

3. Materials for Additive Manufacturing

The AM procedures can employ a wide range of materials, which adds to the adaptability and broad use of this revolutionary technology (Shahrubudina et al.,

2019). Examining materials through different AM methods exposes a variety of materials, such as ceramics, composites, metals, polymers, and biomaterials.

3.1 Ceramics

With 3D printing technology, objects or structures can be created using concrete and ceramics that have small pores or no cracks by setting up the right mechanical qualities (Owen et al., 2018). Ceramic is resistant to fire, durable, and long-lasting. Because they remain liquid until setting, ceramics may be molded into almost any form or geometry, making them perfect for building projects (Baldassarre and Ricciardi, 2017). Ceramics find use in aerospace and dental industries (Owen et al., 2018). Ceramics include substances such as alumina (Owen et al., 2018), bioactive glasses (Gmeiner et al., 2015), and zirconia (Yanko et al., 2018). Alumina powder is processed and used in 3D printing technology and finds extensive use in chemical, microelectronics, adsorbent, catalyst, and aerospace industries among other high-tech industries (Owen et al., 2018). Alumina has a very difficult curing process. Additionally, in a subsequent experiment, glass-ceramic and bioactive glass were processed into dense parts using an SLA machine. It greatly increases bending strength (Gmeiner et al., 2015). Increasing the mechanical strength will allow bioactive glass in clinically relevant structures like bone and scaffolds (Yanko et al., 2018).

3.2 Metals

Across many industries, the production of complex and long-lasting components has been transformed by metal AM (Horst et al., 2018). The commonly used materials include Inconel (Hitzler et al., 2018), titanium (Trevisan et al., 2018), aluminum and its alloy (Martin et al., 2017), and stainless steel (Murr et al., 2012). Each has unique attributes that make it appropriate for a particular use. Because of its corrosion resistance property, stainless steel is frequently employed in the production of specialized tools and functional prototypes (Ashby and Cebon, 2014). Titanium is used in medical implants and aircraft parts because of its excellent strength-to-weight ratio (Ashby and Cebon, 2014). Aluminum is used in the automobile industry to produce components due to its lightweight nature (Ashby and Cebon, 2014). Turbine blades for the aerospace and energy sectors are among the many applications that use Inconel (Murr 2016), which is well-known for its corrosion resistance and high-temperature tolerance (Uhlmann et al., 2015). Directed energy deposition and powder bed fusion have been essential in producing metal components with exact geometries and improved performance (DebRoy et al., 2018).

3.3 Polymers

Polymers are the building blocks of AM because they are adaptable and affordable for a variety of uses (Gibson et al., 2010). In methods like FDM (Caminero et al.,

2018) and SLA, common polymers like polylactic acid (PLA) (Caminero et al., 2018), acrylonitrile butadiene styrene (ABS), and polyethylene terephthalate glycol (PETG) are utilized extensively. Because PLA is easy to use and biodegradable, it is often used in prototyping processes. PLA is created from sustainable resources. ABS is frequently employed in the production of functional components due to its reputation for resilience to impact and strength. PETG (Dizon et al., 2018) is used in industries that need clear components due to its increased strength and transparency. These polymers can create complex designs and precise geometries (Xin et al., 2017). FDM, which extrudes thermoplastic filaments layer by layer, and SLA, which solidifies liquid photopolymer resins with UV light, are essential for prototyping and a variety of manufacturing applications (Caminero et al., 2018).

3.4 Composite

Composites are essential to the advancement of AM because they combine the benefits of various materials to produce improved qualities. Examples of composite materials commonly used in AM procedures like FDM are carbon fiber reinforced polymers (CFRP) (Haoa et al., 2018) and glass fiber reinforced polymers (GFRP) (Sathishkumar et al., 2014). With its remarkable strength-to-weight ratios, CFRP is frequently utilized in the aerospace sector to create strong, lightweight components (Haoa et al., 2018). GFRP offers better resistance to impacts and is frequently used in applications that call for increased durability (Liu et al., 2015). Due to their customized qualities, composite materials are very versatile in AM. They are therefore preferred for generating parts for various industries, including sports equipment, automotive, and aerospace (Tumbleston, 2015).

3.5 Biomaterials

Biomaterials are a new frontier in AM, opening doors to bioprinting, tissue engineering, and healthcare applications (Hutmacher, 2001). Polycaprolactone (PCL) and poly lactic-co-glycolic acid (PLGA) are two notable biomaterials used in techniques such as SLA and FDM (Liu et al., 2015). PCL is frequently utilized in drug delivery systems and medical implants because of its reputation for biocompatibility and gradual disintegration (Tumbleston, 2015). Applications needing the controlled release of medicinal substances use PLGA, a biodegradable polymer (Yigong et al., 2016). The fabrication of personalized and patient-specific implants is made possible by incorporating biomaterials into AM technology, which is revolutionizing the medical and pharmaceutical industries. These materials' inclusion, driven by biocompatibility and bioresorbability (Knowlton et al., 2015), is a prime example of how AM can transform regenerative therapies and personalized medicine (Qian et al., 2018).

4. Applications of Additive Manufacturing

In AM, material properties are vital because they affect the end products' functionality, robustness, and suitability for specific applications. For example, the great strength, heat resistance, and lightweight characteristics of metals like titanium and Inconel make them perfect for utilizing in aircraft applications (Ashby and Cebon, 2014). Polymers with low cost and high usability, such as PLA and ABS, are preferred for quick prototyping (Chua, et al., 2010). High-temperature stable alumina-based ceramics are utilized in industries, while composites like CFRP have a remarkable strength-to-weight ratio that makes them perfect for sports and aircraft equipment (Owen et al., 2018). Medical applications choose biomaterials with biocompatibility and bioresorbability (Murr 2016), such as PCL and PLGA. To guarantee the best possible performance and safety of the manufactured components, factors including mechanical qualities, thermal stability, and regulatory compliance must be carefully considered (Uhlmann et al., 2015). The choice of material is closely linked to the needs of a certain application (Yuanbin et al., 2017). The continuous development and diversification of AM across industries are facilitated by this sophisticated understanding of material properties and careful consideration of application-specific needs (Yuanbin et al., 2017).

a. The aerospace industry has embraced AM for its ability to produce lightweight, intricate, and highly efficient components. Engine components like turbine blades and fuel nozzles are designed, enabling the integration of complex geometries such as cooling channels (Mahesh et al, 2017). This not only reduces overall weight but also improves performance in gas turbines, an essential factor for modern aerospace engines (Liu and Shin 2018). AM is being increasingly used to produce lightweight structural parts such as brackets, contributing to reduced fuel consumption and increased efficiency (Frazier, 2014), also complex and precise tools can be produced faster, reducing downtime and costs in manufacturing processes (Gibson et al., 2015).
b. The automotive industry benefits from AM in various ways, from rapid prototyping to the production of customized (Goh et al., 2019), high-performance parts. It enables rapid design iterations and testing, making it possible to prototype vehicle components much faster than with traditional methods (Thompson et al., 2017).
c. Healthcare and biomedical sectors use AM significantly to manufacture custom implants that are tailored to individual patients, improving both functionality and comfort (Ventola, 2014). AM is being used to develop custom surgical guides that help improve precision during operations to enhance surgical accuracy, reduce operative time, and improve patient recovery (Rengier et al., 2010).
d. The construction and architecture industry is offering new ways to build structures more efficiently and sustainably through AM. Large-scale 3D printers are being used to construct buildings, significantly reducing construction time and material waste which can lower construction costs (Buswell et al., 2007) while allowing for greater design flexibility (Lim et al., 2012).

e. Energy and power generation industry use AM to improve efficiency in the energy sector, particularly in the design and production of components for power generation systems by manufacturing complex turbine blades for gas and wind turbines. These components benefit from weight reductions and optimized designs (Thompson et al., 2017).
f. Museums and heritage (Karg at al., 2018) institutions are using AM to create replicas of valuable artifacts for both display and educational purposes (Reiff et al., 2019).

5. Challenges in Additive Manufacturing

There are still issues in the field of AM which affect how widely it is adopted and integrated into the industrial process. Because the qualities needed for successful AM processes, including powder flowability or melt viscosity, frequently differ from typical manufacturing materials, material compatibility and attributes provide a major challenge. It is still difficult to obtain the right surface finish straight out of the printing process; this calls for post-processing procedures that can compromise efficiency and accuracy. Concerns about scaling and production speed persist, especially in large-scale manufacturing settings where the efficiency of the technology could be compromised (Ligon et al., 2017). AM technologies are often slow in comparison to traditional manufacturing processes such as injection molding or Computer Numerical Control (CNC) machining, making them less suitable for high-volume production (Wang et al., 2021). Additional obstacles arise from legal considerations and the construction of industry standards since the rapid progress of AM technology necessitates the parallel development of thorough guidelines to guarantee product safety, quality, and regulatory compliance (Kruth et al., 1998). Due to the lack of regulatory frameworks, companies hesitate to adopt technologies without clear guidelines ensuring product quality and safety (Herzog et al., 2016). Additionally, concerns related to intellectual property and digital rights management (DRM) have emerged as companies explore the potential for decentralized production using 3D printing (Despeisse et al., 2017). The ability to easily replicate designs raises issues regarding counterfeit goods and unauthorized use of proprietary designs (Khorram Niaki and Nonino, 2017).

6. Future Trends in Additive Manufacturing

Dynamic patterns that highlight ongoing development and incorporation into numerous industries characterize the technology's future. New technologies that have the potential to completely change the manufacturing landscape include bioprinting and metal 3D printing (Thompson et al., 2017). Powder bed fusion and directed energy deposition are two new metal 3D printing techniques that have recently emerged, and they provide the aerospace and medical sectors with enhanced scalability and accuracy (Xibing et al., 2014). Conversely, bioprinting has the potential to completely transform healthcare by making it possible to create customized tissues and organs. Another significant trend is the integration of AM technologies with Industry 4.0 and smart manufacturing, (Ugur et al., 2017) where they are becoming essential elements of network and data-driven production processes (Thompson et al., 2017). This makes it easier to monitor, optimize, and adjust manufacturing processes in real time. Additionally, the application of sustainable and eco-friendly methods is becoming more popular. Industries and researchers are investigating the application of recycled and bio-based materials, and additional ways to optimize energy use in AM processes (Agrawal and Vinodh, 2019). These upcoming patterns show a direction toward AM techniques that are more effective, linked, and sustainable (Agrawal and Vinodh, 2019).

Conclusion

Distinguished by its layer-by-layer construction using digital models, AM stands in contrast to traditional subtractive manufacturing methods, offering unparalleled design freedom and application versatility. In this chapter, all of the AM procedures are described in great detail, including powder bed fusion, sheet lamination, material extrusion, binder jetting, vat photopolymerization, and directed energy deposition. Applications across industries, from aerospace to healthcare, underscore the transformative impact of AM on design, prototyping, and manufacturing. Despite these advancements, challenges persist in material compatibility, surface finish, scalability, and regulatory standards. Looking ahead, the future trends in AM, such as emerging technologies like metal 3D printing and bioprinting, integration with Industry 4.0, and a growing emphasis on sustainability, promise to reshape manufacturing landscapes.

References

Agrawal, R. and Vinodh, S. 2019. State of Art Review on Sustainable Additive Manufacturing. *Rapid Prototyping Journal* 25(6): 1045–1060.

Amanda, S. and J. Al'Aref, S.J. 2018. *3D Printing Applications in Cardiovascular Medicine*. Elsevier.

Ashby, M.C. and Cebon, D. 2014. *Material and Design: The Art and Science of Material Selection in Product Design*. Elsevier.

Baldassarre, F. and Ricciardi, F. 2017. The Additive Manufacturing in the Industry 4.0 Era: The Case of an Italian FabLab. *Journal of Emerging Trends in Marketing and Management* 1(1): 105–115.

Burkhart M, Jan C. Aurich, 2015. Framework to Predict the Environmental Impact of Additive Manufacturing in the Life Cycle of a Commercial Vehicle, 29: 408-413

Buswell, R.A., de Silva, W.R. Leal, Jones S.Z. and Dirrenberger, J. 2007. 3D Printing in Construction. *Building and Environment* 119: 190–207.

Caminero, M.A., Chacon, J.M., Garcia-Moreno, I. and Rodriguez, G.P. 2018. Impact Damage Resistance of 3D Printed Continues Fibre Reinforced Thermoplastic Composites Using Fused Deposition Modelling. *Composite Part B: Engineering* 148: 93–103.

Chaudhary, R., Fabbri, P. and Leoni, E. 2022. Additive Manufacturing by Digital Light Processing: A Review. *Progress in Additive Manufacturing* 8: 331–351.

Chua, C.K., Leong, K.F. and Lim, C.S. 2010. *Rapid Prototyping: Principles and Applications*. World Scientific.

Chaudhary, R., Fabbri, P. and Leoni, E. 2022. Additive Manufacturing by Digital Light Processing: A Review. *Progress in Additive Manufacturing* 8: 331–351.

DebRoy, T., Wei, H.L., Zuback, J.S., Mukherjee, T., Elmer, J.W., Milewski, J.O., Beese, J.O., Wilson-Heid, A. and Zhang, W. 2018. Additive Manufacturing of Metallic Components-Process, Structure and Properties. *Progress in Materials Science* 92: 112–224.

Despeisse, M., Baumers, M., Brown, P., Charnley, F., Ford, S.J., Garmulewicz, A. and Wu, D.T. 2017. Unlocking Value for a Circular Economy through 3D Printing: A Research Agenda. *Technological Forecasting and Social Change* 115: 75–84.

Dizon, J.R.C., Espera, A.H. Jr, Chen, Q. and Advincula, R.C. 2018. Mechanical Characterization of 3D-printed Polymers. *Additive Manufacturing* 20: 44–67.

Frazier, W.E., 2014. Metal Additive Manufacturing: A Review. *Journal of Materials Engineering and Performance* 23(6): 1917–1928.

Gibson, I., Rosen, D.W. and Stucker, B. 2010. *Additive Manufacturing Technologies: Rapid Prototyping to Direct Digital Manufacturing*. New York: Springer. New York, USA.

Gibson, I., Rosen, D. and Stucker, B. 2015. *Additive Manufacturing Technologies: 3D Printing, Rapid Prototyping, and Direct Digital Manufacturing* (2nd edition). Springer.

Gmeiner, R., Deisinger, U., Schönherr, J., Lechner, B., Detsch, R., Boccaccini, A.R. and Stampfl, J. 2015. Additive Manufacturing of Bioactive Glasses and Silicate Bioceramics. *Journal of Ceramic Science and Technology* 6: 75–86.

Goh, G.D., Goh, G.L., Agarwal, S., Sing, S.L., Dikshit V. and Yeong, W.Y. 2019. Additive Manufacturing in Unmanned Aerial Vehicles (UAVs): Challenges and Potential. *Progress in Aerospace Sciences* 105: 222–244.

Haoa, W., Liua, Y., Zhouc, H., Chenb, H. and Fangb, T.D. 2018. Preparation and Characterization of 3D Printed Continuous Carbon Fiber Reinforced Thermosetting Composite. *Polymer Testing* 65: 29–34.

Herzog, D., Seyda, V., Wycisk, E. and Emmelmann, C. 2016. Additive Manufacturing of Metals. *Acta Materialia* 117: 371–392.

Hitzler, L., Alifui-Segbaya, F., William, P., Heine, B., Heitzmann, M., Hall, W., Merkel, M. and Ochner, A. 2018. Additive Manufacturing of Cobalt-based Dental Alloys: Analysis of Microstructure and Physicomechanical Properties. *Advances in Materials Science and Engineering* 8: 1–12.

Horst, D.J., Duvoisin, C.A. and Viera, R.A. 2018. Additive Manufacturing at Industry 4.0: A Review. *International Journal of Engineering and Technical Research* 8(8): 1–8.

Hutmacher, D.W. 2001. Scaffold Design and Fabrication Technologies for Engineering Tissues, State of the Art and Future Perspectives. *Journal of Biomaterials Science, Polymer Edition* 12(1): 107–24.

Karg, M., Becker, K.J. and Jones, P. 2018. Additive Manufacturing in Cultural Heritage Restoration: Challenges and Opportunities. *Restaurator: International Journal for the Preservation of Library and Archival Material* 39(1): 1–25.

Khoshnevis, B. 2016. Contour Crafting: A Revolution in Civil Engineering. *Journal of Sustainable Construction and Design* 7(3): 75–80.

Knowlton, S., Onal, S., Chu, Yu H.S., Zhao, J. and Tasoglu, S. 2015. Bioprinting for Cancer Research. *Trends in Biotechnology* 133(9): 504–513.

Kodama, H., 1981. Automatic Method for Fabricating a Three-dimensional Plastic Model with Photo-hardening Polymer. *Review of Scientific Instruments* 52(11): 1770–1773.

Kruth, J.P., Leu, M.C. and Nakagawa, T. 1998. Progress in Additive Manufacturing and Rapid Prototyping. *CIRP Annals* 47(2): 525–540.

Lang, M., 2017. An Overview of Laser Metal Deposition, A Publication of the Fabricators & Manufacturers Association. Available at https://www.thefabricator.com/article/additive/an-overview-of-laser-metal-deposition (Accessed in November 06, 2023).

Ligon, S.C., Liska, R., Stampfl, J., Gurr, M. and Mülhaupt, R. 2017. Polymers for 3D Printing and Customized Additive Manufacturing. *Chemical Reviews* 117(15): 10212–10290.

Lim, S., Buswell, R.A., Le, T.T., Austin, S.A., Gibb, A.G. and Thorpe, T. 2012. Developments in Construction-scale Additive Manufacturing Processes. *Automation in Construction* 21: 262–268.

Liu, S. and Shin, Y.C. 2018. Additive Manufacturing of Ti6Al4V Alloy: A Review. *Progress in Aerospace Sciences* 105: 111–131.

Low, Z., Chua, Y.T., Ray, B.M., Mattia, D., Metcalfe, I.S. and Patterson, D.A. 2017. Perspective on 3D Printing of Separation Membranes and Comparison to Related Unconventional Fabrication Techniques. *Journal of Membrane Science* 523(1): 596–613.

Mahesh, P., Kumar, N.S. and Bhavikatti, R. 2017. Additive Manufacturing: Defence and Aerospace Applications. *Materials Science and Engineering Conference Series* 330(2): 1–22.

Martin, J.H., Yahata, B.D., Hundley, J.M., Mayer, J.A., Schaedler, T.A. and Pollock, T.M. 2017. 3D Printing of High-strength Aluminium Alloys. *Nature* 549(7672): 356–369.

Muller, A. and Karevska, S. 2016. How Will 3D Printing Make Your Company the Strongest Link in the Value Chain? EY's Global 3D Printing Report. Available at https://www.ey.com/Publication/vwLUAssets/ey-global-3d-printing-report-2016-fullreport/$FILE/ey-global-3d-printing-report-2016-full-report.pdf (accessed May 03 2022).

Murr, L.E. 2016. Frontiers of 3D Printing/Additive Manufacturing: From Human Organs to Aircraft Fabrication. *Journal of Materials Sciences and Technology* 3(10): 987–995.

Murr, L.E., Gaytan, S.M. and Ramirez, D.A. 2012. Metal Fabrication by Additive Manufacturing Using Laser and Electron Beam Melting Technologies. *Journal of Materials Science & Technology* 28(1): 1–14.

Owen, D., Hickey, J., Cusson, A., Ayeni, O.I., Rhoades, J., Yifan, D., Limmin, W., Hye-Yeong, P., Nishant, H., Raikar, P.P, Yeon-Gil, J. and Jing, Z. 2018. 3D Printing of Ceramic Components Using a Customized 3D Ceramic Printer. *Progress in Additive Manufacturing* 3: 3–9.

Qian Y., Hanhua, D., Jin, S., Jianhua, H., Bo, S., Qingsong, W. and Yusheng, S. 2018. A Review of 3D Printing Technology for Medical Applications. *Engineering* 4(5): 729–742.

Reiff, M.R., Barnett, Z.R., de Beer, D.J. and Campbell, R.I. 2019. The Role of Additive Manufacturing in Artifact Replication and Restoration: A Case Study. *Journal of Cultural Heritage* 35: 176–185. https://doi.org/10.1016/j.culher.2018.09.001

Rengier, F., Mehndiratta, A., von Tengg-Kobligk, H., Zechmann, C.M., Unterhinninghofen, R., Kauczor, H.U. and Giesel, F.L. 2010. 3D Printing Based on Imaging Data: Review of Medical Applications. *International Journal of Computer Assisted Radiology and Surgery* 5(4): 335–341.

Sathishkumar, T.P., Satheeshkumar S. and Naveen, J. 2014. Glass Fiber-reinforced Polymer Composites—A Review. *Journal of Reinforced Plastics and Composites* 33(13): 1–14.

Shahrubudina, N., Leea, T.C. and Ramlana, R. 2019. An Overview on 3D Printing Technology: Technological, Materials, and Application. *Procedia Manufacturing* 35: 1286–1296.

Silbernagel, C. 2018. Additive Manufacturing 101-4: What is Material Jetting? Canada Makers. Available at http://canadamakes.ca/what-is-material-jetting/ (accessed on 5 April 2024).

Stansbury, J.W. and Idacavage, M.J. 2016. 3D Printing with Polymers: Challenges among Expanding Options and Opportunities. *Dental Materials* 32: 54–64.

Syed, A.M.T., Elias, P.K., Amit, B., Susmita, B., Lisa, O. and Charitidis, C. 2018. Additive Manufacturing: Scientific and Technological Challenges, Market Uptake and Opportunities. *Materials Today* 21(1): 1–16.

Thompson, M.K., Moroni, G., Vaneker, T., Fadel, G., Campbell, R.I., Gibson, I. and Martina, F. 2017. Design for Additive Manufacturing: Trends, Opportunities, Considerations, and Constraints. *CIRP Annals*, 66(2): 737–760.

Tiwari, S.K., Pande, S., Agrawal, S. and Bobade, S.M. 2015. Selection of Selective Laser Sintering Materials for Different Applications. *Rapid Prototyping Journal* 21(6): 630–648.

Trevisan, F., Calignano, F., Aversa, A., Marchese, G., Lombardi, M., Biamino, S., Ugues, D. and Manfredi, D. 2018. Additive Manufacturing of Titanium Alloys in the Biomedical Field: Processes, Properties and Applications. *Journals Indexing & Metrics* 16(2): 57–67.

Tsukimori, K., Hagura, K. and Yoshida, H. 2018. Wind Turbine Components through Additive Manufacturing: A Survey. *Journal of Renewable and Sustainable Energy* 10(6): 063307.

Tumbleston, J.R. 2015. Continuous Liquid Interface Production of 3D Objects. Sciencexpress 347(6228): 1349–1352.

Ugur, M.D., Gharehpapagh, B., Yaman, U. and Dolen, M. 2017. The Role of Additive Manufacturing in the Era of Industry 4.0. *Procedia Manufacturing* 11: 545–554.

Uhlmann, E., Kersting, R., Klein, T.B., Cruz, M.F. and Borille, A.V. 2015. Additive Manufacturing of Titanium Alloy for Aircraft Components. *Procedia CIRP* 35: 55–60.

Ventola, C.L. 2014. Medical Applications for 3D Printing: Current and Projected Uses. *Pharmacy and Therapeutics* 39(10): 704–711.

Vijayavenkataraman, S., Fuh, J.Y.H. and Lu, W.F. 2017, 3D Printing and 3D Bioprinting in Pediatrics. *Bioengineering (Basel)* 4(3): 63.

Wang, W., Gao, W., Zhang, Y. and Tan, C. 2021. Additive Manufacturing of Metals: From Fundamental Challenges toward Applications. *Engineering* 7(7): 1030–1037.

Xibing, G., Ted, A. and Kevin, C. 2014. Review on Powder-based Electron Beam Additive Manufacturing Technology. *Manufacturing Review 1, 2*. St. Louis, Missouri, US, June 18–20, 507–515.

Xin, W., Man, J., Zuowan, Z., Jihua, G. and David, H. 2017. 3D Printing of Polymer Matrix Composites: A Review and Prospective. *Composites Part B* 110: 442–458.

Yanko, T., Panov, S., Sushchyns'ky, O., Pylypenko, M. and Dmytrenko, O. 2018. Zirconium Alloy Powders for Manufacture of 3D Printed Articles Used in Nuclear Power Industry. *Problems of Atomic Science and Technology* 1: 148–153.

Yee, L.Y., Yong, S.E.T., Heang, K.J.T., Zheng, K.P., Xue, Y.L., Wai, Y.Y., Siang, C.H.T. and Augustinus, L. 2017. 3D Printed Bio-models for Medical Applications. *Rapid Prototyping Journal* 23(2): 227–235.

Yigong, L., Hamid, Q., Snyder, J., Chengyang, W. and Wei, S. 2016. Evaluating Fabrication Feasibility and Biomedical Application Potential of in situ 3D Printing Technology. *Rapid Prototyping Journal* 22(6): 947–955.

Yuanbin, W., Blache, R. and Xun, X. 2017. Selection of Additive Manufacturing Processes. *Rapid Prototyping Journal* 23(2): 434–447.

Ze-Xian, L., Yen, T.C., Ray, M.R., Mattia, D., Metcalfe, I.S. and Patterson, D.A. 2017. Perspective on 3D Printing of Separation Membranes and Comparison to Related Unconventional Fabrication Techniques. *Journal of Membrane Science* 523: 596–613.

4 Multi-Material Vat Photopolymerization 3D Printing: A Brief Review of Technology, Design, and Process

Harshit K. Dave[1], Sandip T. Patel[2], Dumitru Nedelcu[3], Mohd. Khairol Anuar Bin Mohd Ariffin[4] and Keyur P. Desai[5]

1. Vat Photopolymerization

Vat photopolymerization is an additive manufacturing (AM) process in which liquid photopolymer in a vat is selectively cured by light-activated polymerization, as defined by ISO 17296-1 (ISO, 2015). This 3D printing technology involves, layer by layer, curing of a liquid photosensitive polymer by exposing it to a certain wavelength of light. Majority of the commercial vat photopolymerization 3D printing technology make photopolymers react with radiation in the ultraviolet (UV) spectrum. However, several recent innovations have successfully employed infrared radiation (IR) as well as visible light systems (Bártolo, 2011; Gibson et al., 2015). This 3D printing technology has been in existence for over four decades. In 1984, Hull submitted a patent application (Hull, 1986) for 3D printing technique using vat photopolymerization and later established the pioneering additive manufacturing business, 3D Systems Inc. The word stereolithography (SLA) was first used by Hull and has now been widely recognized as a synonym for vat photopolymerization in common parlance. In reality, SLA is subcategory of vat photopolymerization technology. Over time, several more technologies in this field have been created and commercialized (see Figure 1). Other proven and commercialized vat photopolymerization 3D printing techniques include digital light processing (DLP), liquid crystal display (LCD), and continuous liquid interface production (CLIP) (Pagac et al., 2021; Schmidleithner and Kalaskar, 2018).

[1,5] *Professor*, Department of Mechanical Engineering, Sardar Vallabhbhai National Institute of Technology, Surat.

[2] *Assistant Professor*, Department of Mechanical Engineering, Government Engineering College, Valsad.

[3] *Professor*, Department of Manufacturing Engineering, Gheorghe Asachi Technical University of Iasi.

[4] *Professor*, Department of Mechanical and Manufacturing Engineering, Faculty of Engineering, Universiti Putra.

E-mail: stpatel1983@gmail.com[2]; nedelcu1967@yahoo.com[3]; khairol@upm.edu.my[4]; kpd@med.svnit.ac.in[5]

* Corresponding author: harshitkumar@yahoo.com

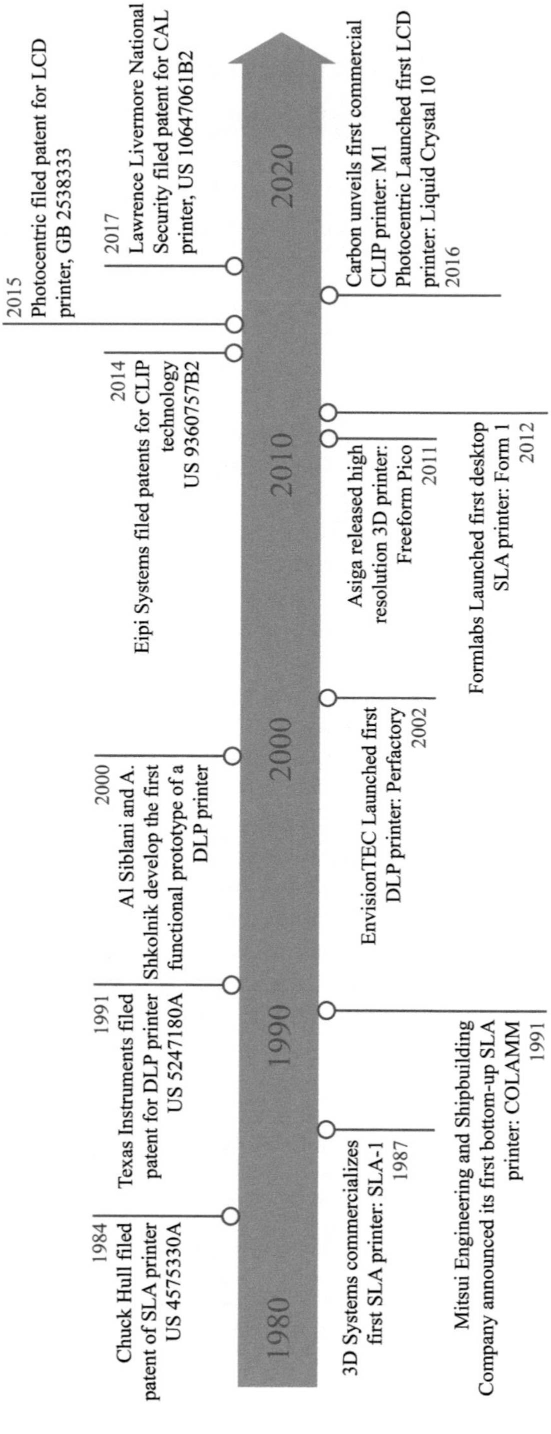

Fig. 1 Key events of vat photopolymerization 3D printing technology

A vat photopolymerization–based 3D printer typically includes a light source that emits a specific wavelength of light, a computer-controlled imaging system that selectively exposes areas of the resin, a movable platform for holding the printed object, and a resin vat filled with liquid photopolymer resin. The key to vat photopolymerization working principle resides in the photopolymerization process, where a photosensitive resin solidifies when exposed to a certain wavelength of light. An imaging system controlled by a computer selectively cures the resin in the desired pattern using a light source, resulting in the formation of a layer. Each layer can be precisely controlled in terms of thickness, usually falling between 0.4 μm and 300 μm (Shah et al., 2023). Moreover, X-Y (lateral) resolution usually ranges between 6.5 μm and 120 μm (Shah et al., 2023). This level of precision enables the creation of high-resolution parts. Once a layer has been cured, the build platform advances by a distance equal to the thickness of one layer. This may involve either the platform ascending or the vat descending, or vice versa, depending upon the architecture of the system. Before the next exposure, the use of a re-coating blade or a similar mechanism ensures the application of a fresh and consistent layer of liquid resin onto the previously cured layer. The entire process is repeated layer by layer until the component is fully completed. After the entire part is constructed, it is carefully taken off the build platform and thoroughly cleaned to eliminate any residual uncured resin. To ensure complete polymerization of the resin and to improve the material qualities of the final component, further UV curing might be used. The chemical designs and printer configuration have the ability to control the size, accuracy, resolution, build speed, and physical qualities of the parts (Bao, 2022; Li et al., 2023; Quan et al., 2020).

The SLA, DLP, CLIP, and LCD are the major commercialized categories of vat photopolymerization according to their light source and the imaging system (see Figure 2). Conventionally, SLA employs a concentrated beam of UV light or a laser to scan and cure each point of desired cross-section. DLP, CLIP, and LCD-based vat photopolymerization employs controlled area light source to simultaneous irradiate desired cross-section, which eliminates shortcomings of point-to-point scanning. These techniques are obviously faster than SLA as they do not need any moving parts and the whole layer solidifies in a single step. Through the use of a digital projector screen, the DLP 3D printing technique is able to simultaneously flash a single image of each layer across the platforms. An improvement over DLP, CLIP technology enables continuous layer-less printing of photopolymeric structures using a same light source and imaging equipment (Tumbleston et al., 2015). The CLIP 3D printer uses a continuous curing process, in which a photosensitive resin is solidified using UV light that passes through an oxygen-permeable window. This makes it one of the fastest AM technology available. The LCD vat photopolymerization technology uses LCD screen to mask off a visible light source.

Vat polymerization 3D printers are categorized into top-down and bottom-up configurations according to moving direction of build platform and elevator, as shown in Figure 3 (Chaudhary et al., 2023; Schmidleithner and Kalaskar, 2018;

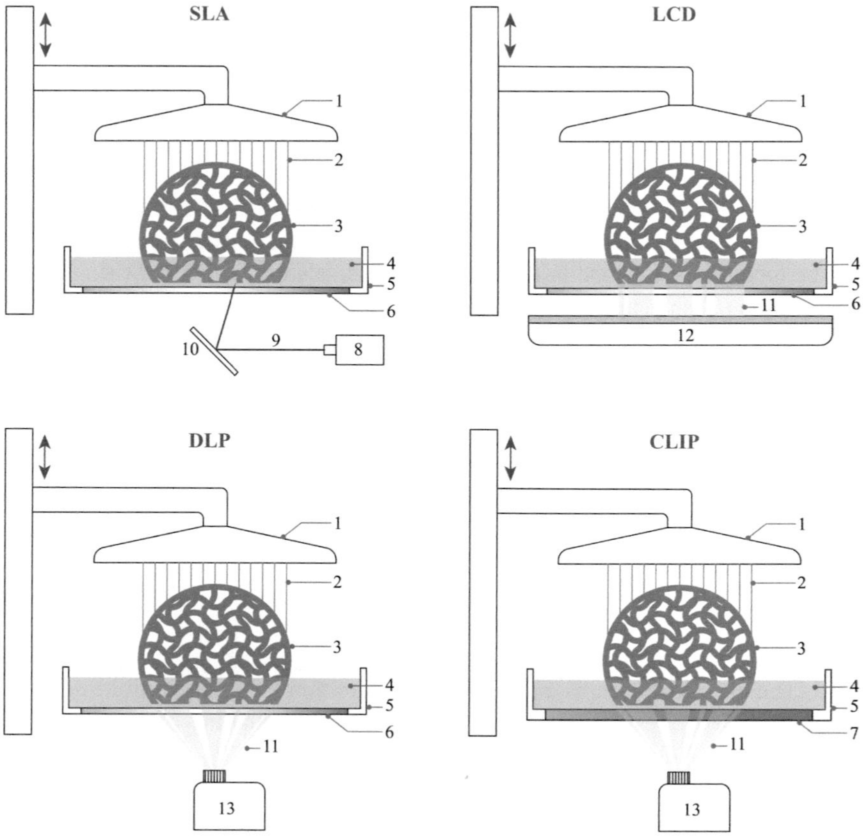

1. Build platform; 2. Support structure; 3. Build part; 4. Liquid resin; 5. Vat; 6. Transparent plate; 7. Oxygen permeable window; 8. Laser unit; 9. Laser beam; 10. Mirror; 11. Light beam matrix; 12. LCD screen; 13. DLP projector

Fig. 2 Schematic of SLA, DLP, CLIP, and LCD printers

Taormina et al., 2018). Top-down printers have the build platform positioned below the light source, and on curing of each successive layer, the build platform descends into vat of liquid photocurable resin by one layer thickness. Bottom-up printers have the build platform positioned above the light source and the part is built in upside down orientation. The resin vat has a transparent bottom which allows light to pass through it. On curing of each successive layer, the build platform ascends from the vat filled with photosensitive liquid resin. The gap kept between part and transparent bottom is one layer thickness. It means that less amount of liquid photocurable resin will be required in vat for bottom-up configuration compared to top-down configuration (Chaudhary et al., 2023). SLA and DLP printers may be found in both configurations for commercial use, although LCD and CLIP printers are typically offered in a bottom-up configuration.

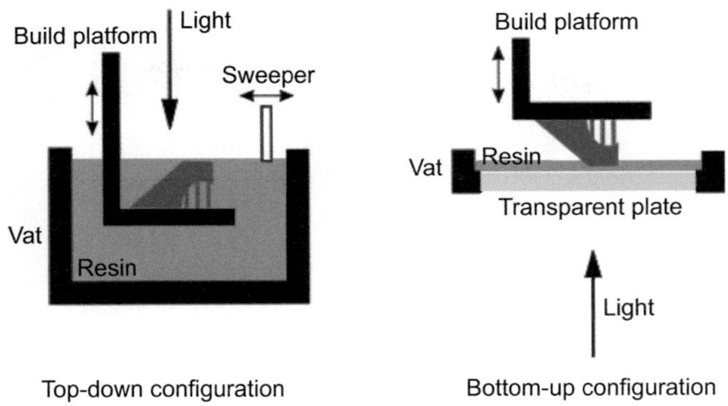

Fig. 3 Schematic of top-down and bottom-up configuration, modified and reprinted from Schmidleithner and Kalaskar (2018). Image permission under the creative commons attribution 3.0 license.

Volumetric printing, the next evolution of vat photopolymerization, distinguishes itself from conventional layer-by-layer techniques by fabricating 3D objects in a single unified process allowing rapid, freeform, layer-less 3D printing. In this developing domain, recently two technologies have emerged, xolography (Garmshausen et al., 2020) and tomographic volumetric additive manufacturing (Loterie et al., 2022) or computed axial lithography (CAL) (Kelly et al., 2020). Xolography is a technique that uses intersecting light beams of various wavelengths to superimpose a light sheet in the printing volume using linear excitation (Regehly et al., 2020; Stüwe et al., 2024). While CAL utilizes the principle of projecting series of 2D light patterns corresponding to the tomographic projections of the desired object from many different angles onto the vial of photocurable resin, usually by means of rotating vial, as depicted in Figure 4 (Kelly et al., 2019; Shusteff et al., 2017). Recent research has dived into further investigation and enhacement of its print fidelity (Bhattacharya et al., 2021; Loterie et al., 2020; Rackson et al., 2021), feedstock (Bernal et al., 2019, 2022; Cook et al., 2020; Kollep et al., 2022a; Qiu et al., 2023; Rizzo et al., 2021; Schwartz et al., 2022; Thijssen et al., 2023; Xie et al., 2023), and speed (Weisgraber et al., 2023). The comparision of major categories pertaining to vat polymerization is given in Table 1.

2. Multi-material Construct

Multi-material vat photopolymerization (MMVP) printed parts found application in a variety of domains ranging from aerospace to biomedical to consumer products (Bandyopadhyay and Heer, 2018; Bhusal et al., 2022; Grix et al., 2018; Hisham et al., 2023; Miri et al., 2018; Murr, 2016; Nazir et al., 2023; Roohani et al., 2023; Wang et al., 2017; B.Zhang et al., 2019). Designing highly customized, intricate,

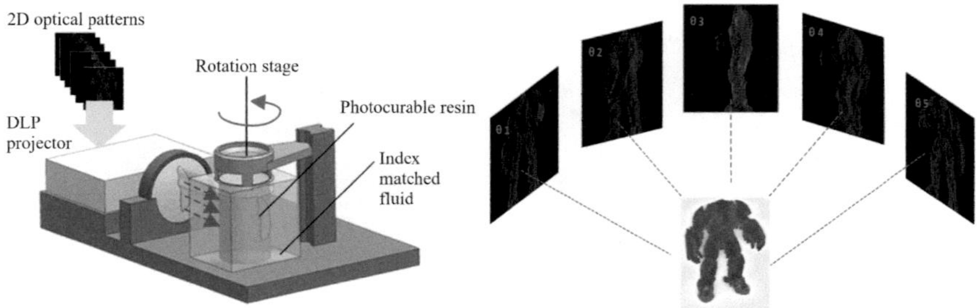

Fig. 4 Computed axial lithography (CAL)–based volumetric printing, reprinted from Zhang et al. (2023). Image permission under the creative commons attribution 4.0 license.

Table 1 Comparison of vat polymerization major categories

Technology	*Scanning Method*	*Resolution [μm]*		*Print Speed [mm/h]*	*Light Source*
		X-Y	*Z*		
SLA	Point-to-Point	25	25–300	4–31	Laser (355–405 nm)
DLP	Area	25	25–175	25–150	DLP projector (385–405 nm)
LCD	Area	50	25–300	15–170	LCD screen (405 nm)
CLIP	Conti. Area	75	1–100	500–1000	DLP projector (385–405 nm)
CAL	Volume	50	50	55 mm^3/s^*	DLP projector (405 nm)

Most of the information is sourced from the websites of commercial suppliers.
*Being a layer-less process, the print speed of CAL is expressed by volume/time.

practical, and valuable articles with enhanced properties is just one of the numerous opportunities that MMVP provides, encompassing but not limited to mechanical, thermal, optical, magnetic, electrical, and multifunctional characteristics. Recent advancements in additive manufacturing of functionally graded materials, programmable materials, metamaterial, and 4D parts calls for promising future of MMVP. Multi-material components can have enhanced physical properties and performance due to varying material compositions (Vaezi et al., 2013). The construct and layer-wise arrangement of multiple materials are responsible for the unique properties of 3D printed part.

General configuration of MMVP part is shown in Figure 5. Here, example for two material construct is depicted along with their layer-wise material sequence. The simplest form is when the entire layer is made of single material. Multi-material fabrication difficulty level increases when there is more than one material in a single layer. In case of vat photopolymerization, there are two types of part configurations when there is more than one material in a single layer. In the first configuration, body of one material within set of layers does not shadow body of the other material. In this case, all layers of the first material are fabricated first and then all layers of the other material are fabricated to reduce frequency of changing the material. In the second configuration, one material shadows another material for every layer. This makes such part fabrication complex as for every layer material changeover is required. Still, this type of construct is most promising as it has multi-oriented structure. For the scope of this chapter, we henceforth call multi-material AM parts depicted in Figures 5a, 5b, and 5c as class-I, class-II, and class-III, respectively. The layer-wise material sequence for class-II part is possible only with top-down printer. But for bottom-up printer, the material sequence for class-II part is according to class-III part.

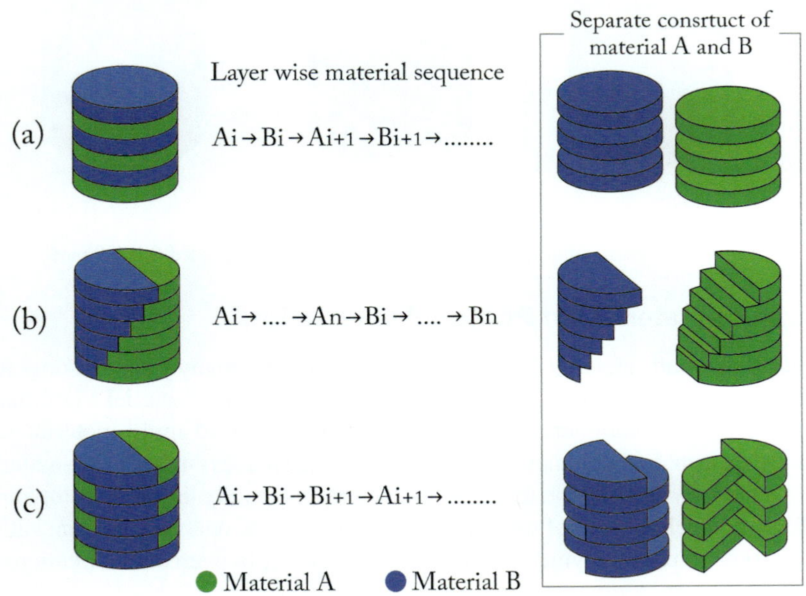

Fig. 5 Typical configuration of multi-material AM part (a) class-I: layer-wise single material; (b) class-II: layer-wise multiple material—all layers of particular material exposed to perpendicular light source; (c) class-III: layer-wise multiple material—one material is in shadow region of another material.

Another category of MMVP part is functionally graded material (FGM), wherein change from one material to another is gradual. In the context of multi-material, FGM exhibits a distinct compositional gradient of two distinct materials, setting it apart from ordinary composite materials (Loh et al., 2018; Mahamood and Akinlabi, 2017a, 2017b; Mirzaali et al., 2020). There are three types of compositional gradients that FGM parts may have: one-dimensional, two-dimensional, and three-dimensional (Muller et al., 2012), as depicted in Figure 6. Similar to class-I part, one dimensional gradient can be created layer wise in MMVP part. In two-dimensional FGM part, material gradient has to be created in single layer, while for three-dimensional gradient, the gradient has to be created in single layer as well layer-wise. This configuration is most complex to be fabricated using vat polymerization. FGMs are excellent choices for applications requiring two distinct properties (Nohut and Schwentenwein, 2022). As an example, the external surface of gear teeth needed greater hardness and the inner material needed higher toughness.

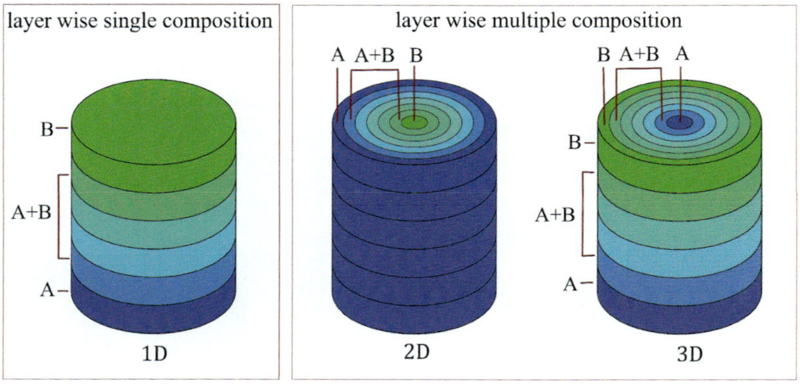

Fig. 6 Types of compositional gradients in functionally graded MMVP part

3. Multi-material Vat Photopolymerization

Research on multi-material AM has been continuing for many years but only few multi-material AM technologies, such as Polyjet and multi-material FDM, have seen the light of commercialization. The commercialized multi-material AM processes are mainly polymer-based and slow, which makes current multi-material AM technology not suitable for mass production and mainly used for prototyping. Still the compelling idea of multi-material AM fuels the research and innovation for various vat photopolymerization processes in order to overcome limitations of the current technology.

Compared to other AM processes, vat photopolymerization offers high dimensional accuracy, high-quality surface finish and printing of bio compatible materials, which impales many researchers select this process for multi-material technology development (Choi et al., 2011; Vaezi et al., 2013; Wicker and MacDonald, 2012).

The viscous liquid state of raw material creates several challenges to MMVP. The major challenges include raw material's changing mechanism, cross-contamination between raw materials, cleaning of build part and build platform between material change, and reducing build time (Choi et al., 2011; Salonitis, 2014). Many researchers have over the years presented several concepts and inversions to overcome these challenges. According to their core concept, different approaches are presented in subsequent topics.

3.1 Multi-vat Techniques

Multi-vat techniques involve 3D printer having more than one vat and each vat contains different material. Thus, the number of materials required to build a part is equal to the number of vats to be employed. The basic strategy in multi-vat technique is to change vat when change of material is required. In between vat changing process, to reduce cross-contamination of materials, cleaning of build part and build platform is required. Different methods and mechanisms to fulfill these basic requirements are proposed by various researchers and inventors in order to make the process faster, reliable, and efficient.

A team of faculties and researchers at W.M. Keck Center for 3D Innovation, University of Texas at El Paso, is working for the last two decades to advance the field of MMVP. Their significant contribution to the domain has resulted in a number of publications and patents. Over the years, they have been engaged in developing the necessary methods and systems to achieve precise control over material placement and structure creation, which resulted in conceiving the technology of multi-material stereolithography (MMSL) machine, hybrid stereolithography, and direct write (Hybrid SL/DW) machine (Medina et al., 2010; Wicker et al., 2009, 2011, 2012). The Hybrid SL/DW technique is explained in section 3.4.

A patent on MMSL apparatus was filled in the United States by Wicker et al. in 2004 based on top-down SLA and multi-vat design. The research group has conceived the idea of MMSL prototype by retrofitting 3D Systems 250/50 stereolithography machine (Lozoya, 2006;Wicker et al., 2004). The apparatus is further developed and stand-alone automated MMSL machine is fabricated in which only laser scanning system is retained from 250/50 stereolithography machine (Choi et al., 2011; Inamdar et al., 2006; Wicker and MacDonald, 2012). The MMSL machine setup and its important mechanisms are depicted in Figure 7(a). There are three to four circularly arranged vat carousel, attached to a rotary mechanism, which can rotate about vertical axis to align various vats beneath the platform. The process of multi-material part creation involves immersing the build platform in a vat filled with the initial resin material and solidifying successive layers of this resin. Then the platform is lifted out of the first vat, and before changing the material vat, cleaning operation is carried out. Cleaning of build platform and part reduces cross-contamination of materials. The build platform is connected to the elevator using an automated rotating stage, allowing it to spin around a horizontal axis. This rotation enables the building, washing, curing, and drying of parts at different angles. The stereolithography build continues with other materials till the complete build of multi-material part in cyclical manner. A custom-made program was developed in LabVIEWfi to manage and control all operations of MMSL machine. An alternate linear configuration of numerous stereolithography machines and/or alternative production technologies was also designed and patented, which allows the platform to be relocated to different sections of the machine by means of a horizontal traverse (Wicker et al., 2009). In top-down MMSL machine, employing a sweeper is not possible for the class-II and class-III geometry due to an existing layer of material that obstructs the movement of the sweeper as it traverses the vat. In order to resolve this problem, a deep-dip coating technique (non-contact recoating) was employed along with low viscosity resins to eliminate the requirement for sweeping. In order to minimize the quantity of material changeovers, Kim et al. (2010) developed a process-planning algorithm that depends on construct of part.

Researchers at University of Southern California, Los Angeles, developed an MMVP technique based on bottom-up DLP and multi-vat design having a rotary table, which rotates the resin vats and cleaning stations relative to build platform (Chen and Zhou, 2015; Zhou et al., 2011, 2013). Part fabrication is faster due to use of DLP projection system, compared to conventional laser SLA. Using a bottom-up design and a combination of brush wiping, ultrasonic washing, and drying, they were able to increase cleaning efficiency and decrease material waste during the resin tank changeover. The material wiped by soft brush is collected in brush tank for further recycling. A custom-made program is developed in C++ to control functions of MMVP machine.

Another MMVP 3D printer based on bottom-up DLP and multi-vat design was developed by researchers from Concordia University, Montreal, Canada (Matte, 2019; Matte et al., 2018, 2019). Drawing inspiration from automated storage and retrieval systems in order to reduce material swapping time, this printer has a vertical storage tower to store and retrieve material vats. Their research was focused on designing scalable and sustainable machine having faster speed of operation. Unlike horizontal traverse or rotary stages to swap different materials, vertical material tower can accommodate more number of vats without increasing machine footprint area. During material exchange, build parts are cleaned by wiping with a soft silicon brush mounted to the vat, washing with cleaning solution sprayed by a sprinkler below the build platform and within the cleaning vat, and drying with fans.

The Centre for Rapid and Sustainable Product Development of the Polytechnic Institute of Leiria, Portugal, developed stereo-thermal-lithographic (STLG), a top-down MMVP process that uses UV and IR radiation to initiate polymerization in a liquid resin with photo and thermal initiators, as depicted in Figure 7(b) (Almeida et al., 2015; Bartolo & Mitchell, 2003). STLG is similar to two DLP systems having UV and IR light sources, whose projections are combined using image overlay system and then radiated to cure resin layer. It has more tunability compared to conventional vat photopolymerization.

Recent research has shown the successful use of vacuum cleaning (Mao et al., 2021), air jet cleaning (Y. Zhang et al., 2019), and centrifugal force–enabled non-contact cleaning (see Figure 7(c)) (Cheng et al., 2022) mechanisms for small build volumes during resin tank changeovers.

3.2 Mono-vat Techniques

Fabricating multi-material part using only one vat is rather challenging. Different strategies attempted by various researchers and inventors are described in this section. So far, there are four main strategies employed by researchers: (i) using external resin dispensing device; (ii) changing material from vat; (iii) employing

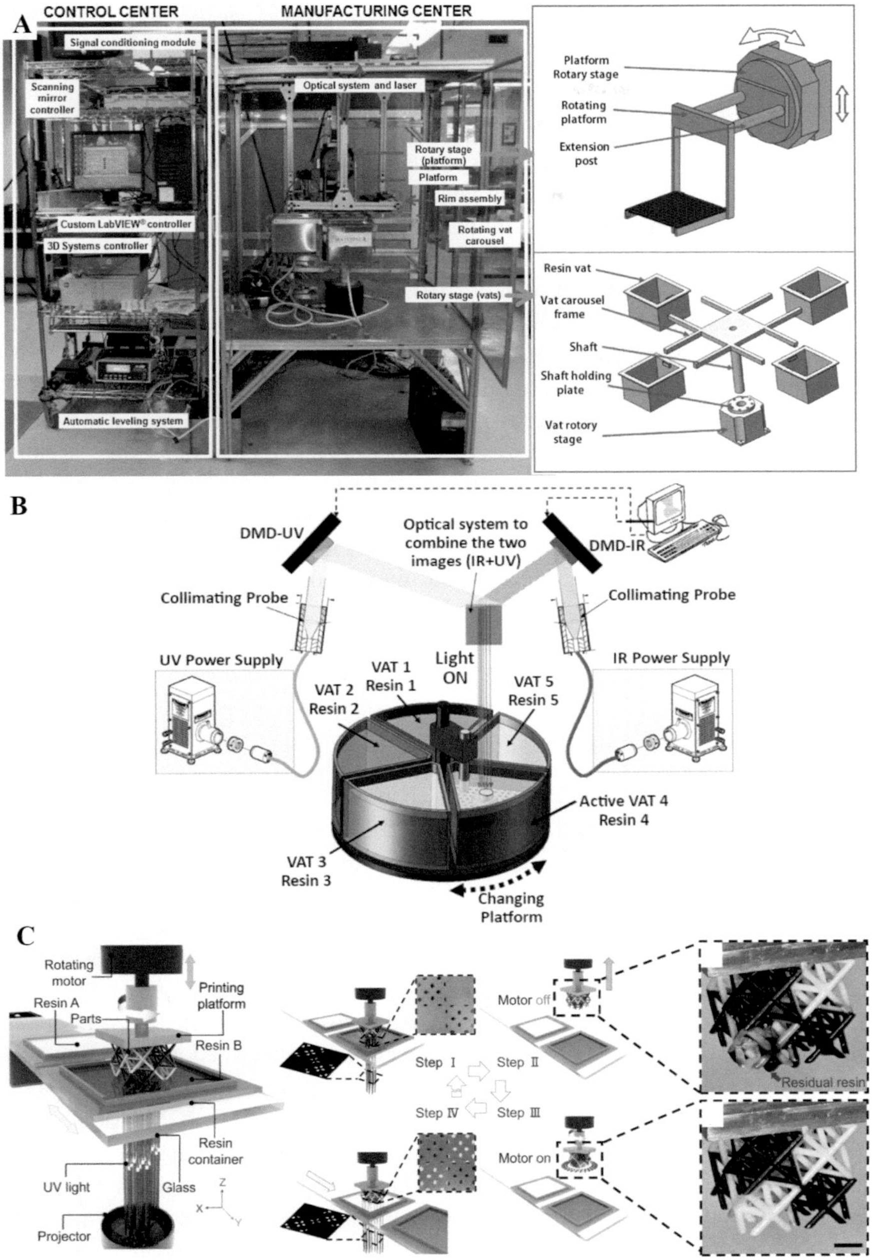

Fig. 7 Examples of multi-vat MMVP systems (a) MMSL technology developed at University of Texas at El Paso, reprinted with permission from Choi et al. (2011); (b) Schematic of MMVP 3D printer based on STGL, reprinted with permission from Bártolo (2011); (c) DLP-based centrifugal MMVP system, image permission under the creative commons attribution 4.0 license, reprinted from Cheng et al. (2022).

wavelength-based polymerization; and (iv) inclusion of particles or fibers into resin.

3.2.1 External Resin Dispensing Method

This approach utilizes a droplet dispenser, like pipette or syringe, to directly dispense several materials layer by layer in top-down design printers. Very few printer concepts were developed in this category. SLA-based position- and time-controlled dispensing is employed to create multicolour parts, but with poor dimensional accuracy (Gong et al., 2010). The printing of multi-material FGM components using top-down DLP technology, as described in a patent application by Zheng and Deotte (Zheng and Deotte, 2017), involves filling a vat with a carrier fluid that is immiscible with the build material. The Zheng and Deotte (2017) droplet dispenser dispenses micro- or nano-liter droplet with a variety of materials on substrate. The build platform was positioned such that the droplet was partially submerged in the carrier fluid while the upper surface of the droplet was not covered by the carrier fluid. Droplet pipetting followed by photo polymerization creates layered part. The same team of researchers had also filed a patent on similar concept for top-down CLIP printer (Deotte and Zheng, 2018) (see Figure 8(a)). In CLIP printer, the carrier fluid is used to hold resin against the transparent window. The small amount of resin is quickly switched using integrated microfluidic channels.

3.2.2 Material Exchange Method

In this method, the basic strategy is to clean the vat and build part whenever material change is required. Researchers at Virginia Tech, Blacksburg, had developed single-vat MMVP technology based on bottom-up DLP configuration, depicted in Figure 8(b) (Chen, 2016; Chen and Zheng, 2018; Hensleigh et al., 2020; Zheng, 2021). The basic strategy employed was to clean and feed the new material in vat when change of build material was required. Three linear actuators were driving X, Y, and Z axis cartridges. The printer featured an automated resin supply system and cleaning system to increase speed and efficiency of material swapping and cleaning. Micro-fluidic channels were integrated in wash unit and polydimethylsiloxane (PDMS) brush to jet clean leftover resin from the X axis PDMS window and build part, respectively, using ethanol. PDMS brush was also cleaned by wiping it onto protruded fins before and after wipe cleaning was performed on PDMS window. During cleaning of build part, an active vibration was added by an integrated vibrator motor in Z axis. After cleaning, a new material was perfused into the vat by Y axis material jethead. The material changeover time for each layer was 10 seconds, which accounted for both the cleaning and drying processes. This MMVP technique is capable of printing class-III parts having high resolution.

Computed axial lithography (CAL) is very compatible with printing material on an intricate, pre-existing 3D structure (Kelly et al., 2019). Because of this,

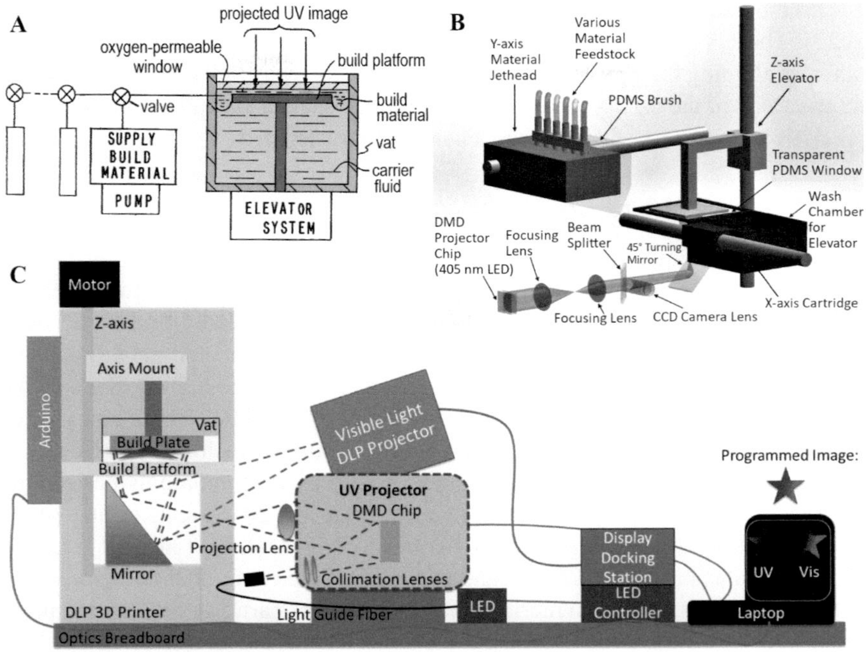

Fig. 8 Examples of mono-vat MMVP systems (a) Schematic of system facilitates external resin dispensing in top-down CLIP printer, modified and reprinted from Deotte and Zheng (2018); (b) Schematic of system based on bottom-up DLP printer with robotic material exchange system, image permission under the creative commons attribution 4.0 license, reprinted from Chen (2016); (c) Schamatic of multi-wavelength MMVP 3D printer, image permission under the creative commons attribution 4.0 license, reprinted from Schwartz and Boydston, (2019).

convex substrates may be quickly printed with any conformal geometries. A multi-material component may be printed onto an existing substrate or without one at all by sequentially swepping the resin from vile (Chansoria et al., 2023; Größbacher et al., 2023; Soliman et al., 2023).

3.2.3 *Wavelength-based Polymerization*

The concept of multi-wavelength MMVP is to allow a mixture of multiple resins to reside in the same resin vat; each resin is then engineered to polymerize when exposed to different wavelengths of light. This concept was utilized by Schwartz and Boydston using UV and visible irradiation (Schwartz and Boydston, 2019). Simultaneous series of sliced heterogeneous images were parallelly projected by visible lightand UV-DLP projector (as depicted in Figure 8(c)). This technique is capable of processing two different materials in a single vat without additional cleaning cycle.

A recent study used a dual colour tomographic volumetric 3D printing approach to demonstrate that photoresins sensitive to different wavelengths may be cured by utilizing both visible and UV light sources concurrently (Wang et al., 2022). The UV and visible light channels were arranged perpendicular to each other by aligning their projection centerlines in such a way that they crossed at the rotation axis of the curing volume. This technique enables the quick production of precise components with diverse mechanical characteristics.

3.2.4 Polymer Matrix Composite Material

Single vat technique is generally employed for fabrication of polymer matrix composite material using vat photopolymerization. Majority of the research done till date in this domain is focused on fabrication feasibility and physical property analysis for composite material. Researchers have attempted inclusion of continuous fiber by manual laying (Karalekas and Antoniou, 2004; Sano et al., 2018) or using fiber laying mechanisms (Charan et al., 1994; Gupta and Ogale, 2002), limited to simple topology and resulting poor dimensional accuracy. Not much research is done for continuous fiber reinforced vat photopolymerization. However, particles (He et al., 2015; Kalsoom et al., 2016; Kim et al., 2014; Kurimoto et al., 2015; Yang et al., 2016) or short fibers (Eng et al., 2017; Lin et al., 2015; Palaganas et al., 2017; Sano et al., 2018; Xu et al., 2020) are premixed with resin for vat photopolymerization. Premixing of particles or discontinuous fibers creates composites having randomly oriented and uniformly dispersed inclusions. Thus, there is a need for controlling fiber orientation in resin to create more complex and functional composite material (Goh et al., 2019; Zindani and Kumar, 2019).

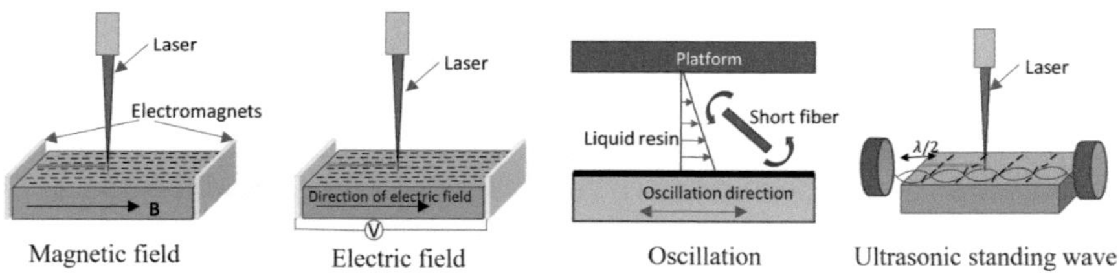

Fig. 9 Various fiber alignment methods in AM composites for vat photopolymerization, reprinted with permission from Goh et al. (2019).

Various techniques demonstrated by researchers to control fiber orientation in liquid resin include employing magnetic field (Lu et al., 2017; Martin et al., 2015; Safaee and Chen, 2019), electric field (Holmes and Riddick, 2014; Nakamoto et al., 2013; Yang et al., 2017, 2019), acoustic fields (Scholz et al., 2014), and shear-induced alignment (Yunus et al., 2016) as depicted in Figure 9. Among these techniques, Martin et al. had specified a DLP-based magnetic printer concept of creating highly programmable discontinuous fiber composites using orientable magnetic field, illustrated in Figure 10 (Erb and Martin, 2020; Martin et al., 2015). Before premixing to resin, reinforcing particles must be magnetized if nonmagnetic reinforcing materials are to be used. The magnetic field was simultaneously applied

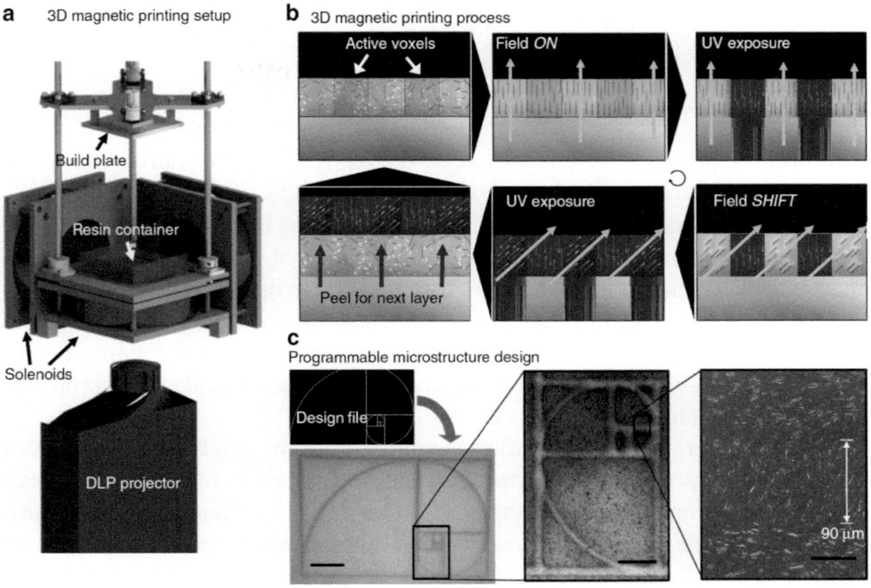

Fig. 10 (a) The 3D magnetic-printer setup; (b) The 3D magnetic printing process; (c) printed sample of the golden rectangle, from design files to micro-structure, reprinted from Martin et al. (2015). Image permission under the creative commons attribution 4.0 license.

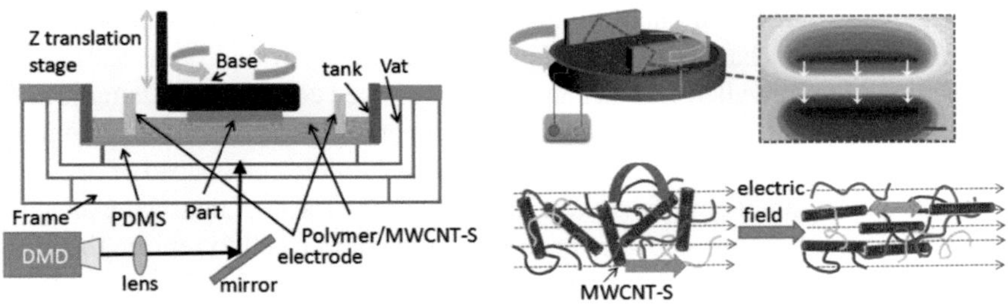

Fig. 11 Schematic diagram of the electrically assisted 3D printer, reprinted with permission from Yang et al. (2017)

using electromagnetic solenoids, which systematically aligns and selectively polymerizes set of voxels programmed to have specific reinforcement orientation within each layer of printed material. This enables to create complex bioinspired reinforcement structures. Similarly, Yang et al. (2017) had specified the technique to create bioinspired anisotropic reinforcement structures using custom-made bottom-up DLP printerClick or tap here to enter text. Using rotating platform, two electrodes with applied DC electric field control orientation of reinforcement, as depicted in Figure 11.

The emerging volumetric printing is well suited for particle and discontinuous fiber reinforcement composite materials. Recently, the strength of these approaches was demonstrated through printing of ceramic (Kollep et al., 2022b) and glass (Barbera et al., 2024; Toombs et al., 2022) reinforced multimaterial.

3.3 *Micro-stereolithography*

The process principle of micro-stereolithography (μSL) is similar to that of conventional vat photopolymerization (SLA, DLP, and LCD) with the difference being in the thickness of print layer. The layer thickness in case of μSL is between 5 μm and 10 μm, while for conventional vat photopolymerization it is typically in the range of 100 μm (Salonitis, 2014). Modification in optical system and resin is need to achieve low vertical and horizontal resolution in printed parts. In case of laser-based μSL, more focused laser beam is used while digital micromirror device (DMD) or LCD-based system employed reduction optics to scale down projected image (Bertsch and Renaud, 2011).

Multi-material μSL fabrication techniques can be fundamentally derived from the techniques discussed in the previous sections. Researchers at W.M. Keck Center for 3D Innovation, University of Texas at El Paso, had employed multi-material μSL fabrication based on single-vat design where material changeover process included manual cleaning and micro pipetting new material in the vat (Arcaute et al., 2006, 2007, 2010; Choi et al., 2010). Kim et al. (2018) employed multi-vat concept by manually pausing the machine and switching the vat in commercial top-down DLP printer. Ge et al. (2016) employed automatic vat changing mechanism in top-down DLP printer having multiple vat for multi-material 4D printing. Similarly, bottom-up DLP printer equipped with multi-vat and automatic vat changing mechanism was employed by Mu et al. (2017) for fabrication of multiwalled carbon nanotubes reinforced composites. Zhang (2017) had employed multi-wavelength, single vat concept using custom-made bottom-up DLP printer. Han et al. (2019) had employed closed fluidic cell instead of open vat to achieve quick and active material exchange. The process of material exchange included the introduction of fresh material into the fluidic cell to replace the prior material present in the printing chamber, as illustrated in Figure 12.

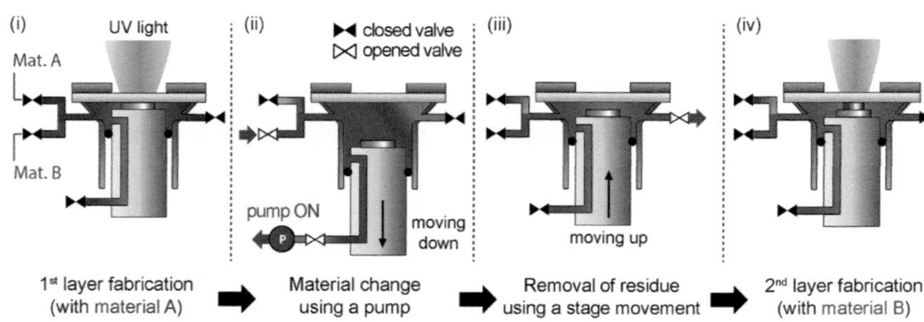

Fig. 12 Schematic of multi-material mSL process proposed by Han et al., reprinted from Han et al. (2019)

3.4 Hybrid Techniques

Hybrid techniques deal with integration of multiple manufacturing techniques including AM as primary technique to fabricate multi-material parts having unique properties or application. For example, in few researches, direct writing technique is integrated with vat photopolymerization to fabricate 3D monolithic structures that consist of integrated electronics and printed interconnect (Joe Lopes et al., 2012; Li et al., 2016). The fabrication process involves the sequential steps of: (i) partially constructing the structural component using vat photopolymerization; (ii) using automated direct writing to create electrical interconnects while simultaneously curing the dispensed material; (iii) embedding the component; and (iv) cleaning the constructed part before steps (ii) and (iii).

4. Summary and Future Trends

Devlopment of MMVP technology began more than a decade ago. Many researchers came up with innovative MMVP techniques as stated earlier. The summary and comparison of these researches is given in Table 2. Nonetheless, MMVP has not yet reached a commercially viable stage of development. Researchers have come a very long way but a few challenges remain in order to maximize this technology. The current challenges of MMVP are listed as follows:

- *Speed of fabrication:* The material exchange method in single vat or mulit-vat system required cleaning and drying of affected components which can cross-contaminate liquid resins. In case of composite material, magnetic or electric field–driven fiber alignment takes time to align the fibers based on strength of field. These added requirements increase build time per layer apart from time for light exposer to cure single layer. When there are more than one materials in single layer (class-III part), multiple material swapping and light irradiation cycle has to be repeated for each layer. It makes layer-based MMVP 3D printing time consuming. The speed of fabrication should be increased by finding more efficient methods. This may be potentially achived by using layer-less volumetric printing techniques.
- *Material combinations:* In previous researches, various combinations of materials had been tested for multiple applications. There are large numbers of material combinations possible and each cannot be tested for their inter material bonding strength. Thus, more research is required to find best material combinations pertaining to specific application.
- *Software:* To cater to the special needs of different MMVP techniques currently available, slicing and machine management software are inadequate. This is backed by previous research work in which machine management software were created in-house. The slicing software for vat photopolymerization are not available for multi-material structures. Thus, separate material constructs are sliced separately and their slice images are merged and sorted according to layer-wise material sequence.

Table 2 Summary of research on MMVP techniques

Organization	*Printer Configuration*	*Multi-material Mechanism*	*Operating Software*	*Build Area*	*Part Class*	*Time*	*Patent*	*Ref.*
University of Texas, USA	Multi-vat, SLA, Top-down	Swapping of vat	Custom make	165′165 mm	I,II,III	NR	US7556490B2, US7658603B2	(Choi et al., 2011; Wicker & MacDonald, 2012)
University of Southern California, USA	Multi-vat, DLP, Bottom-up	Swapping of vat	NR	48′36 mm	I,II,III	~180s*	US9120270B2	(Zhou et al., 2013)
Concordia University, Canada	Multi-vat, DLP, Bottom-up	Swapping of vat	NR	305′152 mm	I,II,III	63s*	NR	(Matte et al., 2019)
Polytechnic Institute of Leiria, Portugal	Multi-vat, STLG, Top-down		NR	NR	I,II,III	NR	PT107096A	(Bartolo & Mitchell, 2003)
Lawrence Livermore National Security LLC, USA	Mono-vat, DLP, Top-down	Pipetting of material on immiscible carrier fluid	Custom make	NR	I,II,III	NR	US11345879B2	(Zheng & Deotte, 2017)
Lawrence Livermore National Security LLC, USA	Mono-vat, CLIP, Top-down	Micro-fluidic dispensing of materials on immiscible carrier fluid	Custom make	NR	I	NR	US10118338B2	(Deotte & Zheng, 2018)
VirginiaTech, Blacksburg, USA	Mono-vat, DLP, Bottom-up	Material swapping from vat	Custom make	20′15 mm	I,II,III	10s*	US10987865B2	(Chen & Zheng, 2018)
University of Washington, USA	Mono-vat, DLP, Bottom-up	Irradiating different wavelength on resin	Creation Workshop	NR	I,II,III	NR	NR	(Schwartz & Boydston, 2019)
Northeastern University, Boston, USA	Mono-vat, DLP, Bottom-up	Reinforcement alignment control by magnetic field	Creation Workshop	NR	3D FGM	12s**	US10703052B2	(Martin et al., 2015)

(*Contd.*)

Table 2 *Contd.*

Organization	*Printer Configuration*	*Multi-material Mechanism*	*Operating Software*	*Build Area*	*Part Class*	*Time*	*Patent*	*Ref.*
University of Southern California, USA	Mono-vat, DLP, Bottom-up	Reinforcement alignment control by electric field	Custom make	NR	2D FGM	60s**	NR	(Yang et al., 2017)
Lawrence Livermore National Security LLC, USA	CAL	Material swapping from vat	Custom make	NR	I,II	NR	US11446861B2	(Kelly et al., 2019)

* Time per layer for material swapping, cleaning, and drying (irradiation time not included).
** Time per layer for fiber alignment (irradiation time not included).
NR = Not Reported

In wake of the fourth industrial revolution, 3D printing holds key to the disruptive manufacturing technology of future. Multi-material printing can enable this disruption as it can bring products on table which we have never even heard of. Future trends for MMVP will be fueled by paradigm shift in design and manufacturing practices.

A major future trend for MMVP is to increase speed of 3D printing. Robotic-powered 3D printing is an emerging field to speed up the process with the aid of automation. Another option is to employ CLIP technique in MMVP as it is the fastest 3D vat photopolymerization process. But due to its continuous layering operation, material swapping is difficult. Researchers have to find efficient and practical ways to print multi-material parts using CLIP technology. Volumetric printing recently emerged as a paradigm shift in MMVP. When it comes to printing speed and material variety, volumetric printing is exceptional. Nevertheless, a resolution greater than 50 μm remains unachievable. The future direction for volumetric printing based on MMVP should prioritize improving print quality, resolution, and the ability to print on a larger scale.

Accurate analytical models exist for vat photopolymerization due to intelligible chemistry of photopolymerization (Gibson et al., 2015). Similar analytical models can be developed for MMVP to predict inter-material bonding strength and their curing mechanism. In future, specialized computer-aided design (CAD) and slicing software has to be developed for MMVP, and especially for volumetric printing, to cater to the vivid requirements of the process.

References

Almeida, H., Bártolo, P.J., Mateus, A.J., Ferreira, N.M. and Alves, N.M. 2015. *Stereo System Term Multi-material Lithography and Appropriate Process* (Patent PT107096A).

Arcaute, K., Mann, B.K. and Wicker, R.B. 2006. Stereolithography of Three-dimensional Bioactive Poly(Ethylene Glycol) Constructs with Encapsulated Cells. *Annals of Biomedical Engineering* 34(9): 1429–1441.

Arcaute, K., Mann, B. and Wicker, R. 2010. Stereolithography of Spatially Controlled Multi-material Bioactive Poly(Ethylene Glycol) Scaffolds. *Acta Biomaterialia* 6(3): 1047–1054.

Arcaute, K., Zuverza, N., Mann, B. and Wicker, R. 2007. Multi-material Stereolithography: Spatially-controlled Bioactive Poly(Ethylene Glycol) Scaffolds for Tissue Engineering. *Proceedings of 18th Annual Solid Freeform Fabrication Symposium*, 458–469.

Bandyopadhyay, A. and Heer, B. 2018. Additive Manufacturing of Multi-material Structures. *Materials Science and Engineering: R: Reports* 129: 1–16.

Bao, Y. 2022. Recent Trends in Advanced Photoinitiators for Vat Photopolymerization 3D Printing. *Macromolecular Rapid Communications* 43(14): 2200202.

Barbera, L., Madrid-Wolff, J., Emma, R., Masania, K., Boniface, A., Loterie, D., Delrot, P., Moser, C. and Studart, A.R. 2024. Multimaterial Volumetric Printing of Silica-based Glasses. *Advanced Materials Technologies* 9(7): 2202117.

Bártolo, P.J. 2011. Stereolithographic Processes. In Bártolo, P. (ed.), *Stereolithography*, 1–36. Springer US.

Bartolo, P.J. and Mitchell, G. 2003. Stereo-thermal-lithography: A New Principle for Rapid Prototyping. *Rapid Prototyping Journal* 9(3): 150–156.

Bernal, P.N., Bouwmeester, M., Madrid-Wolff, J., Falandt, M., Florczak, S., Rodriguez, N.G., Li, Y., Größbacher, G., Samsom, R., van Wolferen, M., van der Laan, L.J.W., Delrot, P., Loterie, D., Malda, J., Moser, C., Spee, B. and Levato, R. 2022. Volumetric Bioprinting of Organoids and Optically Tuned Hydrogels to Build Liver-like Metabolic Biofactories. *Advanced Materials* 34(15): 2110054.

Bernal, P.N., Delrot, P., Loterie, D., Li, Y., Malda, J., Moser, C. and Levato, R. 2019. Volumetric Bioprinting of Complex Living-tissue Constructs within Seconds. *Advanced Materials* 31(42): 1904209.

Bertsch, A. and Renaud, P. 2011. Microstereolithography. In Bártolo, P. (ed.), *Stereolithography*, 81–112. Springer US.

Bhattacharya, I., Toombs, J. and Taylor, H. 2021. High Fidelity Volumetric Additive Manufacturing. *Additive Manufacturing* 47: 102299.

Bhusal, A., Dogan, E., Nguyen, H.-A., Labutina, O., Nieto, D., Khademhosseini, A. and Miri, A.K. 2022. Multi-material Digital Light Processing Bioprinting of Hydrogel-based Microfluidic Chips. *Biofabrication* 14(1): 014103.

Chansoria, P., Rütsche, D., Wang, A., Liu, H., D'Angella, D., Rizzo, R., Hasenauer, A., Weber, P., Qiu, W., Ibrahim, N.B.M., Korshunova, N., Qin, X. and Zenobi-Wong, M. 2023. Synergizing Algorithmic Design, Photoclick Chemistry and Multi-Material Volumetric Printing for Accelerating Complex Shape Engineering. *Advanced Science* 10(26): 2300912.

Charan, R., Renault, T., Ogale, A.A. and Bagchi, A. 1994. Automated Fiber-reinforced Composite Prototypes. *Fifth International Conference on Rapid Prototyping*, 91–97.

Chaudhary, R., Fabbri, P., Leoni, E., Mazzanti, F., Akbari, R. and Antonini, C. 2023. Additive Manufacturing by Digital Light Processing: A Review. *Progress in Additive Manufacturing* 8(2): 331–351.

Chen, D. 2016. *A Multi-Material Projection Stereolithography System for Manufacturing Programmable Negative Poisson's Ratio Structures*. Virginia Polytechnic Institute.

Chen, D. and Zheng, X. 2018. Multi-material Additive Manufacturing of Metamaterials with Giant, Tailorable Negative Poisson's Ratios. *Scientific Reports* 8(1): 9139.

Chen, Y. and Zhou, C. 2015. *Digital Mask-Image-Projection-based Additive Manufacturing That Applies Shearing Force to Detach Each Added Layer* (Patent US9120270B2). United States Patent and Trademark Office (USPTO).

Cheng, J., Wang, R., Sun, Z., Liu, Q., He, X., Li, H., Ye, H., Yang, X., Wei, X., Li, Z., Jian, B., Deng, W. and Ge, Q. 2022. Centrifugal Multimaterial 3D Printing of Multifunctional Heterogeneous Objects. *Nature Communications* 13(1): 7931.

Choi, J.-W., Kim, H.-C. and Wicker, R. 2011. Multi-material stereolithography. *Journal of Materials Processing Technology*, *211*(3), 318–328. https://doi.org/10.1016/j.jmatprotec.2010.10.003.

Choi, J.-W., MacDonald, E. and Wicker, R. (2010). Multi-material Microstereolithography. *The International Journal of Advanced Manufacturing Technology* 49(5–8): 543–551.

Cook, C.C., Fong, E.J., Schwartz, J.J., Porcincula, D.H., Kaczmarek, A.C., Oakdale, J.S., Moran, B.D., Champley, K.M., Rackson, C.M., Muralidharan, A., McLeod, R.R. and Shusteff, M. 2020. Highly Tunable Thiol-Ene Photoresins for Volumetric Additive Manufacturing. *Advanced Materials* 32(47): 2003376.

Deotte, J.R. and Zheng, X. 2018. *Additive Manufacturing with Integrated Microliter Resin Delivery* (Patent US10118338B2). USPTO.

Eng, H., Maleksaeedi, S., Yu, S., Choong, Y.Y.C., Wiria, F.E., Tan, C.L.C., Su, P.C. and Wei, J. 2017. 3D Stereolithography of Polymer Composites Reinforced with Orientated Nanoclay. *Procedia Engineering* 216: 1–7.

Erb, R. and Martin, J.J. 2020. *Additive Manufacturing of Discontinuous Fiber Composites Using Magnetic Fields* (Patent US10703052B2).

Garmshausen, Y., Reuter, M., Regehly, M. and Radzinski, D. 2020. *Process and Apparatus for Locally Polymerizing a Starting Material by Dual Color Photopolymerization and Method for Volumetric Printing of a Shaped Body* (Patent Application No. EP2020065818W).

Ge, Q., Sakhaei, A.H., Lee, H., Dunn, C.K., Fang, N.X. and Dunn, M.L. 2016. Multimaterial 4D Printing with Tailorable Shape Memory Polymers. *Scientific Reports* 6(1): 31110.

Gibson, I., Rosen, D. and Stucker, B. 2015. Vat Photopolymerization Processes. In *Additive Manufacturing Technologies*, 63–106. New York: Springer.

Goh, G.D., Yap, Y.L., Agarwala, S. and Yeong, W.Y. 2019. Recent Progress in Additive Manufacturing of Fiber Reinforced Polymer Composite. *Advanced Materials Technologies* 4(1): 1800271.

Gong, H., Wang, M. and Wang, Y. 2010. Color Stereolithography Based on Time-Pressure Dispensing Process. *2010 IEEE International Conference on Mechatronics and Automation*, 99–103.

Grix, T., Ruppelt, A., Thomas, A., Amler, A.-K., Noichl, B., Lauster, R. and Kloke, L. 2018. Bioprinting Perfusion-enabled Liver Equivalents for Advanced Organ-on-a-Chip Applications. *Genes* 9(4): 176.

Größbacher, G., Bartolf-Kopp, M., Gergely, C., Bernal, P.N., Florczak, S., de Ruijter, M., Rodriguez, N.G., Groll, J., Malda, J., Jungst, T. and Levato, R. 2023. Volumetric Printing across Melt Electrowritten Scaffolds Fabricates Multi-material Living Constructs with Tunable Architecture and Mechanics. *Advanced Materials* 35(32): 2300756.

Gupta, A. and Ogale, A.A. 2002. Dual Curing of Carbon Fiber Reinforced Photoresins for Rapid Prototyping. *Polymer Composites* 23(6): 1162–1170.

Han, D., Yang, C., Fang, N.X. and Lee, H. 2019. Rapid Multi-material 3D Printing with Projection Micro-stereolithography Using Dynamic Fluidic Control. *Additive Manufacturing* 27: 606–615.

He, M., Zhao, Y., Wang, B., Xi, Q., Zhou, J. and Liang, Z. 2015. 3D Printing Fabrication of Amorphous Thermoelectric Materials with Ultralow Thermal Conductivity. *Small* 11(44): 5889–5894.

Hensleigh, R., Cui, H., Xu, Z., Massman, J., Yao, D., Berrigan, J. and Zheng, X. 2020. Charge-programmed Three-dimensional Printing for Multi-material Electronic Devices. *Nature Electronics* 3(4): 216–224.

Hisham, M., Salih, A.E. and Butt, H. 2023. 3D Printing of Multimaterial Contact Lenses. *ACS Biomaterials Science & Engineering* 9(7): 4381–4391.

Holmes, L.R. and Riddick, J.C. 2014. Research Summary of an Additive Manufacturing Technology for the Fabrication of 3D Composites with Tailored Internal Structure. *JOM* 66(2): 270–274.

Hull, C.W. 1986. *Apparatus for Production of Three-dimensional Objects by Stereolithography* (Patent US 4,575,330 A). USPTO.

Inamdar, A., Magana, M., Medina, F., Grajeda, Y. and Wicker, R. 2006. Development of an Automated Multiple Material Stereolithography Machine. *Proceedings of 17th Annual Solid Freeform Fabrication Symposium*, 624–635.

ISO. 2015. *ISO 17296-1:2015, Additive Manufacturing–General Principles–Part 1:* International Organization for Standardization (ISO).

Joe Lopes, A., MacDonald, E. and Wicker, R.B. 2012. Integrating Stereolithography and Direct Print Technologies for 3D Structural Electronics Fabrication. *Rapid Prototyping Journal* 18(2): 129–143.

Kalsoom, U., Peristyy, A., Nesterenko, P.N. and Paull, B. 2016. A 3D Printable Diamond Polymer Composite: A Novel Material for Fabrication of Low Cost Thermally Conducting Devices. *RSC Advances* 6(44): 38140–38147.

Karalekas, D. and Antoniou, K. 2004. Composite Rapid Prototyping: Overcoming the Drawback of Poor Mechanical Properties. *Journal of Materials Processing Technology* 153–154: 526–530.

Kelly, B., Panas, R., Shusteff, M., Spadaccini, C., Taylor, H. and Bhattacharya, I. 2020. *System and Method for Computed Axial Lithography (CAL) for 3D Additive Manufacturing* (Patent US10647061B2). USPTO.

Kelly, B.E., Bhattacharya, I., Heidari, H., Shusteff, M., Spadaccini, C.M. and Taylor, H.K. 2019. Volumetric Additive Manufacturing via Tomographic Reconstruction. *Science* 363(6431): 1075–1079.

Kim, H., Choi, J. and Wicker, R. 2010. Scheduling and Process Planning for Multiple Material Stereolithography. *Rapid Prototyping Journal* 16(4): 232–240.

Kim, K., Zhu, W., Qu, X., Aaronson, C., McCall, W.R., Chen, S. and Sirbuly, D.J. 2014. 3D Optical Printing of Piezoelectric Nanoparticle–Polymer Composite Materials. *ACS Nano* 8(10): 9799–9806.

Kim, Y., Castro, K., Bhattacharjee, N. and Folch, A. 2018. Digital Manufacturing of Selective Porous Barriers in Microchannels Using Multi-material Stereolithography. *Micromachines* 9(3): 125.

Kollep, M., Konstantinou, G., Madrid-Wolff, J., Boniface, A., Hagelüken, L., Sasikumar, P.V.W., Blugan, G., Delrot, P., Loterie, D., Brugger, J. and Moser, C. 2022. Tomographic Volumetric Additive Manufacturing of Silicon Oxycarbide Ceramics. *Advanced Engineering Materials* 24(7): 2101345.

Kurimoto, M., Yamashita, Y., Ozaki, H., Kato, T., Funabashi, T. and Suzuoki, Y. 2015. 3D Printing of Conical Insulating Spacer Using Alumina/UV-cured-Resin Composite. *2015 IEEE Conference on Electrical Insulation and Dielectric Phenomena (CEIDP)*, 463–466.

Li, J., Wasley, T., Nguyen, T.T., Ta, V.D., Shephard, J.D., Stringer, J., Smith, P., Esenturk, E., Connaughton, C. and Kay, R. 2016. Hybrid Additive Manufacturing of 3D Electronic Systems. *Journal of Micromechanics and Microengineering* 26(10): 105005.

Li, Y., Zhang, X., Zhang, X., Zhang, Y. and Hou, D. 2023. Recent Progress of the Vat Photopolymerization Technique in Tissue Engineering: A Brief Review of Mechanisms, Methods, Materials and Applications. *Polymers* 15(19): 3940.

Lin, D., Jin, S., Zhang, F., Wang, C., Wang, Y., Zhou, C. and Cheng, G.J. 2015. 3D Stereolithography Printing of Graphene Oxide Reinforced Complex Architectures. *Nanotechnology* 26(43): 434003.

Loh, G.H., Pei, E., Harrison, D. and Monzón, M.D. 2018. An Overview of Functionally Graded Additive Manufacturing. *Additive Manufacturing* 23: 34–44.

Loterie, D., Delrot, P. and Moser, C. 2020. High-resolution Tomographic Volumetric Additive Manufacturing. *Nature Communications* 11(1): 852.

Loterie, D., Delrot, P. and Moser, C. 2022. *Methods and Apparatus for Three-dimensional Fabrication by Tomographic Back Projections* (Patent US11446861B2). USPTO.

Lozoya, O.A. 2006. *Development and Demonstration of a Multiple Material Stereolithography System*. University of Texas at El Paso. Available at https://www.proquest.com/docview/304958803 . It is last accessed on 20-FEB-2025.

Lu, L., Guo, P. and Pan, Y. 2017. Magnetic-field-assisted Projection Stereolithography for Three-dimensional Printing of Smart Structures. *Journal of Manufacturing Science and Engineering* 139(7): 071008.

Mahamood, R.M. and Akinlabi, E.T. 2017. Introduction to Functionally Graded Materials. *In Functionally Graded Materials*, 1–8. Springer International Publishing.

Mao, H., Jia, W., Leung, Y.-S., Jin, J. and Chen, Y. 2021. Multi-material Stereolithography Using Curing-on-Demand Printheads. *Rapid Prototyping Journal* 27(5): 861–871.

Martin, J.J., Fiore, B.E. and Erb, R.M. 2015. Designing Bioinspired Composite Reinforcement Architectures via 3D Magnetic Printing. *Nature Communications* 6(1): 8641.

Matte, C.-D. 2019. *Geometric Simulation and Additive Manufacturing of Multi-material Soft Robots*. Concordia University. Available at https://spectrum.library.concordia.ca/985953/ (last accessed on 20-FEB-2025)

Matte, C.-D., Pearson, M., Trottier-Cournoyer, F., Dafoe, A. and Kwok, T.-H. 2018. Multi-material Digital Light Processing Printer with Material Tower and Spray Cleaning. *Proceedings of the ASME 2018 13th International Manufacturing Science and Engineering Conference, Volume 4: Processes*, 11.

Matte, C.-D., Pearson, M., Trottier-Cournoyer, F., Dafoe, A. and Kwok, T.H. 2019. Automated Storage and Active Cleaning for Multi-material Digital-Light-Processing Printer. *Rapid Prototyping Journal* 25(5): 864–874.

Medina, F., Wicker, R., Palmer, J.A., Davis, D.W., Chavez, B.D. and Gallegos, P.L. 2010. *Methods and Systems for Integrating Fluid Dispensing Technology with Stereolithography* (Patent US7658603B2). USPTO.

Miri, A.K., Nieto, D., Iglesias, L., Goodarzi Hosseinabadi, H., Maharjan, S., Ruiz-Esparza, G.U., Khoshakhlagh, P., Manbachi, A., Dokmeci, M.R., Chen, S., Shin, S.R., Zhang, Y.S. and Khademhosseini, A. 2018. Microfluidics-enabled Multimaterial Maskless Stereolithographic Bioprinting. *Advanced Materials* 30(27): 1800242.

Mirzaali, M.J., Herranz de la Nava, A., Gunashekar, D., Nouri-Goushki, M., Veeger, R.P.E., Grossman, Q., Angeloni, L., Ghatkesar, M.K., Fratila-Apachitei, L.E., Ruffoni, D., Doubrovski, E.L. and Zadpoor, A.A. 2020. Mechanics of Bioinspired Functionally Graded Soft-Hard Composites Made by Multi-material 3D Printing. *Composite Structures* 237: 111867.

Mu, Q., Wang, L., Dunn, C.K., Kuang, X., Duan, F., Zhang, Z., Qi, H.J. and Wang, T. 2017. Digital Light Processing 3D Printing of Conductive Complex Structures. *Additive Manufacturing* 18: 74–83.

Muller, P., Mognol, P. and Hascoet, J.-Y. 2012. Functionally Graded Material (FGM) Parts: From Design to the Manufacturing Simulation. *Proceedings of the ASME 2012 11th Biennial Conference on Engineering Systems Design and Analysis* 4: 123–131.

Murr, L.E. 2016. Frontiers of 3D Printing/Additive Manufacturing: From Human Organs to Aircraft Fabrication. *Journal of Materials Science & Technology* 32(10): 987–995.

Nakamoto, T., Kanehisa, O. and Sugawa, Y. 2013. Whisker Alignment in Microparts Using Laser Stereolithography with Applied Electric Field. *Journal of Advanced Mechanical Design, Systems, and Manufacturing* 7(6): 888–902.

Nazir, A., Gokcekaya, O., Md Masum Billah, K., Ertugrul, O., Jiang, J., Sun, J. and Hussain, S. 2023. Multi-material Additive Manufacturing: A Systematic Review of Design, Properties, Applications, Challenges and 3D Printing of Materials and Cellular Metamaterials. *Materials & Design* 226: 111661.

Nohut, S. and Schwentenwein, M. 2022. Vat Photopolymerization Additive Manufacturing of Functionally Graded Materials: A Review. *Journal of Manufacturing and Materials Processing* 6(1): 17.

Pagac, M., Hajnys, J., Ma, Q.-P., Jancar, L., Jansa, J., Stefek, P. and Mesicek, J. 2021. A Review of Vat Photopolymerization Technology: Materials, Applications, Challenges and Future Trends of 3D Printing. *Polymers* 13(4): 598.

Palaganas, N.B., Mangadlao, J.D., de Leon, A.C.C., Palaganas, J.O., Pangilinan, K.D., Lee, Y.J. and Advincula, R.C. 2017. 3D Printing of Photocurable Cellulose Nanocrystal Composite for Fabrication of Complex Architectures via Stereolithography. *ACS Applied Materials & Interfaces* 9(39): 34314–34324.

Qiu, W., Gehlen, J., Bernero, M., Gehre, C., Schädli, G.N., Müller, R. and Qin, X. 2023. A Synthetic Dynamic Polyvinyl Alcohol Photoresin for Fast Volumetric Bioprinting of Functional Ultrasoft Hydrogel Constructs. *Advanced Functional Materials* 33(20): 2214393.

Quan, H., Zhang, T., Xu, H., Luo, S., Nie, J. and Zhu, X. 2020. Photo-curing 3D Printing Technique and Its Challenges. *Bioactive Materials* 5(1): 110–115.

Rackson, C.M., Champley, K.M., Toombs, J.T., Fong, E.J., Bansal, V., Taylor, H.K., Shusteff, M. and McLeod, R.R. 2021. Object-space Optimization of Tomographic Reconstructions for Additive Manufacturing. *Additive Manufacturing* 48: 102367.

Regehly, M., Garmshausen, Y., Reuter, M., König, N.F., Israel, E., Kelly, D.P., Chou, C.-Y., Koch, K., Asfari, B. and Hecht, S. 2020. Xolography for Linear Volumetric 3D Printing. *Nature* 588(7839): 620–624.

Rizzo, R., Ruetsche, D., Liu, H. and Zenobi-Wong, M. 2021. Optimized Photoclick (Bio) Resins for Fast Volumetric Bioprinting. *Advanced Materials* 33(49): 2102900.

Roohani, I., Newsom, E. and Zreiqat, H. 2023. High-resolution Vat-Photopolymerization of Personalized Bioceramic Implants: New Advances, Regulatory Hurdles and Key Recommendations. *International Materials Reviews* 68(8): 1075–1097.

Safaee, S. and Chen, R. 2019. Investigation of a Magnetic Field-assisted Digital-Light-Processing Stereolithography for Functionally Graded Materials. *Procedia Manufacturing* 34: 731–737.

Salonitis, K. 2014. Stereolithography. *Comprehensive Materials Processing* 10: 19–67.

Sano, Y., Matsuzaki, R., Ueda, M., Todoroki, A. and Hirano, Y. 2018. 3D Printing of Discontinuous and Continuous Fibre Composites Using Stereolithography. *Additive Manufacturing* 24: 521–527.

Schmidleithner, C. and Kalaskar, D.M. 2018. Stereolithography. In D. Cvetkovic (ed.), *3D Printing*, 1–22. InTech.

Scholz, M.-S., Drinkwater, B.W. and Trask, R.S. 2014. Ultrasonic Assembly of Anisotropic Short Fibre Reinforced Composites. *Ultrasonics* 54(4): 1015–1019.

Schwartz, J.J. and Boydston, A.J. 2019. Multimaterial Actinic Spatial Control 3D and 4D Printing. *Nature Communications* 10(1): 791.

Schwartz, J.J., Porcincula, D.H., Cook, C.C., Fong, E.J. and Shusteff, M. 2022. Volumetric Additive Manufacturing of Shape Memory Polymers. *Polymer Chemistry* 13(13): 1813–1817.

Shah, M., Ullah, A., Azher, K., Rehman, A.U., Juan, W., Aktürk, N., Tüfekci, C.S. and Salamci, M.U. 2023. Vat Photopolymerization-based 3D Printing of Polymer Nanocomposites: Current Trends and Applications. *RSC Advances* 13(2): 1456–1496.

Shusteff, M., Browar, A.E.M., Kelly, B.E., Henriksson, J., Weisgraber, T.H., Panas, R.M., Fang, N.X. and Spadaccini, C.M. 2017. One-step Volumetric Additive Manufacturing of Complex Polymer Structures. *Science Advances* 3(12): eaao5496.

Soliman, B.G., Longoni, A., Wang, M., Li, W., Bernal, P.N., Cianciosi, A., Lindberg, G.C.J., Malda, J., Groll, J., Jungst, T., Levato, R., Rnjak-Kovacina, J., Woodfield, T.B.F., Zhang, Y.S. and Lim, K.S. 2023. Programming Delayed Dissolution into Sacrificial Bioinks for Dynamic Temporal Control of Architecture within 3D-Bioprinted Constructs. *Advanced Functional Materials* 33(8): 2210521.

Stüwe, L., Geiger, M., Röllgen, F., Heinze, T., Reuter, M., Wessling, M., Hecht, S. and Linkhorst, J. 2024. Continuous Volumetric 3D Printing: Xolography in Flow. *Advanced Materials* 36(4): 2306716.

Taormina, G., Sciancalepore, C., Messori, M. and Bondioli, F. 2018. 3D Printing Processes for Photocurable Polymeric Materials: Technologies, Materials and Future Trends. *Journal of Applied Biomaterials & Functional Materials* 16(3): 151–160.

Thijssen, Q., Quaak, A., Toombs, J., De Vlieghere, E., Parmentier, L., Taylor, H. and Van Vlierberghe, S. 2023. Volumetric Printing of Thiol-Ene Photo-Cross-Linkable Poly(ε-caprolactone): A Tunable Material Platform Serving Biomedical Applications. *Advanced Materials* 35(19): 2210136.

Toombs, J.T., Luitz, M., Cook, C.C., Jenne, S., Li, C.C., Rapp, B.E., Kotz-Helmer, F. and Taylor, H.K. 2022. Volumetric Additive Manufacturing of Silica Glass with Microscale Computed Axial Lithography. *Science* 376(6590): 308–312.

Tumbleston, J.R., Shirvanyants, D., Ermoshkin, N., Janusziewicz, R., Johnson, A.R., Kelly, D., Chen, K., Pinschmidt, R., Rolland, J.P., Ermoshkin, A., Samulski, E.T. and DeSimone, J.M. (2015). Continuous Liquid Interface Production of 3D Objects. *Science* 347(6228): 1349–1352.

Vaezi, M., Chianrabutra, S., Mellor, B. and Yang, S. 2013. Multiple Material Additive Manufacturing—Part 1: A Review. *Virtual and Physical Prototyping* 8(1): 19–50.

Wang, B., Engay, E., Stubbe, P.R., Moghaddam, S.Z., Thormann, E., Almdal, K., Islam, A. and Yang, Y. 2022. Stiffness Control in Dual Color Tomographic Volumetric 3D Printing. *Nature Communications* 13(1): 367.

Wang, X., Jiang, M., Zhou, Z., Gou, J. and Hui, D. 2017. 3D Printing of Polymer Matrix Composites: A Review and Prospective. *Composites Part B: Engineering* 110: 442–458.

Weisgraber, T.H., de Beer, M.P., Huang, S., Karnes, J.J., Cook, C.C. and Shusteff, M. 2023. Virtual Volumetric Additive Manufacturing (VirtualVAM). *Advanced Materials Technologies* 8(23): 2301054.

Wicker, R., Medina, F., Arcaute, K., Ochoa, L., Elkins, C. and Mann, B. 2012. *Hydrogel Constructs Using Stereolithography* (Patent US8197743B2). USPTO.

Wicker, R., Medina, F. and Elkins, C. 2004. Multiple Material Micro-fabrication: Extending Stereolithography to Tissue Engineering and Other Novel Applications. *Proceedings of 15th Annual Solid Freeform Fabrication Symposium*, 754–764.

Wicker, R., Medina, F. and Elkins, C. 2009. *Multi-material Stereolithography* (Patent US7556490B2). USPTO.

Wicker, R., Medina, F. and Elkins, C. 2011. *Methods for Multi-material Stereolithography* (Patent US7959847B2). USPTO.

Wicker, R.B. and MacDonald, E.W. 2012. Multi-material, Multi-technology Stereolithography. *Virtual and Physical Prototyping* 7(3): 181–194.

Xie, M., Lian, L., Mu, X., Luo, Z., Garciamendez-Mijares, C.E., Zhang, Z., López, A., Manríquez, J., Kuang, X., Wu, J., Sahoo, J.K., González, F.Z., Li, G., Tang, G., Maharjan, S., Guo, J., Kaplan, D.L. and Zhang, Y.S. 2023. Volumetric Additive Manufacturing of Pristine Silk-based (Bio)Inks. *Nature Communications* 14(1): 210.

Xu, Z., Ha, C.S., Kadam, R., Lindahl, J., Kim, S., Wu, H.F., Kunc, V. and Zheng, X. 2020. Additive Manufacturing of Two-phase Lightweight, Stiff and High Damping Carbon Fiber Reinforced Polymer Microlattices. *Additive Manufacturing* 32: 101106.

Yang, Y., Chen, Z., Song, X., Zhang, Z., Zhang, J., Shung, K.K., Zhou, Q. and Chen, Y. 2017. Biomimetic Anisotropic Reinforcement Architectures by Electrically Assisted Nanocomposite 3D Printing. *Advanced Materials* 29(11): 1605750.

Yang, Y., Chen, Z., Song, X., Zhu, B., Hsiai, T., Wu, P.-I., Xiong, R., Shi, J., Chen, Y., Zhou, Q. and Shung, K.K. 2016. Three Dimensional Printing of High Dielectric Capacitor Using Projection based Stereolithography Method. *Nano Energy* 22: 414–421.

Yang, Y., Li, X., Chu, M., Sun, H., Jin, J., Yu, K., Wang, Q., Zhou, Q. and Chen, Y. 2019. Electrically Assisted 3D Printing of Nacre-inspired Structures with Self-sensing Capability. *Science Advances* 5(4): eaau9490.

Yunus, D.E., Shi, W., Sohrabi, S. and Liu, Y. 2016. Shear Induced Alignment of Short Nanofibers in 3D Printed Polymer Composites. *Nanotechnology* 27(49): 495302.

Zhang, B., Gao, L., Ma, L., Luo, Y., Yang, H. and Cui, Z. 2019. 3D Bioprinting: A Novel Avenue for Manufacturing Tissues and Organs. *Engineering* 5(4): 777–794.

Zhang, R. 2017. *Stereolithographic Hydrogel Printing of 3D Microfluidic Cell Culture Chips*. DTU Nanotech.

Zhang, Y., Liu, M., Liu, H., Gao, C., Jia, Z. and Zhai, R. 2023. Edge-enhanced Object-Space Model Optimization of Tomographic Reconstructions for Additive Manufacturing. *Micromachines* 14(7): 1362.

Zhang, Y., Ng, C.J., Chen, Z., Zhang, W., Panjwani, S., Kowsari, K., Yang, H.Y. and Ge, Q. 2019. Miniature Pneumatic Actuators for Soft Robots by High-resolution Multimaterial 3D Printing. *Advanced Materials Technologies* 4(10): 1900427.

Zheng, X. 2021. *3D Printing Systems and Methods Thereof* (Patent US10987865B2). USPTO.

Zheng, X. and Deotte, J.R. 2017. *Emulsion Stereolithography and 3D Printing of Multimaterials and Nanoscale Material Gradients* (Patent US11345879B2). USPTO.

Zhou, C., Chen, Y., Yang, Z. and Khoshnevis, B. 2011. Development of Multi-material Mask-image-projection-based Stereolithography for the Fabrication of Digital Materials. *Annual Solid Freeform Fabrication Symposium*, 65–80.

Zhou, C., Chen, Y., Yang, Z. and Khoshnevis, B. 2013. Digital Material Fabrication Using Mask-image-projection-based Stereolithography. *Rapid Prototyping Journal* 19(3): 153–165.

Zindani, D. and Kumar, K. 2019. An Insight into Additive Manufacturing of Fiber Reinforced Polymer Composite. *International Journal of Lightweight Materials and Manufacture* 2(4): 267–278.

5 Sustainable Production Planning for Additive Manufacturing in a Circular Economy Environment Setting

Mihaela-Elena Ulmeanu[1], Cristian-Vasile Doicin[2*], Giuseppe Lamanna[3], Huseyin Kursat Celik[4] and Constantin Gheorghe Opran[5]

1. Introduction

In today's manufacturing landscape, there's a paramount need to minimize the environmental, economic, and social impact of the production and consumption cycle. Both manufacturers and suppliers are gravitating towards circular production models and processes. Within this transition, additive manufacturing (AM) offers a transformative pathway to achieving the objectives of the circular economy (CE), a model aiming to reduce waste, promote recycling, and ensure sustainable resource use (Bal and Badurdeen, 2022; Sauerwein et al., 2019).

Contrary to the traditional linear economy's "take-make-dispose" approach, CE emphasizes maximizing resource and energy circularity. The shift to a more restorative cycle relies on six fundamental concepts: Reduce, Recover, Reuse, Remanufacture, Recycle, and Redesign (Jawahir and Bradley, 2016). In this context, AM plays a pivotal role. It not only reduces production waste but also promotes the use of recycled materials, enabling the creation of on-demand, customized products and facilitating repair and remanufacturing (Priyadarshini et al., 2022; Verma et al., 2022; Enyoghasi and Badurdeen, 2021). Another key advantage of AM is its ability to create products from recycled materials, whether metal or polymeric (Romanova et al., 2022). This opens up new possibilities for businesses (Krishnendu et al., 2023) to use waste as a resource, turning what would have been discarded as waste into valuable products. In addition to reducing waste and promoting recycling, AM also enables the creation of on-demand, customized products (Kerin and Pham, 2020). This has the potential to reduce overproduction

[1] *Associate Professor*, Department of Manufacturing Engineering, The National University of Science and Technology Politehnica Bucharest.

[2,5] *Professor*, Department of Manufacturing Engineering, The National University of Science and Technology Politehnica Bucharest.

[3] *Professor*, Department of Engineering, University of Campania Luigi Vanvitelli.

[4] *Professor*, Department of Agricultural Machinery and Technology Engineering, Faculty of Agriculture, Akdeniz University.

E-mail: mihaela.ulmeanu@upb.ro[1]; giuseppe.lamanna@unicampania.it[3]; hkcelik@akdeniz.edu.tr[4]; constantin.opran@upb.ro[5]

* Corresponding author: cristian.doicin@upb.ro

and stock surplus, which is one of the main drivers of waste in the traditional linear economy. With 3D printing, businesses can produce only the products they need, when they need them, reducing the risk of overproduction and surplus (Zhu et al., 2021). The decentralized and versatile nature of AM also enables repair and remanufacturing of products. This is particularly important in the context of the CE, as it helps to extend the lifespan of products and reduce the need for raw materials (Gao et al., 2021). In many cases, it is possible to repair or upgrade products using 3D printing, reducing the need to dispose of them and replace them with new products (Prashar et al., 2022).

From a circular and sustainable point of view, AM not only provides opportunities related to the process, it must also be considered alongside the thinking of the product, i.e., effecting the design of it, crucial for the implementation of sustainable cycles of production and consumption of goods. Design freedoms of material, geometry, and topology have extremely expanded and conventional manufacturing constraints (e.g., material difference and tooling accessibility) have largely been removed due to the introduction of AM (Sousa and Pham, 2023).

There are several AM technologies that apply CE principles and bring about their advantages (Ferreira et al., 2023; Cruz Sanchez et al., 2020). Some of the most common technologies are fused deposition modelling (FDM), stereolithography (SLA), and selective laser sintering (SLS). FDM can use thermoplastic filament which can be made from recycled materials, making it a sustainable option for businesses looking to reduce their environmental footprint (Üreten et al., 2019). The photosensitive resin used in SLA technologies can be made from bio-based materials, making it a viable alternative to traditional materials (Hui, et al., 2022). The plastic or metal powders used in SLS can be made from recycled materials, reducing the need for virgin raw materials. In addition to these technologies, many businesses are also exploring the use of closed-loop systems for material recovery, recycling, and reuse in AM. This involves collecting used or waste materials from the production process, recycling them, and using the recycled materials to produce new products, creating a closed-loop system for material usage (Nascimento et al., 2022). These technologies, along with the use of closed-loop systems for material recovery, demonstrate the potential for AM to play a critical role in the transition toward a CE. By reducing waste, promoting recycling, and enabling the use of sustainable materials, these technologies offer businesses a way to reduce their environmental footprint while also improving their production processes.

Despite the current and potential benefits that this technology provides, especially when corelated alongside CE, there are many difficulties and application inertia. Some of the barriers to the use of AM materials include: the relationship between material properties and powder properties that is limited (Romanova et al., 2022); a characterization of AM materials, including input material characteristics such as powder size, shape, and chemistry, not always available (Benedetti et al., 2021); no standardized AM-specific methods for conducting interlaboratory studies of AM materials (Slotwinski and Garboczi, 2015); the underlying factors responsible for machine-to-machine variability (McGregor et al., 2021); and the qualification of raw materials that is limited (Chen et al., 2022). These limitations affect business strategies, management styles, and manufacturing goals for a variety of companies, leading to a lower adoption rate of AM for large-scale production facilities. The same difficulties are faced by research AM facilities and educational institutions.

This chapter proposes a different approach on a tertiary educational AM facility aiming at lowering some of the limitations by optimizing manufacturing parameters to plan the production of 3D printed complex parts, within a CE setting. Optimizing 3D printing parameters within a CE environment is critical for reducing waste, minimizing the environmental impact of the technology, and maximizing the economic efficiency of the production process. Cost-effective parts that meet the requirements of CE can be produced by balancing the trade-offs between printing time, material usage, and high quality. The focus is on a new hypothesis through which the AM environment of a higher education institution (HEI) is approached as a small and medium enterprise (SME) in an effort to optimize quality, time, and production costs, whilst generating new and important implications for managerial practices.

2. Methods and Materials

The current research is aimed at optimizing the production planning and parameters for a 3D printed part. This is a crucial step toward creating a more sustainable and cost-effective manufacturing process within a CE environment (Jayasekara, et al., 2022). In a CE, the focus is on conserving resources, reducing waste, and promoting sustainability (Zeba et al., 2021). The use of 3D printing in the production of parts offers numerous advantages, including versatility, speed, and affordability, making it an ideal technology to use in a CE context (Rojas et al., 2021). This research has three main targets for a 3D printing application, which mainly concern the manufacturing process, i.e., the first phase of a product's life cycle: reduce overall costs; efficient assembly (reduce assembly and fit operations of components); and end-of-life strategy (reuse, recycle, or discard the final part after main function is completed).

The first target of the research, reducing costs, is fundamental for business strategy alignment with the CE. This target emphasizes less material consumption without sacrificing product functionality. Strategies here encompass optimization of infill density, removal of unnecessary support structures, shortening print durations, and automation of post-processing (Papacharalampopoulos et al., 2018; Qamar Tanveer et al., 2022; Golab et al., 2022). Additionally, reducing printing time and labour costs for post-processing will also have a significant impact on the overall cost of production.

The second target, efficient assembly, aims to streamline the manufacturing process, reducing both economic and environmental costs. By innovatively redesigning parts for minimal assembly and ensuring precision in component fits, there's a potential reduction in labour costs and time (Daneshmand et al., 2023; Javaid et al., 2021).

Finally, the third target, reusing the final part after the main function is completed, is an integral component of CE. Designing for reuse, incorporating recyclable materials, and environmentally conscious discarding methods epitomize this target (Camilleri, 2019; Corona et al., 2019). Elements such as water-soluble support structures and filters for nanoparticle and odour management further contribute to a sustainable production cycle.

To realize these targets, the research was based on a three-direction strategy group. Each group contains specific methods and materials tailored for the respective target. For cost reduction, techniques like part design optimization and employment of faster printing technologies were examined. Automated post-processing techniques and streamlined manual processes were also explored for potential labour savings. The efficient assembly group concentrated on part design improvements for fewer components and better component fit. Methods investigated include precision tolerances during printing and the exploration of snap-fit designs. In addressing the end-of-life strategy, the research considered the material selection, notably biodegradable thermoplastics like Polylactic Acid (PLA). The emphasis was also on materials that promote reuse, are easily recyclable, and diminish waste.

Building on these strategy groups, the authors develop a methodology to plan the production of additive manufactured parts within a CE environment setting. The chapter further proposes an integrative hypothesis to approach the academic working environment as an SME and apply optimization principles based on three main economic indicators, namely quality, time, and cost. The need for conducting the current research stemmed from working in a tertiary education environment, implementing 3D printing laboratories and projects for undergraduate and postgraduate courses. To manifest the aforementioned hypothesis, an integrative framework was used by constructing an AM academic ecosystem consisting of a network of academic experts, researchers, and industry specialists who collaborate to advance the field of 3D printing. The ecosystem was divided into four main categories: hardware, materials, post-processing, and services (Figure 1). The hardware component targets academic research focused on improving the speed, accuracy, and reliability of 3D printing technology, based on CE principles. Exploration of new materials, as well as the development of new material formulations and processing techniques are aimed at improving the mechanical properties, durability, and functionality of 3D printed parts. Post-processing involves the development of techniques to finish and refine 3D printed parts after they are printed and includes removing of support structures, sanding, polishing, painting or coating, and other finishing techniques. Academic research in this area focuses on improving the quality and accuracy of post-processing methods to achieve high-quality finished parts.

The final integrative category of the AM ecosystem are the services which include the development of design, prototyping, and testing, and even product development research on specific topics. This area is focused on improving the efficiency and effectiveness of services, as well as developing new ones to meet

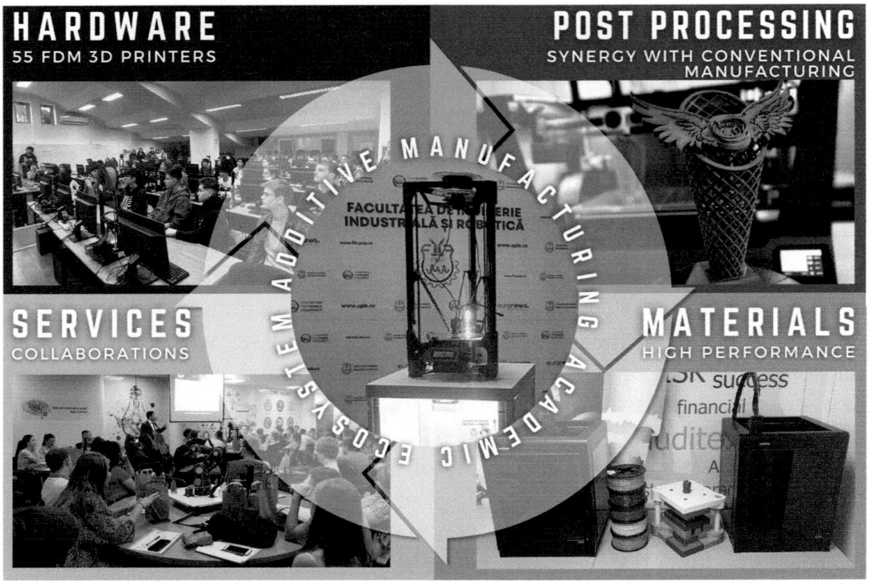

Fig. 1 Additive manufacturing academic ecosystem for undergraduate and postgraduate courses

the needs of specific academic partners. The proposed AM academic ecosystem is a dynamic and collaborative environment that fosters innovation and advancement among academic institutions, researchers, and industry partners, driving the development of new technologies, materials, and services that will continue to transform the manufacturing landscape.

Within the proposed AM academic ecosystem, a total of four disciplines from six program studies were selected for the analysis, as follows: product design and development (laboratory and project), additive manufacturing equipment and technologies (laboratory and project), additive manufacturing technologies (laboratory), and additive manufacturing (laboratory and project). The disciplines are taught during one semester (14 weeks) in the third and fourth years for undergraduate studies and the first year of masters' studies in a technical university. In total, a number of 240 students attend the laboratory and project activities provided within the structures of the four disciplines. Per week, disciplines have a two-hour laboratory session and two-hour project session, totalling up at 28 hours per semester for disciplines with laboratories and 56 hours per semester for disciplines with laboratory and project activities. For every application session (laboratory or project) each student is required to design and 3D print specific parts in accordance with a given task. FDM is the main technology used for the manufacturing of various parts within all the aforementioned disciplines within the analysed tertiary education establishment.

The average material consumption estimation in the last five years for these disciplines is presented in Table 1. A total of 111.74 kg of plastic filament was obtained from direct scrap material. Indirect scrap material is generated in two main ways: irreversible damage to the 3D printed part, rendering it unusable; parts that completed their function, thus ending the life cycle with the discarding phase. More than 50% of 3D printed parts end up as indirect scrap because of damage or end of life cycle. Even though all scrap material from non-conforming parts, support structures, and other deburring materials is usually collected and recycled as regular plastic material, it is crucial to identify more efficient methods for recycling, reusing, or discarding of indirect scrap parts. Usually, all discarded 3D printed products that are not contaminated with other substances are collected within the faculty and they either get recycled at local community recycling centres as other regular plastic components, or they are shredded and made up into new filament. This practice is used in order to align with CE goals in Europe set on 2 December 2015 in the Circular Economy Package (Mikula et al., 2021). Community recycling is quite tedious due to local legislation, which accepts only certain types of filaments to go into the recycling process, like PLA which is known to be biodegradable.

Table 1 Material consumption for additive manufacturing disciplines from a tertiary education institution

Academic Year	*No. of FDM 3D Printers*	*Material Consumption (kg)*	*Scrap Parts (%)**	*Support Structures (%)**
2017–2018	20	38	35	8
2018–2019	35	50	31	8
2019–2020	48	85	23	7
2020–2021	55	100	19	5
2021–2022	55	120	15	7

*Percentage of total material consumption per academic year.

The research aims at lowering the overall material consumption and percentages of scrap parts generated during the process, including support structures, as well as identifying viable solutions for indirect scrap parts. An important part of 3D printed products during the 14-week activities is represented by disposable tools, fixtures, and jigs, as the AM disciplines are included in three main study areas: mechatronics and robotics, industrial engineering, and applied sciences. Thus, the current research follows the design and parameter optimization of 3D printed complex molds for resin casting, considering CE principles applied in an academic environment. 3D printed molds for resin casting are bespoke designed and manufactured as secondary products. Once the resin is casted and demolded, the 3D printed parts are discarded. A common application undertaken with all students is design, manufacture and testing of 3D printed molds for different variations of socket covers. The Computer Aided Design (CAD) model for this application is designed with two components, the top mold cavity and bottom mold cavity (Figure 2).

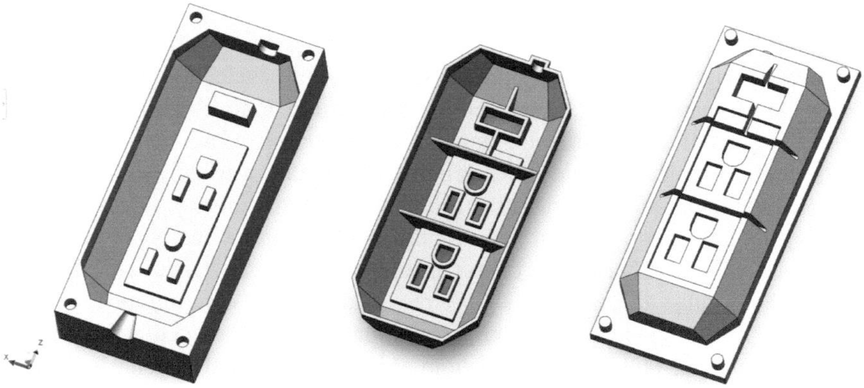

Fig. 2 Complex mold CAD model for resin casting of a socket cover component (left to right): top mold cavity; resin casted part; bottom mold cavity

For this application, students work in groups of three and the manufacturing time frame must fit in one week, from one laboratory session to the subsequent one. Thus, using 20 Zortrax 3D printers, 80 complex molds should be printed in a seven-day time frame, considering that the machines can print overnight. To accurately analyze the production planning options for the selected complex casting mold within a CE environment setting, both common application 3D printing materials and specialized thermoplastics were considered. Thus, six materials were selected, namely: polylactic acid (Z-PLA), acrylonitrile butadiene styrene copolymer (Z-ABS, Z-ULTRAT), high impact polystyrene (Z-HIPS), polyethylene terephthalate glycol modified (Z-PETG), and acrylonitrile styrene acrylate copolymer (Z-ASA Pro). Amongst the plethora of available printing filaments, the materials were selected due to their properties (Table 2) in relation to the function of the proposed part. The working environment of the 3D printed molds involves, amongst others, withstanding temperatures of 80°C due to the exothermic reactivity of the casting resin for approximately 3 minutes. Z-PLA was the first choice for this analysis as it is an ecological material which is derived from natural, biodegradable resources with advantages that fit the CE principles, a very low shrinkage rate, and a high degree of hardness (Table 2). This material is suitable for models with superior details and exceptionally smooth surface, as well as delivering 3D printing high-quality parts, and consumer goods. Z-ABS was the second natural choice as it is a universal solution for economical prototyping and manufacturing due to the low-cost and versatility of the material. This material also withstands high temperatures and it is tough enough for basic functional testing, and is also efficient for economical short-series production of casings, accessories, and educational props. Parts manufactured from Z-ABS are easy to post-process and can be treated chemically or mechanically.

Z-HIPS was also included in the research due to its high level of hardness and durability to withstand chemical agents. Compared to standard high impact

Table 2 Material properties for analyzed 3D printing filaments (adapted from https://zortrax.com/filaments)

Mechanical Properties	*Z-PLA*	*Z-ABS*	*Z-HIPS*	*Z-PETG*	*Z-ULTRAT*	*Z-ASA Pro*	*Test Method*
Tensile strength	47.95 MPa	30.46 MPa	16.90 MPa	40.18 MPa	32.60 MPa	24.21 MPa	ISO 527:1998
Breaking stress	46.53 MPa	25.89 MPa	13.02 MPa	37.27 MPa	30.70 MPa	21.94 MPa	ISO 527:1998
Elongation at max tensile stress	3.80%	4.52%	1.87%	5.83%	3.78%	2.64%	ISO 527:1998
Elongation at break	4.32%	11.08%	7.75%	6.39%	4.87%	2.76%	ISO 527:1998
bending stress	56.80 MPa	46.30 MPa	29.30 MPa	55.30 MPa	54.00 MPa	45.60 MPa	ISO 178:2011
Flexural modulus	1.47 GPa	1.08 GPa	1.18 GPa	1.39 GPa	1.85 GPa	1.36 GPa	ISO 178:2011
Izod impact, notched	3.14 kJ/m^2	8.93 kJ/m^2	4.82 kJ/m^2	3.27 kJ/m^2	5.26 kJ/m^2	3.81 kJ/m^2	ISO 180:2004
Thermal Properties							
Glass transition temperature	57.06°C	107.89° C	96.68°C	77.65°C	106.40°C	80.99°C	ISO 11357-3:2014
Other Properties							
Melt flow rate	14.48 g/10 min Load 2.16 kg Temperature 190°C	11.75 g/10 min Load 10 kg Temperature 220°C	7.14 g/10 min Load 5 kg Temperature 200°C	20.02 g/10 min Load 2.16 kg Temperature 230°C	43.88 g/10 min Load 5 kg Temperature 260°C	74.91 g/10 min Load 10 kg Temperature 220°C	ISO 1133:2006
Specific density	1.292 g/cm3	1.195 g/cm^3	1.136 g/cm^3	1.295 g/cm^3	1.179 g/cm^3	1.176 g/cm^3	ISO 1183-3:2003
Shore hardness (D)	79.8	69.2	73.2	71.4	73.4	68	ISO 868:1998

polystyrene, Z-HIPS has been modified to provide a unique semi-mat finish, making it a good candidate for resin casting as it prevents leakage of the unsolidified fluid through the pores and gaps of the contact surface. The material also has a low-shrinkage rate and allows the manufacture of prototype casings and industrial parts with no post-processing needed. Z-PETG is an industrial-grade material which offers outstanding resistance to impact of time, most acids and alcohols, and weak alkalies. Its properties include durability and tensile strength, as it combines glycol with standard PET material to eliminate brittleness. Because the material is immune to ultraviolet (UV) light, it is a good candidate for manufacturing complex mold models which allow reusage for several resin casting product batches. Z-ULTRAT is an ABS plastic blend designed for durability and excellent surface quality and works great whenever high quality is a requirement. This 3D printing material is resistant to high temperatures and impacts. Based on ABS, the material can be easily post-processed with acetone and mechanical treatment. This thermoplastic allows to reliably achieve properties found in items manufactured with injection molding technology. Z-ULTRAT is a very well suited candidate for cost-effective low-series production. The last selected material is Z-ASA Pro, which is durable and exhibits superior resistance to external factors, including UV radiation and extreme weather conditions, extending the life span of models and allowing reuse of cast molds for several manufacturing cycles. Z-ASA Pro is a suitable choice for prototypes of fixtures, jigs, and specialized tools, displaying a high temperature resistance, no deformation, no bleaching or no changing of physical characteristics even after several extensive uses of the manufactured part.

The scrap percentage of 15% must be taken into account when designing the manufacturing program. Post-processing tasks account for 0.5% of the AM time of each component.

3. Results

The purpose of this research was to highlight the fact that production planning for AM within a CE environment involves the careful selection of materials, processes, equipment, post-processing methods, and manufacturing parameters to minimize waste and reduce the environmental impact of AM while maximizing the economic efficiency of the production process. For the current research, the academic setting sets the process and equipment, namely: FDM is used on the Zortrax equipment available within the university labs. Thus, the materials and 3D printing parameters are analyzed in order to optimize the manufacturing results targeting the completion of a CE environment. The optimization of 3D printing parameters for FDM technology is critical in a CE environment as it helps to reduce resource consumption, to improve part quality, reduce costs, and increase sustainability. For example, optimizing the infill density can result in stronger, more functional parts that are less likely to fail. By optimizing the 3D printing parameters, FDM technology was used to support the principles of CE by reducing waste and minimizing the environmental impact of AM. In this case, production planning was undertaken based on the previously mentioned specifications. The main 3D printing parameters that were considered constant are presented in Table 3, for all six materials.

Table 3 Constant 3D printing parameters for manufacturing of complex molds

Parameter	*Value*
Support structure angle [deg]	30
Smart bridges	Yes
Support lite	Yes
Nozzle diameter [mm]	0.4
Platform temperature [o]	80
Print quality	High
Print type	Normal
Infill density [%]	50
Contour-infill gap [mm]	0.4
Contour-top gap [mm]	0.25
Top surface layers	8
Bottom surface layers	4
Seam	Random
Max. wall thickness [mm]	3.13
First layer gap [mm]	0.31
Raft layers [No.]	7

Two main parameters were varied for this study. The layer height and infill print pattern are important parameters that influence the optimization of production planning of a 3D printed complex mold. Layer height, which refers to the thickness of each layer of material that is deposited during the 3D printing process, has a direct impact on the quality and accuracy of the final part, as well as on the overall production time. A smaller layer height results in a smoother surface finish, but also increases the overall printing time. In the case of complex molds, it may be necessary to use a smaller layer height to ensure that the final part is accurate and has the desired surface finish. Infill refers to the internal structure of a 3D printed part that provides support and strength. The infill print pattern refers to the pattern used to create the internal structure of the part. There are several different infill patterns available in Z-Suite, each with its own advantages and disadvantages. For example, a grid pattern provides good support and strength, while a honeycomb pattern is more efficient in terms of material usage. Choosing the right infill print pattern is important for optimizing the production planning of complex molds, as it can have a significant impact on the overall cost, time, and quality of the final part. By carefully balancing the trade-offs between printing time, material usage, and part quality, it is possible to optimize the 3D printing process and produce high-quality, cost-effective parts that meet the requirements of the CE. The analysis is undertaken for all six materials in order to identify the best fit for this application (Table 4).

Table 4 Variable 3D printing parameters for FDM of complex mold—1 top and 1 bottom mold cavity

Material	*Layer Thickness [mm]*	*Print Pattern*	*Estimated Print Time [h min]*	*Material Usage [m]/[g]*	*Material Cost [€] **	*Energy Cost [€]*	*Print Cost [€]*
Z-PLA	0.14	PATT.0	16h 25min	74.76m / 213g	8.87	1.21	10.08
		PATT.1	21h 25min	72.73m / 207g	8.62	1.57	10.20
		PATT.2	23h 03min	84.32m / 240g	10.00	1.69	11.69
		PATT.3	17h 55min	69.02m / 197g	8.21	1.32	9.52
	0.19	PATT.0	14h 44min	76.22m / 217g	9.04	1.08	10.12
		PATT.1	18h 12min	74.32m / 212g	8.83	1.34	10.17
		PATT.2	19h 22min	84.92m / 242g	10.08	1.42	11.50
		PATT.3	15h 47min	71.11m/ 203g	8.45	1.16	9.61
	0.29	PATT.0	11h 25min	81.67m / 233g	9.70	0.84	10.54
		PATT.1	13h 55min	79.74m / 227g	9.45	1.02	10.48
		PATT.2	13h 38min	89.63m / 256g	10.66	1.00	11.66
		PATT.3	11h 43min	76.93m / 219g	9.12	0.86	9.98
Z-ABS	0.14	PATT.0	16h 38min	74.53m / 177g	9.22	1.22	10.44
		PATT.1	21h 43min	72.43m / 172g	8.95	1.60	10.55
		PATT.2	23h 17min	84.07m / 200g	10.41	1.71	12.12
		PATT.3	18h 09min	68.79m / 164g	8.54	1.33	9.87
	0.19	PATT.0	13h 58min	76.93m / 183g	9.53	1.03	10.55
		PATT.1	17h 27min	74.91m / 178g	9.27	1.28	10.55
		PATT.2	18h 35min	85.30m / 203g	10.57	1.37	11.93
		PATT.3	15h 01min	71.79m / 171g	8.90	1.10	10.01
	0.29	PATT.0	11h 10min	80.00m / 190g	9.89	0.82	10.71
		PATT.1	13h 19min	78.03m / 186g	9.68	0.98	10.66
		PATT.2	13h 56min	87.71m / 209g	10.88	1.02	11.90
		PATT.3	11h 48min	75.46m / 180g	9.37	0.87	10.24

(Contd.)

Table 4 (*Contd.*)

Material	*Layer Thickness [mm]*	*Print Pattern*	*Estimated Print Time [h min]*	*Material Usage [m]/[g]*	*Material Cost [€] **	*Energy Cost [€]*	*Print Cost [€]*
Z-HIPS	0.14	PATT.0	14h 13min	73.85m / 183g	11.98	1.04	13.02
		PATT.1	18h 44min	71.75m / 178g	11.65	1.38	13.03
		PATT.2	21h 25min	83.38m / 206g	13.48	1.57	15.06
		PATT.3	16h 23min	68.13m / 169g	11.06	1.20	12.26
	0.19	PATT.0	12h 27min	76.47m / 189g	12.37	0.91	13.28
		PATT.1	15h 34min	74.50m / 184g	12.04	1.14	13.19
		PATT.2	17h 28min	85.14m / 211g	13.81	1.28	15.09
		PATT.3	13h 58min	71.38m / 177g	11.58	1.03	12.61
	0.29	PATT.0	10h 43min	77.25m / 191g	12.50	0.79	13.29
		PATT.1	12h 53min	75.41m / 187g	12.24	0.95	13.19
		PATT.2	13h 28min	84.55m / 209g	13.68	0.99	14.67
		PATT.3	11h 20min	72.84m / 180g	11.78	0.83	12.61
Z-PETG	0.14	PATT.0	16h 56min	75.04m / 227g	14.86	1.24	16.10
		PATT.1	22h 06min	73.01m / 221g	14.46	1.62	16.09
		PATT.2	23h 34min	84.19m / 254g	16.62	1.73	18.36
		PATT.3	18h 27min	69.32m / 210g	13.74	1.36	15.10
	0.19	PATT.0	14h 26min	75.66m / 229g	14.99	1.06	16.05
		PATT.1	18h 03min	73.71m / 223g	14.60	1.33	15.92
		PATT.2	19h 10min	84.22m / 255g	16.69	1.41	18.10
		PATT.3	15h 30min	70.41m / 213g	13.94	1.14	15.08
		PATT.0	12h 02min	78.12m / 236g	15.45	0.88	16.33
	0.29	PATT.1	14h 11min	76.31m / 231g	15.12	1.04	16.16
		PATT.2	14h 47min	85.38m / 258g	16.89	1.09	17.97
		PATT.3	12h 38min	73.75m / 223g	14.60	0.93	15.52
Z-ULTRAT	0.14	PATT.0	16h 33min	76.67m / 182g	11.91	1.22	13.13
		PATT.1	21h 34min	74.44m / 177g	11.58	1.58	13.17
		PATT.2	23h 11min	86.70m / 206g	13.48	1.70	15.19
		PATT.3	18h 03min	70.72m / 168g	11.00	1.33	12.32
	0.19	PATT.0	14h 39min	77.50m / 184g	12.04	1.08	13.12
		PATT.1	18h 08min	75.40m / 179g	11.72	1.33	13.05
		PATT.2	19h 18min	86.51m / 206g	13.48	1.42	14.90
		PATT.3	15h 42min	72.15m / 172g	11.26	1.15	12.41
	0.29	PATT.0	11h 30min	77.60m / 185g	12.11	0.84	12.95
		PATT.1	13h 38min	75.70m / 180g	11.78	1.00	12.78
		PATT.2	14h 14min	84.93m / 202g	13.22	1.05	14.27
		PATT.3	12h 07min	73.16m / 174g	11.39	0.89	12.28

(*Contd.*)

Table 4 (*Contd.*)

Material	*Layer Thickness [mm]*	*Print Pattern*	*Estimated Print Time [h min]*	*Material Usage [m]/[g]*	*Material Cost [€] **	*Energy Cost [€]*	*Print Cost [€]*
Z-ASA Pro	0.14	PATT.0	16h 31min	74.36m / 207g	13.55	1.21	14.76
		PATT.1	21h 31min	72.25m / 202g	13.22	1.58	14.80
		PATT.2	23h 09min	83.91m / 234g	15.32	1.70	17.02
		PATT.3	18h 00min	68.62m / 191g	12.50	1.32	13.82
	0.19	PATT.0	14h 39min	74.99m / 209g	13.68	1.08	14.76
		PATT.1	18h 08min	73.09m / 204g	13.35	1.33	14.68
		PATT.2	19h 17min	83.29m / 232g	15.18	1.42	16.60
		PATT.3	15h 41min	69.98m / 195g	12.76	1.15	13.91
	0.29	PATT.0	11h 11min	77.10m / 215g	14.07	0.82	14.89
		PATT.1	13h 21min	75.34m / 210g	13.74	0.98	14.73
		PATT.2	13h 57min	84.29m / 235g	15.38	1.02	16.41
		PATT.3	11h 49min	72.91m / 203g	13.29	0.87	14.15

*Costs are estimated from https://store.zortrax.com/ for 800g material spools of black standard materials: Z-PLA—€33.32, Z-ABS—€41.65, Z-HIPS—€52.36, Z-PETG—€52.36, Z-ULTRAT—€52.36, Z-ASA Pro—€52.36.

Energy costs calculated in Table 4 were based on the fact that a Zortrax M300+ 3D printer has a maximum power consumption of 360 W per hour of functioning and the average electricity cost is 0.2041 €/kwh. Parameter analysis was done in Z-Suite for a pair of one top mold cavity and one bottom mold cavity positioned in the centre of the 3D printing platform of a Zortrax M300+ equipment (Figure 3).

Based on the data in Figure 4, the fastest print times are achieved with a 0.29 mm layer thickness and a PATT.0 infill pattern for all materials. Z-HIPS has the quickest print time with a 10 hours and 43 minutes time frame for manufacturing of a complex mold for a socket cover. The second fastest is PATT.3, also for 0.29 mm layer thickness.

It is somewhat expected that the time frame reduces with the increase in layer thickness, and in general, all the selected materials had this evolution. Nevertheless, Figure 4 shows that for all materials a layer thickness of 0.14 mm results in faster printing times than some patterns from the 0.19 mm and 0.29 mm analyzed groups, thus ensuring for a better overall structural integrity and surface finish of the printed mold. This is of particular importance for the selected application because the mold active surfaces and details will not need any further post-processing, and it will also allow reusage of one mould for several casting cycles, promoting CE principles within the proposed AM academic ecosystem for undergraduate and postgraduate courses.

It is observed that PATT.0 has also the fastest printing time amongst all four patterns that were studied, in each layer thickness category, followed by PATT.3.

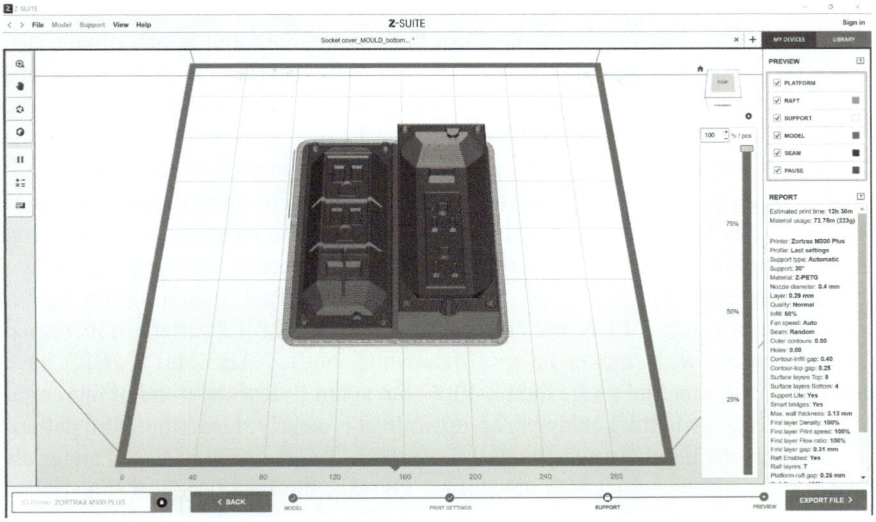

Fig. 3 Z-Suite parameter simulation for a complex mold of a socket cover component: material—PETG, layer thickness—0.29 mm, infill print pattern—PATT. 3

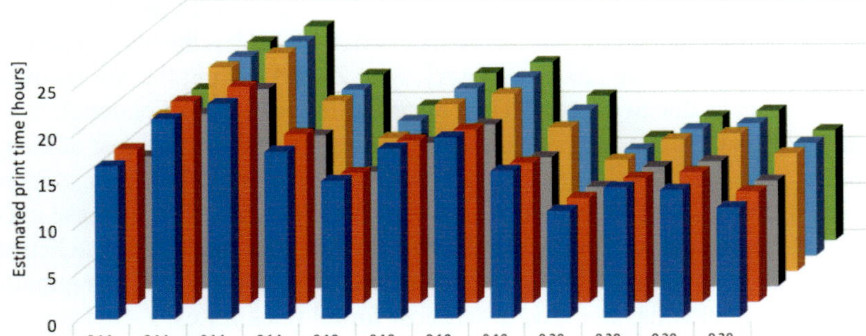

	0.14, PATT. 0	0.14, PATT. 1	0.14, PATT. 2	0.14, PATT. 3	0.19, PATT. 0	0.19, PATT. 1	0.19, PATT. 2	0.19, PATT. 3	0.29, PATT. 0	0.29, PATT. 1	0.29, PATT. 2	0.29, PATT. 3
PLA	16,42	21,42	23,05	17,92	14,73	18,20	19,37	15,78	11,42	13,92	13,63	11,72
ABS	16,63	21,72	23,28	18,15	13,97	17,45	18,58	15,02	11,17	13,32	13,93	11,80
HIPS	14,22	18,73	21,42	16,38	12,45	15,57	17,47	13,97	10,72	12,88	13,47	11,33
PETG	16,93	22,10	23,57	18,45	14,43	18,05	19,17	15,50	12,03	14,18	14,78	12,63
ULTRAT	16,55	21,57	23,18	18,05	14,65	18,13	19,30	15,70	11,50	13,63	14,23	12,12
ASA Pro	16,52	21,52	23,15	18,00	14,65	18,13	19,28	15,68	11,18	13,35	13,95	11,82

Fig. 4 Estimated print time for a complex mold manufactured using FDM

Overall, the best choice for 3D printing of the mold would depend on the desired trade-off between the print time and material properties such as strength and thermal resistance. In this case, thermal resistance is important, as the used resin is usually exothermal. To make a proper selection of the simulated parameters, material usage is also analyzed for each material type, layer thickness, and infill pattern type (Figure 5).

Based on the results, the highest material usage is offered by Z-PETG (85.38m/258g) with a layer thickness of 0.29 mm. Compared to other materials, Z-PETG has the highest material usage, followed by Z-PLA (89.63m/256g) with a layer thickness of 0.29 mm. Z-ABS has the lowest material usage of all materials, with a maximum usage of 68.79m/164g at a layer thickness of 0.14 mm. It is observed that for all the materials, there is a trend of increased material usage with higher layer thicknesses. This is due to the need to build up the same volume of material with fewer, but thicker, layers. The thicker the layer, the more material is required to fill the layer, leading to increased material consumption. When comparing the different materials, it can be seen that Z-PETG has a slightly higher material usage than Z-PLA, but both materials show a similar pattern of increased material usage with thicker layers. Meanwhile, both Z-ABS and Z-HIPS have slightly lower material usage than Z-PLA, but again follow the trend of increased material usage with thicker layers. Meanwhile, it is clearly shown that infill pattern PATT.2 has the highest rate of material consumption amongst all tested materials and across all layer thickness categories. Overall, the best usage of material in terms of consumption is given by pattern PATT.3.

Printing costs are important when taking into consideration CE goals; thus, an analysis based on material and energy consumption was undertaken for all materials (Figure 6). Z-PLA and Z-ABS values are quite similar, even though the material consumption is lower for Z-ABS. This similarity in cost is given by the price of Z-ABS and Z-PLA filaments, the latter being cheaper. When comparing the cost of different materials, it can be seen that Z-PETG has the highest part cost, followed by Z-ASA Pro and Z-ULTRAT, with Z-PLA having the lowest overall cost. This can be due to the cost of raw materials for each type of material, as well as any additional processing costs. In terms of layer thickness, the cost of the part does not increase as the layer thickness rises for all materials, but change in costs were observed in each layer thickness group across patterns. The increase in cost is more pronounced for Z-PETG and Z-HIPS, while the cost difference between 0.14 mm and 0.29 mm thickness is smaller for Z-ABS and Z-PLA. It is important to note that Z-PLA has the lowest part cost for all thicknesses. Z-PETG has the highest part cost among all materials and thicknesses, mainly due to the fact that is has the highest cost (52.36 €/800g) per spool of filament, but also because it has the highest material consumption out of all six materials (Figure 6). The most effective pattern in terms of costs is PATT.3 across all materials and layer thicknesses.

In order to further plan the production of the complex mold for casting a socket cover, with consideration to the CE targets set in the incipient stages of the research, an optimum material and set of 3D printing parameters were chosen. When choosing the appropriate material and the set of parameters for 3D printing of the complex mold, the considered factors included cost, material properties such as ease of 3D printing, dimensional stability, and thermal resistance, and the specific requirements of the complex resin casting mold. It was important to strike a balance between cost and performance to make the best choice for this research. Based on the results obtained and on the study requirements, Z-PLA material was

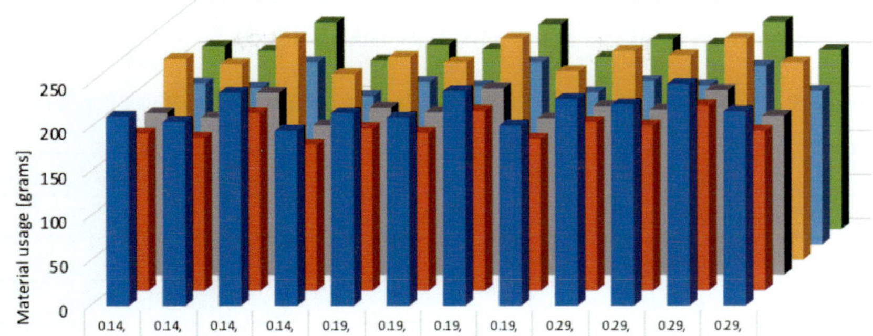

	0.14, PATT. 0	0.14, PATT. 1	0.14, PATT. 2	0.14, PATT. 3	0.19, PATT. 0	0.19, PATT. 1	0.19, PATT. 2	0.19, PATT. 3	0.29, PATT. 0	0.29, PATT. 1	0.29, PATT. 2	0.29, PATT. 3
PLA	213	207	240	197	217	212	242	203	233	227	256	219
ABS	177	172	200	164	183	178	203	171	190	186	209	180
HIPS	183	178	206	169	189	184	211	177	191	187	209	180
PETG	227	221	254	210	229	223	255	213	236	231	258	223
ULTRAT	182	177	206	168	184	179	206	172	185	180	202	174
ASA Pro	207	202	234	191	209	204	232	195	215	210	235	203

Fig. 5 Material usage for a complex mold manufactured using FDM

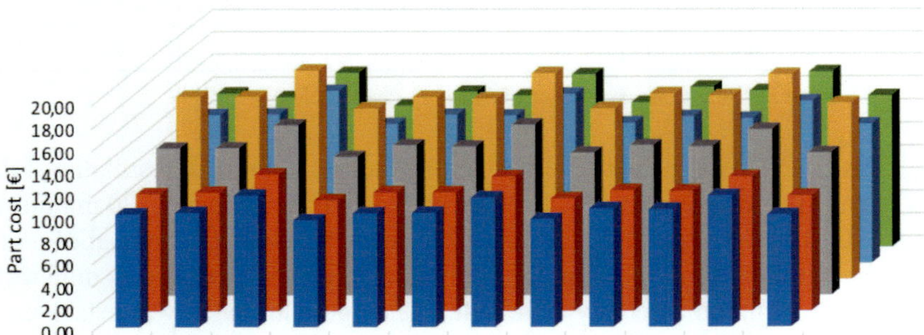

	0.14, PATT. 0	0.14, PATT. 1	0.14, PATT. 2	0.14, PATT. 3	0.19, PATT. 0	0.19, PATT. 1	0.19, PATT. 2	0.19, PATT. 3	0.29, PATT. 0	0.29, PATT. 1	0.29, PATT. 2	0.29, PATT. 3
PLA	10,08	10,20	11,69	9,52	10,12	10,17	11,50	9,61	10,54	10,48	11,66	9,98
ABS	10,44	10,55	12,12	9,87	10,55	10,55	11,93	10,01	10,71	10,66	11,90	10,24
HIPS	13,02	13,03	15,06	12,26	13,28	13,19	15,09	12,61	13,29	13,19	14,67	12,61
PETG	16,10	16,09	18,36	15,10	16,05	15,92	18,10	15,08	16,33	16,16	17,97	15,52
ULTRAT	13,13	13,17	15,19	12,32	13,12	13,05	14,90	12,41	12,95	12,78	14,27	12,28
ASA Pro	13,55	13,22	15,32	12,50	13,68	13,35	15,18	12,76	14,07	13,74	15,38	13,29

Fig. 6 Print cost for a complex mold manufactured using FDM

selected with the PATT.3 infill pattern and a layer thickness of 0.14 mm. This configuration has the lowest cost, €9.52, out of all other tested. It also provides a lower end material consumption with 197 g, the lowest being 177 g for Z-HIPS and the highest being 258 g for Z-PETG. This selection also took into consideration the printing time, which is in the middle range, adding up to 17 hours and 55 minutes, with the highest printing time registered by PETG at a 0.14 mm layer thickness and a PATT.2 infill type. This configuration was further used to plan the optimum batch in order to manufacture all 80 molds within the one-week time frame using the 20 3D printers available in the university labs (Table 5).

Table 5 Production planning of 80 complex molds using an M300 Plus 3D printing equipment

Characteristics	*Values*
Overall dimensions of set* (L × l × h) [mm × mm × mm]	125 × 148 × 26
Optimum batch [no. of sets*]	4
Print time [d h min] / batch	2d 23h 17 min
Post-processing time [d h min] / batch (0.5%)	21.4 min
Manufacturing time [d h min] / batch	2d 23h 38.4 min
Material consumption [m]/[g] / batch	272.86 m/778g
Scrap material (15%) [m]/[g] / batch	40.929 m/116.7g
Manufacturing cost [€] / batch	42.5014
No. of necessary batches to achieve production volume [no.]	20
Total production time [d h min] / production volume	20 printers x 2d 23h 17 min
Total material consumption [m] /[g] / production volume	6275.78 m/17894g
Total manufacturing cost [€] / production volume	850.028

*One set is a pair of one bottom cavity mold and one top cavity mold

This time frame included designing (1.5 hours), Standard Tesselation File (STL) and G-Code preparation (30 minutes), machine maintenance (30 minutes), actual 3D printing using FDM (2d 23h 17 min/one 3D printer), removal of parts from the build platform (6.4 minutes), eliminating raft by peeling (15 minutes), and testing of fit and assembly (1 minute). The optimum batch was established using Z-Suite's option of Replicate the Z-Codex at a 5 mm offset; after all printing parameters have been configured, the print job was estimated at 2 days 23 hours and 17 minutes, with a consumption of 778 grams of filament.

4. Discussions

The primary contribution of the current research lies in its successful optimization of production planning for a complex 3D printed part within a CE environment. This not only enhances the sustainability, efficiency, and cost-effectiveness of the manufacturing process but also has implications for academia and the environment. Previous studies have explored the role of AM in promoting CE principles (Priyadarshini et al., 2022; Verma et al., 2022; Bal and Badurdeen, 2022); however, the current research adds new insights by focusing on a specific academic setting and a bespoke product application. One of the key findings is the importance of material and parameter selection in optimizing the 3D printing process. The conducted research demonstrated, in an SME-like setting, that by selecting appropriate materials (Z-PLA) and optimizing parameters (PATT.3 infill pattern and 0.14 mm layer thickness), it is possible to reduce overall costs, minimize material consumption, and shorten production time, whilst maintaining quality indicators. This aligns with previous research that has emphasized the importance of material and parameter selection in AM (Papacharalampopoulos et al., 2018; Qamar Tanveer et al., 2022), whilst creating a closed-loop system for material usage (Nascimento et al., 2022). In line with these findings, the presented research provides a step-by-step structural approach for academia-based manufacturing facilities, where CE principles have not been a clear priority (Paletta et al., 2019; Nunes et al., 2018). It is important to acknowledge that academic institutions play a key role in steering new business and consumer habits into a CE mindset (Serrano-Bedia and Perez-Perez, 2022); thus, the current study provides substantial evidence on how to efficiently apply optimization principles based on three main economic indicators—quality, time, and cost—with the final goal of proving an integrative hypothesis of approaching the academic working environment as an SME. This organizational strategy, constructing and testing an AM academic ecosystem for undergraduate and postgraduate courses, can be further developed and implemented for AM spin-offs and start-ups, creating a basis for the future business environment.

Comparing the methodologies applied by the authors with previous research, it is evident that the authors' approach focuses on a more holistic optimization process. While some studies have investigated material and parameter selection in isolation (Qamar Tanveer et al., 2022), the presented research considers additional factors such as post-processing, assembly, recycling, waste reduction, and time management, whilst promoting a CE approach on AM. This comprehensive approach has led to the generation of new theoretical insights that can be applied to manufacturing management, in general.

This research helps to reconcile some of the conflicting findings in the literature by demonstrating that material and parameter selection can be optimized to achieve a balance among cost, material consumption, and production time for bespoke products, but in a business factory-like setting. For example, while previous studies have reported varying results regarding the most appropriate materials and parameters for 3D printing (Golab et al., 2022), the authors' work shows that a combination of Z-PLA, PATT.3 infill pattern, and 0.14 mm layer thickness provides the best balance for the specific application of a complex resin casting mold. This demonstration is of particular importance, considering that previous research has identified that personalized manufacturing is in high demand (Kerin and Pham, 2020), with the potential for lowering overproduction and stock surplus (Zhu et al., 2021).

In addition, the proposed research adds to the existing body of literature by investigating the impact of AM on advancing CE principles within the environment of academia. Previous research has primarily focused either on the broader applications of AM and CE (Cruz Sanchez et al., 2020; Ferreira et al., 2023;

Priyadarshini et al., 2022) or on the role of HEIs in transitioning toward a CE attitude (Paletta et al., 2019; Serrano-Bedia and Perez-Perez, 2022). This study highlights the specific benefits of integrating the CE principles within an HEI, which can lead to a more sustainable and efficient production process, as well as contribute to the overall sustainability goals of the institution, by creating synergic competitive advantages. The successful integration of AM and CE principles within an academic context highlights the potential for similar applications in other industries. By recognizing the potential synergies among CE, AM, and HEIs, concepts can be applied into related fields. For example, managers can make more informed decisions about how to incorporate these concepts and their benefits into their operations, leading to more sustainable and efficient production processes that align with the growing emphasis on environmental responsibility in the manufacturing sector. Moreover, the findings emphasize the importance of context-specific optimization. Decisional factors should be aware of the unique requirements of their specific products and industry contexts when implementing AM and CE principles, acknowledging that a tailored approach can help maximize the benefits of these technologies while minimizing potential drawbacks, as demonstrated in this research.

A new avenue investigated by the authors was to approach the academic environment as an SME. This approach offers the potential for new business models to emerge, providing valuable insights into the economic and social implications of such a paradigm shift. By treating HEIs as SMEs, we initiated the exploration of ways in which universities and colleges can leverage the adoption of AM and CE principles to drive innovation and enhance their sustainability efforts. There is a need to differentiate between spin-offs and start-ups and the current proposed academic ecosystem. The first two have an independent legal status and can react very rapidly to technological changes (Krishnendu et al., 2023), whilst the proposed model is a part of the HEI legal status, being obliged to adhere to the HEI regulatory standards and regulations. Although a part of an academic environment, which can prove to be challenging in terms of keeping up with rapid technological changes in the market (Serrano-Bedia and Perez-Perez, 2022), it was demonstrated that the proposed AM academic ecosystem can mimic the behaviour of these two business models, whilst having economic stability given by its base HEI. Through the proposed hypothesis the AM ecosystem can rapidly adapt to any bespoke manufacturing requirement, within an academic setting and CE strategy.

The SME approach also opened the way to a different investigation. Traditionally, a manufacturing environment is based on human, material (capital), and financial resources. Optimization of material or capital resources is usually undertaken within certain technical requirements given by the used technology, in this case AM. One of the main limitations mentioned by Chen (Chen et al., 2022), and also encountered during the conducted research, was the variability of the 3D printers, correlated with low rates of repeatability for each machine. To overcome these limitations, we applied the anthropomorphism principle (Huang et al., 2019) to the SME-AM ecosystem and to the manufacturing process. This new management approach led to a deeper understanding of the proposed organizational structure and its behaviour. Because of their uniqueness, each 3D printer was considered an "individual" with different skills and competences. We found that the manufacturing system should be considered as a "living" entity which can evolve and adapt to specific environmental conditions. Similar to a team within which each individual serves tasks in correspondence with their levels of competence, the AM ecosystem was evaluated with characterization charts for each machine, creating distinct machine profiles, each focused at best completing different targets, amongst which we mention: dimensional accuracy, surface roughness, complex overhanging structures, thin walls, embossed and engraved details, material consumption, speed, mechanical properties through geometry optimization. The authors' proposed approach showed that management strategies for AM should include in their performance improvements and optimizations, the "personality" of each machine if desired to obtain competitive advantages.

Conclusions

Additive manufacturing and CE are a natural fit, with the former offering several key benefits that support the principles of the latter. By reducing waste, promoting recycling, enabling on-demand production, and facilitating repair and reuse, 3D printing has the potential to play a critical role in the transition toward a more sustainable and efficient future. Institutions that embrace the combination of AM and the CE are in a position to embrace benefits of a more sustainable and efficient production process, while also contributing to a more sustainable future.

The current research proposed an integrative hypothesis to approach the academic working environment as an SME and applied optimization principles based on three main economic indicators—quality, time, and cost. A complex casting mold was designed to ensure an efficient use of resources within a 3D printing laboratory. The overall costs were reduced for the 3D printed mold by reducing material consumption through optimization of the infill pattern to PATT.3, which gave the best material usage at 197 g for Z-PLA. The mold was designed so that the need for support structures was eliminated, thus leading to material reduction and therefore to lower costs. Also, the printing time was optimized through the geometry of the part and by selecting optimum 3D printing parameters like infill pattern PATT.3, layer height 0.14 mm, and Z-PLA filament. Labour costs were reduced for post-processing by eliminating deburring of support structures. The mold only needed to be removed from the printing platform and the raft peeling. The CAD model was optimized to minimize the number of components to only two, eliminating any cores or inserts, thus reducing the time and labour required for assembly. The fit between the two mould cavities was adjusted for FDM 3D printing with a 0.35 mm gap between the guiding pins, thus the need for fasteners was eliminated. Recycling was facilitated through the use of PLA, a biodegradable thermoplastic. Future research involves the testing of water-soluble materials, such as Polyvinyl Alcohol (PVA), not only for support structure manufacturing, but also for 3D printing of the entire part, thus combating waste.

The economic implications of the proposed hypothesis are multifaceted. An SME-like framework for manufacturing laboratories (but not limited to them) could lead to increased resource efficiency, waste reduction, and associated cost savings for HEIs. This can help academic institutions optimize their use of resources and reduce operational costs, making them more competitive in the marketplace, this being the next step into adaptive and new business models (Rojas et al. 2021; Nunes et al., 2018). We conclude that the emergence of new business models within the academic environment can lead to the creation of new revenue streams. For instance, universities and colleges could potentially offer specialized courses, workshops, or consultancy services related to AM and CE strategies. This diversification of income sources can contribute to the financial stability and growth of academic institutions (Paletta et al., 2019).

Adoption, development, and implementation of the proposed AM ecosystem is expected to have implications for academic stakeholders as well. For professors, this new approach may require them to adapt their teaching methodologies and curricula to incorporate new topics, leading to the development of new interdisciplinary courses and research opportunities, fostering greater collaboration among faculty members from different departments. For students, this approach can offer a more practical and hands-on learning experience, exposing them to real-world applications of AM and CE principles in an academic setting. This can help students develop valuable skills and knowledge, enhancing their employability and career prospects upon graduation. Since they are also the next generation of managers and company employees, their training and involvement in an interactive learning environment can potentially foster creativity and problem-solving competences for the development of future adaptive business models.

Aside from the economic and educational implications, this new approach also has the potential to promote greater sustainability within academic institutions. By adopting AM and CE principles in the demonstrated ecosystem, universities and colleges can reduce their environmental impact, aligning with global sustainability goals (Priyadarshini et al, 2022). This can not only improve the reputation of HEIs but also serve as a powerful example for other organizations and industries seeking to adopt more sustainable practices (Verma et al., 2022; Krishnendu et al., 2023). By adhering to AM and CE strategies, academic institutions can optimize their use of resources, reduce operational costs, and diversify their revenue streams, ultimately benefiting professors, students, and the wider community. This study contributes to the existing literature on sustainable business models and highlights the importance of embracing new paradigms in the pursuit of a more sustainable future.

The aim to optimize the production planning of a complex 3D printed part within a CE environment was a crucial step toward creating a more sustainable, efficient, and cost-effective manufacturing process within an HEI. By focusing on reducing costs, reducing assembly and fit of components, and recycling the final part, it was possible to outline a more sustainable future strategy for both academia and the environment.

Acknowledgments

This work was supported by a grant from the National Program for Research of the National Association of Technical Universities—GNAC ARUT 2023, Grant No. 14/06.10.2023.

References

Bal, A. and Badurdeen, F. 2022. A Simulation-based Optimization Approach for Network Design: The Circular Economy Perspective. *Sustainable Production and Consumption* 30: 761–775.

Benedetti, M., Plessis, A., Ritchie, R.O., Dallago, M., Razavi, S.M.J. and Berto, F. 2021. Architected Cellular Materials: A Review on Their Mechanical Properties towards Fatigue-tolerant Design and Fabrication. *Materials Science and Engineering: R: Report* 144: 100606.

Camilleri, M.A. 2019. Closing the Loop for Resource Efficiency, Sustainable Consumption and Production: A Critical Review of the Circular Economy. *International Journal of Sustainable Development* 21: 1–17.

Chen, Z., Han, C., Gao, M., Kandukuri, S.Y. and Zhou, K. 2022. A Review on Qualification and Certification for Metal Additive Manufacturing. *Virtual and Physical Prototyping* 17(2): 382–405.

Corona, B., Shen, L., Reike, D., Carreón, J.R. and Worrell, E. 2019. Towards Sustainable Development through the Circular Economy: A Review and Critical Assessment on Current Circularity Metrics. *Resources, Conservation and Recycling* 151: 1–15.

Cruz Sanchez, F.A., Boudaoud, H., Camargo, M. and Pearce, J.M. 2020. Plastic Recycling in Additive Manufacturing: A Systematic Literature Review and Opportunities for the Circular Economy. *Journal of Cleaner Production* 264: 121602.

Daneshmand, M., Noroozi, F., Corneanu, C., Mafakheri, F. and Fiorini, P. 2023. Industry 4.0 and Prospects of Circular Economy: A Survey of Robotic Assembly and Disassembly. *The International Journal of Advanced Manufacturing Technology* 124: 2973–3000.

Enyoghasi, C. and Badurdeen, F. 2021. Industry 4.0 for Sustainable Manufacturing: Opportunities at the Product, Process, and System Levels. *Resources, Conservation & Recycling* 166: 105362.

Ferreira, I.A., Godina, R., Pinto, A., Pinto, P. and Carvalho, H. 2023. Boosting Additive Circular Economy Ecosystems Using Blockchain: An Exploratory Case Study. *Computers & Industrial Engineering* 175: 108916.

Gao, C., Wolff, S. and Wang, S. 2021. Eco-friendly Additive Manufacturing of Metals: Energy Efficiency and Life Cycle Analysis. *Journal of Manufacturing Systems* 60: 459–472.

Golab, M., Massey, S. and Moultrie, J. 2022. How Generalisable Are Material Extrusion Additive Manufacturing Parameter Optimisation Studies? A Systematic Review. *Heliyon* 8(11): e11592.

Huang, F., Wong, V.C. and Wan, E.W. 2019. The Influence of Product Anthropomorphism on Comparative Judgment. *Journal of Consumer Research* 46(5): 936–955.

Hui, J., Yan, Z., Lv, J., Liu, Y., Ding, K. and Chan, F.T.S. 2022. An Investigation on Energy Consumption and Part Quality of Stereolithography Apparatus Manufactured Parts. *Rapid Prototyping Journal* 28(1): 52–67.

Javaid, M., Haleem, A., Singh, R.P., Suman, R. and Rab, S. 2021. Role of Additive Manufacturing Applications towards Environmental Sustainability. *Advanced Industrial and Engineering Polymer Research* 4(4): 312–322.

Jawahir, I.S. and Bradley, R. 2016. Technological Elements of Circular Economy and the Principles of 6R-based Closed-loop Material Flow. In G. Seliger, H. Kohl, and J. Mallon (eds), *13th Global Conference on Sustainable Manufacturing—Decoupling Growth from Resource Use. Procedia CIRP* 40: 103–108.

Jayasekara, D., Lai, N.Y.G., Wong, K.H., Pawar, K. and Zhu, Y. 2022. Level of Automation (LOA) in Aerospace Composite Manufacturing: Present Status and Future Directions towards Industry 4.0. *Journal of Manufacturing Systems* 62: 44–61.

Kerin, M. and Pham, D.T. 2020. Smart Remanufacturing: A Review and Research Framework. *Journal of Manufacturing Technology Management* 31(6): 1205–1235.

Krishnendu, S., Chrisovaladis, M., Soumyadeb, C. and Prasanta Kumar, D. 2023. Impact of Institutional Voids on the Performance of Small and Medium-sized Enterprises. *Journal of Strategy and Management* 16(1): 1–27.

McGregor, D.J., Rylowicz, S., Brenzel, A., Baker, D., Wood, C., Pick, D., Deutchman, H., Shao, C., Tawfick, S. and King, W.P. 2021. Analyzing Part Accuracy and Sources of Variability for Additively Manufactured Lattice Parts Made on Multiple Printers. *Additive Manufacturing* 40: 101924.

Mikula, K., Skrzypczak, D., Izydorczyk, G., Warchoł, J., Moustakas, K., Chojnacka, K. and Witek-Krowiak, A. 2021. 3D Printing Filament as a Second Life of Waste Plastics—A Review. *Environmental Science and Pollution Research* 28(10): 12321–12333.

Nascimento, D.L. de M., Nepomuceno, R.M., Caiado, R.G.G., Maqueira, J.M., Moyano-Fuentes, J. and Garza-Reyes, J.A. (2022). A sustainable circular 3D printing model for recycling metal scrap in the automotive industry. *Journal of Manufacturing Technology Management*, 35(5), 876–892. https://doi.org/10.1108/JMTM-10-2021-0391

Nunes, B.T., Pollard, S.J.T., Burgess, P.J., Ellis, G., De los Rios, I.C. and Charnley, F. 2018. University Contributions to the Circular Economy: Professing the Hidden Curriculum. *Sustainability* 10: 2719.

Paletta, A., Fava, F., Ubertini, F., Bastioli, C., Gregori, G., La Camera, F. and Douvan, A.R. 2019. Universities, Industries and Sustainable Development: Outcomes of the 2017 G7 Environment Ministerial Meeting. *Sustainable Production and Consumption* 19: 1–10.

Papacharalampopoulos, A., Bikas, H. and Stavropoulos, P. 2018. Path Planning for the Infill of 3D Printed Parts Utilizing Hilbert Curves. *Procedia Manufacturing* 21: 757–764.

Prashar, G., Vasudev, H. and Bhuddhi, D. 2022. Additive Manufacturing: Expanding 3D Printing Horizon in Industry 4.0. *International Journal of Interactive Design and Manufacturing* 16(4): 1259–1272.

Priyadarshini, A., Singh, R.K., Mishra, R. and Kamal, M.M. 2022. Adoption of Additive Manufacturing for Sustainable Operations in the Era of Circular Economy: Self-assessment Framework with Case Illustration. *Computers & Industrial Engineering* 171: 108514.

Qamar Tanveer, Md., Mishra, G., Mishra, S. and Sharma, R. 2022. Effect of Infill Pattern and Infill Density on Mechanical Behaviour of FDM 3D Printed Parts—A Current Review. *Materials Today: Proceedings* 62(1): 100–108.

Rojas, R., Rauch, E. and Matt, D.T. 2021. Research Fields and Challenges to Implement Cyber-Physical Production Systems in SMEs: A Literature Review. *Chiang Mai University Journal of Natural Sciences* 20(2): 1–19.

Romanova, V., Mohebbi, M.S., Dymnich, E., Balokhonov, R. and Ploshikhin, V. 2022. A Physically-based Computational Approach for Processing-Microstructure-Property Linkage of Materials Additively Manufactured by Laser Powder Bed Fusion. *International Journal of Mechanical Sciences* 219: 107103.

Sauerwein, M., Doubrovski, E., Balkenende, R. and Bakker, C. 2019. Exploring the Potential of Additive Manufacturing for Product Design in a Circular Economy. *Journal of Cleaner Production* 226: 1138–1149.

Serrano-Bedia, A.-M. and Perez-Perez, M. 2022. Transition towards a Circular Economy: A Review of the Role of Higher Education as a Key Supporting Stakeholder. *Sustainable Production and Consumption* 31: 82–96.

Slotwinski, J.A. and Garboczi, E.J. 2015. Metrology Needs for Metal Additive Manufacturing Powders. *JOM* 67: 538–543.

Sousa, S.D. and Pham, D.T. 2023. Quality Control in Remanufacturing: Distinguishing Features and Techniques. In K. Kim, J. Rickli, and L. Monplaisir (eds), *31st International Conference on Flexible Automation and Intelligent Manufacturing, FAIM 2022*. Detroit, USA, Code 285199.

Üreten, E., Willett, T., Vlasea, M. and Daim, T. 2019. Additive Manufacturing (AM) Technology Assessment for Titanium Hip Implant Fabrication. In T. Daim, M. Dabić, N. Başoğlu, J.R. Lavoie, and B.J. Galli (eds), *R&D Management in the Knowledge Era*, 237–259. London: Springer.

Verma, P., Kumar, V., Daim, T., Kumar Sharma, N. and Mittal, A. 2022. Identifying and Prioritizing Impediments of Industry 4.0 to Sustainable Digital Manufacturing: A Mixed Method Approach. *Journal of Cleaner Production* 356: 131639.

Zeba, G., Dabić, M., Čičak, M., Daim, T. and Yalcin, H. 2021. Technology Mining: Artificial Intelligence in Manufacturing. *Technological Forecasting and Social Change* 171: 120971.

Zhu, C., Li, T., Mohideen, M.M., Hu, P., Gupta, R., Ramakrishna, S. and Liu, Y. 2021. Realization of Circular Economy of 3D Printed Plastics: A Review. *Polymers* 13(5): 1–16.

6 Development of a 3D-printed, Modularized, and Low-cost USV

Anastasios Tzotzis[1*], Athanasios Manavis[2], Nikolaos Efkolidis[3] and Panagiotis Kyratsis[4]

1. Introduction

The need for unmanned surface vessels (USVs) is continuously growing since the past few years due to the increased demand for automated inspection, surveillance, and monitoring platforms. Most commercially available solutions usually cost thousands of euros, depending on their size and functionalities. Therefore, there is an increasing focus on the development of low-cost solutions by the scientific community.

Blaich et al. (2013) worked on the development of a twin-hull autonomous vessel for the study of control algorithms, sensor data fusion as well as path planning and collision avoidance. The purpose of the USV focuses mainly on measuring and surveillance tasks. Therefore, it can be equipped with sonars, echolot, or water samplers. Similarly, a high performance USV was developed by Grenestedt et al. (2016). High speed and the self-righting ability under all sea states are the key characteristics of the specific design. Jo et al. (2019) developed a compact and low-cost platform for water quality monitoring by utilizing easy-to-access materials and components. The platform was equipped with both a PH and turbidity sensor, as well as a temperature sensor. Moreover, the authors designed their own control interface via a smartphone. Johnston and Poole (2017) developed a wave propulsion USV with real-time monitoring capabilities. The development of the USV focused mainly on surveillance purposes, aiming to become a long-term and cost-effective solution. Underwater drones (Binugroho et al., 2018) constitute a similar topic with increasing interest to the scientific community, which demands low-cost and effective solutions. Wang et al. (2019) worked on the development of Roboat, an autonomous USV intended for urban

[1,2] *Postdoctoral Researcher*, Department of Product and Systems Design Engineering, University of Western Macedonia.

[3] *Assistant Professor*, Department of Product and Systems Design Engineering, University of Western Macedonia.

[4] *Professor*, Department of Product and Systems Design Engineering, University of Western Macedonia.

E-mail: a.manavis@uowm.gr[2]; nefkolidis@uowm.gr[3]; pkyratsis@uowm.gr[4]

* Corresponding author: a.tzotzis@uowm.gr

waterways. Roboat is equipped with obstacle avoidance systems, as well as object identification sensors. A similar work was presented by Chaysri et al. (2024). The authors worked on the development of an autonomous, low-cost USV, focusing on robotic contests. The USV was equipped with object recognition systems and the authors utilized machine learning for the realization of the recognition process. Most of the time, low-cost and compact size are the two main factors taken into consideration during design. Sanfilippo et al. (2021) presented AnSweR, a low-cost and compact surface vessel for research purposes. Shi et al. (2016) developed a trimaran-shaped USV with the aid of appropriate software. Similarly, Kolev et al. (2021) presented a small USV design with autonomous navigation. The authors partially used 3D-printed materials for the hull cover. In most designs, the next aspects are not evident. The first is the ability to modularize the hull, and the second is an alternate method of manufacturing, other than the standard methods such as woodworking and molding. Typical materials include polyester fibers, wood, mold thermoplastic and sometimes aluminum. Doğru et al. (2023) recently studied the possibility of using polyethylene terephthalate glycol (PET-G) for 3D-printed USVs. Earlier, Niu et al. (2021) designed a combination of aerial and surface platforms, which were partially manufactured with the 3D-printed method. Despite the usability of PET-G, an increased number of 3D-printing materials have not yet been tested for this purpose. Moreover, the majority of the developed designs favor the classic manufacturing methods.

With the aforementioned features in mind, this chapter proposes a design approach for an autonomous platform, namely, e-AIOLOS, which focuses on three main parameters: cost, alternative manufacturing, and modularity. First, the whole project was processed with a low budget in mind. Next, the hull was designed in such way that both the manufacturing and the assembly processes can be facilitated. In addition, the design intent enabled the modularization of the vessel. Moreover, the platform utilizes a unique propulsion system.

2. Materials and Methods

2.1 USV Description

The aim of the design was the development of a compact, lightweight, and low-cost USV that can be used for safety purposes. 3D-printing was chosen as the manufacturing method, instead of the traditional methods, to take advantage of the fast processing times and increased accuracy that this method offers. The vessel utilizes a displacement hull, trimaran-shaped, since it focuses on the increased buoyancy and energy efficiency. In addition, the vessel utilizes a non-conventional propulsion system; meaning that no water propellers are used, rather it utilizes a thrust vectoring propulsion system. Since one purpose of the vessel is to navigate autonomously around a body of water, Geographical Positioning System (GPS) was used, based on the M10 and QMC5883 chips for the GPS and magnetometer respectively. The installed controller (STM32F405), which is based on open navigation-based flight controller software, is responsible for the monitoring of the various electronic components and handling of the instructions given to each individual system. Furthermore, it can be remotely controlled with multichannel radio system.

The vessel is intended for use in calm waters such as small lakes, ponds, and reservoirs, especially in waters with dense vegetation, for inspecting water platforms, floaters, and other water structures. Moreover, it can be used as a safety drone for accompanying swimmers and water athletes. It is possible to utilize it in search-and-rescue missions as well, providing extra coverage due to the extended operation time compared to unmanned aerial vehicles (UAVs). For this purpose, it carries a payload of analog live-stream camera, high-definition (HD) recording camera, upgradable to digital system. Despite the fact that it is intended mainly for calm inland waters, it is capable to withstand light winds due to its design and low center of gravity (CoG), allowing it to operate at sea as well, in conditions similar to the ones found at small closed bays. Figure 1 illustrates the basic parts of the vessel. Part numbers 1 to 11 correspond to the fore part, the mid part, the aft part, the side floaters, the extendable support tubes, the rudder assembly, the propulsion system, the radio mast, the foreship compartment cap, the midship compartment cap, and the aftship compartment cap, respectively. Table 1 includes the most basic design parameters of the hull. Besides the dimensions, water line represents the length of the hull that is in contact with the water surface, draft denotes the depth of the vessel below the water line, and the displacement is essentially the weight.

Table 1 Basic design parameters

Parameter	*Value*
Length	1180 mm
Width	630 mm
Water line	1020 mm
Draft	60 mm
Displacement	10 kg

Figure 2 illustrates the connections between the electronic components required for the vessel's operation, as well as the communication setup. In specific, the main control unit (MCU) monitors the status and operation of every component, handles the power distribution among the components, and overlays data related to the status of the USV to the monitor. In addition, it processes the waypoint coordinates during the autonomous navigation. The electronic speed controllers (ESCs) are responsible for the efficient power output fed to the motors (M1 and M2). Servos (S1 and S2) are responsible for the steering and the folding-unfolding of the mast. The GPS unit is critical for the autonomous navigation, by receiving accurate positioning coordinates from satellites. The magnetometer is a

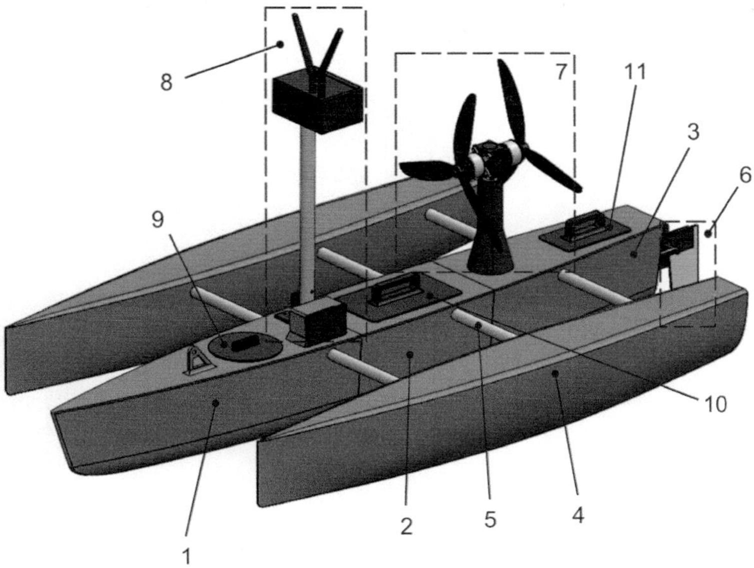

Fig. 1 The basic parts of e-AIOLOS

component that is not visible and therefore is not shown in the diagram of Figure 2. The camera captures live footage from the surroundings and the signal is fed to the video transmitter (VTX), which is responsible to transmit this signal. A ground station comprising of a video receiver (VRX) and a monitor allows the signal to be received and visualized to a monitor. Therefore, the operator is able to see through the vessel's camera. Consequently, the radio receiver (RX) is used to communicate with the radio controller (RC) for manual operation of the vessel. As a future upgrade, the telemetry (T/M) radio will be installed to enable long-range communication between the MCU and a personal computer (PC), in order to track the USV and monitor its status during autonomous navigation.

2.2 *Hull Design*

The hull was designed by utilizing the CAD-based application described in the work by Tzotzis, Efkolidis et al. (2024), with respect to the specifications found in the relevant literature (Guan et al., 2021).

2.2.1 *Manufacturing and Modularization*

One of the most important aspects of the design is the modularization of both the main hull and the side floaters. The hull was designed in such a way that both the manufacturing and the assembly processes could be facilitated. It should be noted that the hull was manufactured in three core pieces (fore, mid, and aft) by employing the fused filament fabrication (FFF) method. Depending on the operation of each part and the mechanical properties requirements, materials such

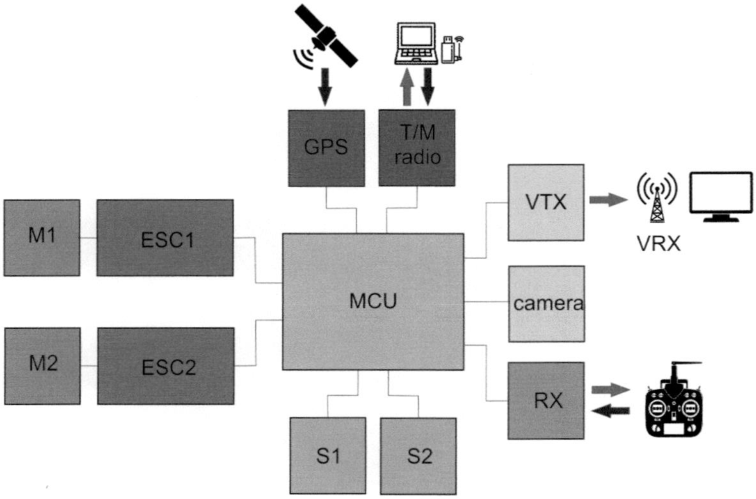

Fig. 2 The connection and communication diagram

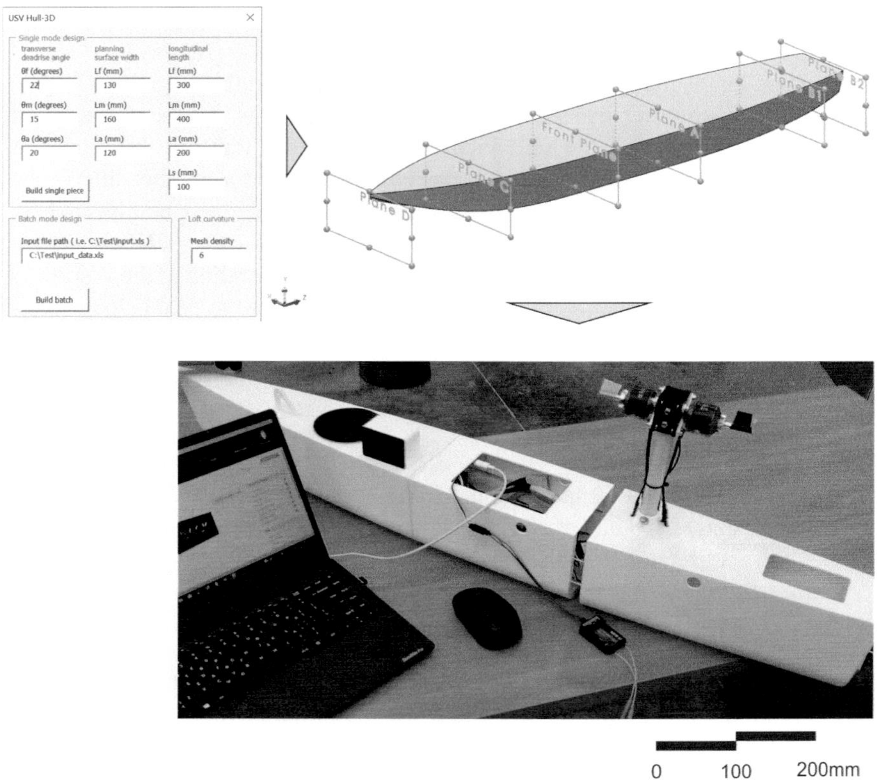

Fig. 3 During the design process of the USV

as Acrylonitrile Butadiene Styrene (ABS), Polylactic Acid (PLA, nylon, Polyvinyl Chloride (PVC) were used. The printing parameters were selected according to the guidelines found in relevant optimization studies such as Tzotzis, Manavis et al. (2024). Despite the fact that all hull pieces were waterproof due to the filament material properties, they were further protected against water by applying layers of varnish. Pieces of hard wire running across the sections of the hull parts were used to secure the parts in place and reinforce the hull. Besides reinforcing, the hard wire acted as guides for the hull parts to slide, facilitating this way the assembly process and enabling the modularization of the main hull. On the other hand, the modularization of the side floaters was achieved by using extendable tubes (see Figure 1). Figure 3 illustrates a stage of the design process.

Regarding the manufacturing process, the most common ways of building unmanned vessels today, regardless of size, are: by extrusion of mechanical thermoplastic materials in a mold (injection molding), as well as molding resins in a mold. In addition, the method of fastening wood sheets to a frame can be used, while in more reinforced constructions, welding aluminum sheets to a frame is an option.

In the first two methods, it is necessary to inject the material into a mold, while in the last two it is necessary to use a frame to fix the linings. In the present case, construction with 3D-printing, neither the use of a mold nor the manufacturing of a fixing frame is required. Therefore, both the time and cost of building are reduced, while at the same time modularization of the hull is achieved. By modularizing the hull, the assembly of the USV is facilitated to an extended degree. Moreover, a modularized hull provides easy access to all of its compartments during the assembly process. In the event that a repair is required, it allows the replacement of the damaged or worn part, without dismantling the whole structure and without damaging any fastening components. Finally, the modularized hull enables the immediate modification of its dimensions (e.g. increase the overall width by replacing the supporting tubes of the side floats with longer ones).

2.2.2 Bill of Materials and Cost

The hull of the USV and its parts were built exclusively with the FFF method. A total of three filament spools were consumed. The electronic components required for the realization of the propulsion and the steering include brushless motors, propellers, ESCs, and servos. The MCU and the GPS unit are necessary for the control and stabilization of the vessel, as well as for its autonomous navigation. Finally, the RC is required to manually control the vessel within a reasonable range, supplementing the autonomous operation of the vessel. The batteries are obviously required to power the USV and can be included in either component group. However, they were added to the control and navigation group since they are directly plugged to the MCU. The surveillance system includes the analog signal VTX, the real-time camera, the analog signal VRX, monitor, HD camera and antennas. The hardware group includes both mechanical and electrical material such as bolts, nuts, paints, varnishes, wires, and connectors. Table 2 presents the required components for the build, in addition to their costs, excluding the cost of a standalone telemetry system. Moreover, Figure 4 visualizes the percentage contribution of each component group to the total build cost.

Table 2 Bill of materials and cost of components

Component	*Group*	*Quantity*	*Cost (€)*
Motors	Propulsion & steering	2	30
Propellers	Propulsion & steering	2	20
Motor speed controllers (ESC)	Propulsion & steering	2	30
Servos	Propulsion & steering	2	20
Main control unit (MCU)	Control & navigation	1	50
Batteries	Control & navigation	2	80
GPS unit	Control & navigation	1	20
Radio control (RC)	Control & navigation	1	80
Video signal transmitter (VTX)	Surveillance system	1	20
Real-time camera	Surveillance system	1	30
Video signal receiver and monitor	Surveillance system	1	100
Antennas	Surveillance system	2	20
HD camera	Surveillance system	1	60
Filament	Auxiliary hardware	3	120
Electrical hardware	Auxiliary hardware	n/a	30
Mechanical hardware	Auxiliary hardware	n/a	50
Paints and varnishes	Auxiliary hardware	n/a	30
		Total cost:	790€

3. Results

3.1 Testing

e-AIOLOS was tested in normal conditions during summer period at lake Polyfytos in Kozani, Greece. Besides buoyancy, it was tested for speed, stability, and maneuverability, as well as operation time and automated navigation. Each test was performed three times under the same conditions, during three randomly selected days. Therefore, each test was carried out nine times in total. The findings pointed out that the vessel was able to maneuver promptly while maintaining its

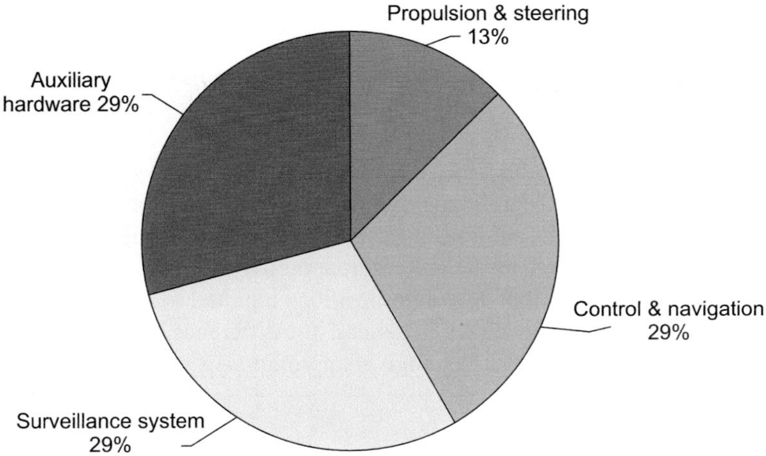

Fig. 4 Percentage contribution of the component costs

stability at the speed of approximately 30 km/h (16 knots). Moreover, it was found that the vessel was able to reach higher speeds; however, it was deemed unsafe due to the increased spraying of water to the motors. Therefore, future work will include design corrections that will allow for higher speeds with safety. The design corrections regard the repositioning of the CoG, the redesign of the main hull's angles, and the use of propellers of different specifications. Figure 5 shows a view from the onboard HD camera of the vessel during the trials. While operating the USV manually, a range of 500 m was achieved by maintaining a relatively low-latency control and video feed, in an unobstructed environment. It should be noted that the equipment used is common in the RC model hobby and therefore fall into the low-budget category.

Fig. 5 View from the onboard HD camera

The vessel's mass, fully loaded, was measured at approximately 7 kg. It was calculated that it can carry another 3 kg of load, such as first-aid material or scientific equipment, with safety. Furthermore, it was estimated that it can easily tow a floating container with material such as a pair of life-jackets, flares, fresh water, and typical pharmaceutical supplies.

Regarding the operation time, a lithium-polymer (LiPo) battery with 8Ah capacity was utilized for the testing, providing about 1 hour of operation at cruising speed (5 km/h). Finally, the automated waypoint navigation testing was successful, indicating that the USV can navigate autonomously with adequate accuracy. Figure 6 compares the initial waypoint positions inputted to the controller with the actual positioning of the vessel. In general, the difference between positioning coordinates was found to be in the range of approximately ±2m.

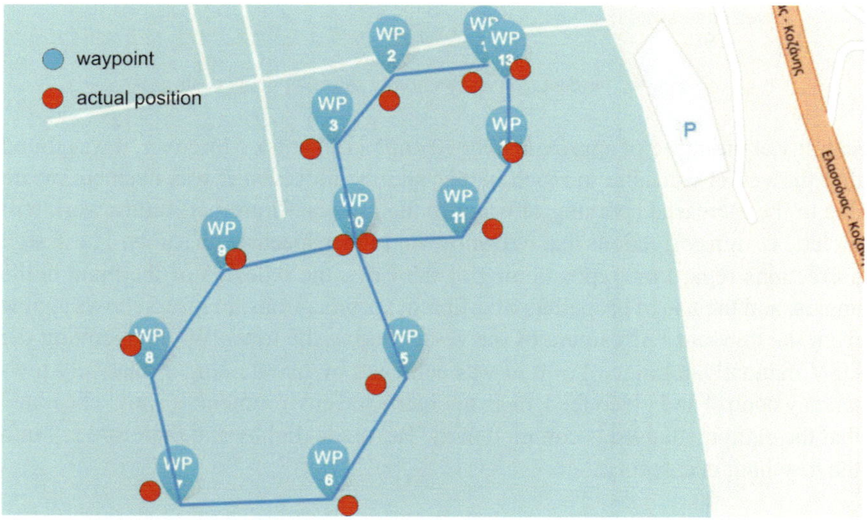

Fig. 6 Comparison between the waypoint coordinates and the actual position of the vessel

3.2 Discussion

The CAD-based parametric design and the FFF method used for the design and fabrication of e-AIOLOS seem promising. The effective and budget-friendly manufacturing process and convenient assembly, in addition to the ability to easily repair or modify the vessel, are the most important benefits that derive from the employment of the aforementioned methods. In terms of the vessel's usability, after the testing, it is evident that the USV is capable of operating under conditions similar to the ones found at small lakes, ponds, reservoirs, and small bays for both surveillance and safety purposes. The performance of the developed design is comparable to the ones referenced in the introduction, in terms of the basic parameters such as speed, maneuverability, buoyancy, and operation. It is, however, difficult to generalize these results and compare the performance between different designs due to the increased number of design aspects. The size, the hull type, the propulsion and steering systems, as well as the purpose of the vessel affect greatly the overall expected performance of the vessel. For example, a trimaran design will always have better stability compared to other types of vessels, whereas a deep vee hull design will always be faster under the same conditions. In general, the present design is based on a fully 3D-printable hull in contrast to most of the designs found in the literature, utilizes a non-conventional propulsion system, and allows for the replacement of the hull's compartments, thanks to its modularized design.

Conclusions

Summarizing the test results, the next parameters were recorded. Maximum safe surface speed of the vessel was measured to be approximately 16 knots, operation range at 500 m, while operation time at 1h when cruising at 4 knots speed. The positioning accuracy provided was measured to be equal to ± 2 m.

In conclusion, potential future upgrades both for video feed and RC, by utilizing modern HD digital video systems and open communication protocols, can increase range and penetration, allowing for significantly farther control and video feed.

Finally, it should be noted that the specific design has received a patent from the Hellenic Industrial Organization, with national number 1010721, further supporting its novel design qualities.

References

Binugroho, E.H., Sanggar Dewanto, R. and Pramadihanto, D. 2018. eROV: Preliminary Design of 5 DOF ROV Using 6 Thrusters Configuration. *2018 International Electronics Symposium on Engineering Technology and Applications (IES-ETA)*, 281–287.

Blaich, M., Wirtensohn, S., Oswald, M., Hamburger, O. and Reuter, J. 2013. Design of a Twin Hull based USV with Enhanced Maneuverability. In *IFAC Proceedings Volumes (IFAC-PapersOnline)* (Vol. 9, Issue PART 1). IFAC, 1–6.

Chaysri, P., Spatharis, C., Vlachos, K. and Blekas, K. 2024. Design and Implementation of a Low-cost Intelligent Unmanned Surface Vehicle. *Sensors* 24(10): 1–20.

Doğru, A., Sözen, A. and Seydibeyoğlu, M.Ö. 2023. Use of (PET-G) Material in Mini-scale Unmanned Surface Vehicle (USV) for Additive Manufacturing BT. In T.H. Karakoc, S. Le Clainche, X. Chen, A. Dalkiran, and A.H. Ercan (eds), *New Technologies and Developments in Unmanned Systems*, 9–14. Springer International Publishing.

Grenestedt, J., Keller, J., Larson, S., Patterson, J., Spletzer, J. and Trephan, T. 2016. LORCA: A High Performance USV with Applications to Surveillance and Monitoring. *SSRR 2015–2015 IEEE International Symposium on Safety, Security, and Rescue Robotics*, 1–6.

Guan, G., Wang, L., Geng, J., Zhuang, Z. and Yang, Q. 2021. Parametric Automatic Optimal Design of USV Hull Form with Respect to Wave Resistance and Seakeeping. *Ocean Engineering* 235: 109462.

Jo, W., Hoashi, Y., Paredes Aguilar, L.L., Postigo-Malaga, M., Garcia-Bravo, J.M. and Min, B.C. 2019. A Low-cost and Small USV Platform for Water Quality Monitoring. *HardwareX* 6: e00076.

Johnston, P. and Poole, M. 2017. Marine Surveillance Capabilities of the AutoNaut Wave-propelled Unmanned Surface Vessel (USV). *OCEANS 2017—Aberdeen*, 3–6.

Kolev, G., Bathaie, S.N.T., Rybin, V., Kulagin, M. and Karimov, T. 2021. Design of Small Unmanned Surface Vehicle with Autonomous Navigation System. *Inventions* 6(4): 1–15.

Niu, H., Ji, Z., Liguori, P., Yin, H. and Carrasco, J. 2021. Design, Integration and Sea Trials of 3D Printed Unmanned Aerial Vehicle and Unmanned Surface Vehicle for Cooperative Missions. *2021 IEEE/SICE International Symposium on System Integration (SII)*, 590–591.

Sanfilippo, F., Tang, M. and Steyaert, S. 2021. The Aquatic Surface Robot (AnSweR), a Lightweight, Low Cost, Multipurpose Unmanned Research Vessel. In S. Yildirim Yayilgan, I.S. Bajwa, and F. Sanfilippo (eds), *Intelligent Technologies and Applications*, 251–265. Springer International Publishing.

Shi, Y., Liu, L., Zhao, J., Long, Y., Chen, J. and Cui, Y. 2016. Mechanical System Design of a Trimaran USV. *Revista de La Facultad de Ingeniería, December 2016*, 31(5): 36–48.

Tzotzis, A., Efkolidis, N., García-Hernández, C. and Kyratsis, P. 2024. CAD-based Automated USV Hull Design. In G. Guxho, T. Kosova Spahiu, V. Prifti, A. Gjeta, E. Xhafka, and A. Sulejmani (eds), *Proceedings of the Joint International Conference: 10th Textile Conference and 4th Conference on Engineering and Entrepreneurship*, 354–362. Springer Nature Switzerland.

Tzotzis, A., Manavis, A. and Efkolidis, N. 2024. Analysis of the Influence of Structural Characteristics on the Tensile Properties of Fused Filament Fabricated ABS Polymer Using Central Composite Design. *Applied Mechanics* 5: 20–35.

Wang, W., Gheneti, B., Mateos, L.A., Duarte, F., Ratti, C. and Rus, D. 2019. Roboat: An Autonomous Surface Vehicle for Urban Waterways. *2019 IEEE/RSJ International Conference on Intelligent Robots and Systems (IROS)*, 6340–6347.

7 Selective Laser Melting: From Research to Industrial Applications

Nicolae Balc[1] and Cosmin Cosma[2*]

1. Introduction

The intricate thermo-physical process of selective laser melting (SLM) is dependent on the laser, material, and process parameters. With the title "Shaped body especially prototype or replacement part production", DE19649865 is the patent for the SLM technique, which was created in 1996 at the Fraunhofer Institute for Laser Technology (Aachen, Germany). The laser, which is categorized as a solid Nd:YAG laser is the fundamental part of SLM systems. It scans the powder bed by emitting light with a wavelength of 1064 nm in the infrared spectrum.

1.1 The Need for and Importance of SLM, in Producing Fully Dense Metal Parts

Metallic items with a relative density of 99.9% may be directly produced using the SLM technology. Tool steel and stainless steel, pure titanium (Ti) or titanium alloy (Ti6Al4V), Co-Cr, aluminum, and nickel-based alloys (Inconel 625, Inconel 718) are the powders that are the focus of multiple SLM research (Yap et al., 2015). Other special powders, including copper, magnesium, tungsten, gold, alumina, and silica (glass) have also been studied using SLM (Fateri et al., 2014). Various applications use the SLM process because it offers a high degree of design freedom, the optimization and integration of functional features, the ability to manufacture small batch sizes at affordable unit costs, and a high degree of product customization even in serial production (3DHubs 2022). The main benefits of SLM parts include (Aniwaa 2019):

- *On-demand production:* SLM manufacturing provides greater flexibility and control over the production line.
- *Complex designs are possible:* SLM technology enables the creation of intricate and highly detailed parts that would otherwise require multiple pieces to be assembled using traditional techniques (i.e., CNC is abbreviation for Computer Numerical Control machining or casting).

[1] *Professor*, Department of Manufacturing Engineering, Technical University of Cluj-Napoca.
[2] *Senior Lecturer*, Department of Manufacturing Engineering, Technical University of Cluj-Napoca. E-mail: nicolae.balc@tcm.utcluj.ro[1]
* Corresponding author: cosmin.cosma@tcm.utcluj.ro

- *Waste reduction:* Because SLM production only uses the material required for a particular item, it generates less waste than CNC machining. Additionally, the un-melted powder can be recycled and reused.
- *Lighter components:* SLM printing enables you to make topological optimized design of parts without compromising their mechanical behavior.

An important aspect in SLM manufacturing is to fabricate 100% fully dense metal parts. SLM parameters including laser power, scanning speed, hatch space, and layer thickness affect the density of produced components. A systematic SLM optimization to reduce the porosity of SLM parts for titanium alloy (Ti6Al4V) is presented in previous publications (Wang et al., 2019). The porosity percentage showed a tendency to steadily rise with increasing scan speed when the laser power is between 50 W and 150 W (Wang et al., 2019). On the other hand, if the scanning speed is set between 850 mm/s and 1750 mm/s, the porosity decreases to 1% when the laser power grows to 200 W–250 W.

Furthermore, recent research explains that adjusting the scanning speed and using a laser power of up to 300 W can decrease the porosity of SLM components. The main conclusions show that the relationships between SLM parameters and tensile strength are not as obvious as anticipated (Majumdar et al., 2019). They also show that relatively small changes in process parameters can result in a wide range of porosity (0.03% to 32%), suggesting that the energy density is not a good indicator of SLM-manufactured parts. More details can be found in previous publications, on how the porosity is influenced by the laser power and laser scanning speed (Majumdar et al., 2019).

For each material, the SLM parameters should be optimized to obtain 100% dense parts, high mechanical strength, reduced surface roughness, and limited manufacturing defects. Heat-affected zones, gas voids, macrocracks, and residual stress phenomena are the main SLM faults (Cosma et al., 2020). To reduce these technological faults, which are caused by fast melting and cooling cycles, optimal SLM process parameters are required. Cherry et al. (2015) studied the influence of SLM parameters on porosity and hardness of 316L stainless steel samples. They observed that if the porosity of SLM specimens is between 0.5% and 2%, then the surface hardness is limited influence, ranging from 225 HV10 to 215 HV10. If the porosity percent increases between 2% and 4%, the hardness decreases to 190 HV10. A dependent down trend was established between hardness and porosity. More details can be found in previous publications, regarding the hardness versus porosity percentage (Cherry et al., 2015).

An example showing the impact of SLM parameters' physical–mechanical properties is listed in Table 1. The tensile tests were carried out on SLM samples made of 316L stainless steel. Young modulus, elongation at fracture, yield strength, ultimate tensile strength, and relative density were among the characteristics that were determined. Because of the generative manufacturing principle, which has also been noted in previous investigations, the tensile tests demonstrated the anisotropic behavior of SLM samples (Cosma et al., 2020; Wang et al., 2019).

Table 1 Effect of the SLM parameters (Cosma et al., 2020)

Laser Power (W)	*Scanning Speed (mm/s)*	*Energy Density (J/mm³)*	*YS (MPa)*	*UTS (MPa)*	*Elongation at Fracture (%)*	*Young Modulus (GPa)*	*Relative Density (prel)*
130	1034	44.98	630	672	6	182	98.3
170		58.82	654	689	8	173	98.5
160	400	142.85	590	648	7	178	99.6
110	500	78.57	774	823	12	192	99.3
150	750	71.42	703	786	10	184	99.3
180	950	67.66	783	841	13	194	99.1

Note: YS—yield tensile strength; UTS—ultimate tensile strength

1.2 Types of Equipment and Manufactures

The size of the work platform and the installed laser determine how much SLM equipment costs, which typically ranges from €130,000 to €600,000. Arcam, Concept Laser, Eos, Matsuura, Realizer, Renishaw, Sisma, and Trumpf are the primary producers of SLM equipment or related technologies such as direct metal laser sintering (DMLS) and electron beam melting (EBM) (Cosma et al., 2018). Some available SLM machines are Arcam Q20plus, Concept Laser M2 Cusing, Eos Eosint M 280, Matsuura Lumex Avance-25, Realizer SLM 250, Renishaw AM 400, Sisma MySint100, and Trumpf TruPrint 1000 LMF.

Due to the expiration of several SLM patents since 2017, low-cost Chinese machine makers appeared. A review of the Chinese suppliers reveals that 21 businesses produce SLM equipment.

In 2016, the Fraunhofer Institute for Laser Technology and FH Aachen collaborated to create a low-cost SLM machine for small and medium-sized enterprises, which was priced at €30,000. Moreover, Deckel Maho Gildemeister (DMG) Company has created a hybrid system that includes a complete 5-axis milling machine and a 5-axis material deposition using a coaxial nozzle for uniform powder distribution. By combining the SLM method with CNC milling, flexible systems were created, and new components could be produced because of alternating machining and material deposition techniques. This hybrid system costs more than €1,000,000.

The available SLM equipment and manufactures are presented on many websites (Aniwaa 2019). The average SLM product costs between €900 and €5,000 to manufacture (including post-processing). The typical costs of the main SLM technological manufacturing stages are listed in Table 2. It can be observed that the cost of materials and post-processing add significantly to the total cost (Hubs 2024).

On the other hand, the production rate of SLM systems varies between 10 cm^3/h and 40 cm^3/h. To provide an increased build rate, the latest SLM systems are equipped with two or four laser sources up to 400 W. Practically, this technology

Table 2 Estimative costs of SLM manufacturing

Production Step	*Operation*	*Cost*
Manufacturing	Material cost	€50–€680 per kilogram*
	SLM cost	€500–€6,000 per build** Or €50–€80 per hour
Post-processing	Stress relief heat treatment	€300–€600 per build
	Part finishing (support removal)	€80–€200 per part
	Hot isostatic pressing heat treatment (HIP parameters: 1185°C, 150 MPa pressure, 4h holding time)	€95–€160 per cycle
	CNC machining (turning, drilling, 3-axis or 5-axis milling, etc.)	€300–€2,000 per part Or €50–€120 per hour
	Surface treatments	€150–€400 per part

*More specific details about the cost in Section 3.1, "Realizer 250 equipment, working principle and available powders"; **A detailed example is shown in Section 6.1, Case study: Mold manufacturing for injection of plastic parts".

allows lasers to work simultaneously, but each one scans a different part of the powder bed.

2. Steps and Procedures in the SLM Process

To achieve the best outcome and reasonable cost of the parts, the models should always go through an optimization process. The general guidelines for designing the SLM-manufactured parts are presented by different authors and open sources.

The basic phases in the SLM technique's procedure are:

- Step 1: Designing the virtual model of the part considering the rules detailed in Figure 3. Save the Computer Aided Design (CAD) and Standard Triangle Language (STL).
- Step 2: Slice the 3D CAD model of the part into individual layers. Every layer depicts the part's cross-section at a certain Z-height.
- Step 3: The first layer of metal powder is spread on the SLM platform when the data preparation is complete.
- Step 4: According to the slice information, a laser beam that travels via deflection mirrors melts the powder material and reveals the first layer's cross-section.
- Step 5: The next powder layer is spread once the SLM platform has been reduced by one layer thickness. Common values of layer thickness are 20, 30, 40, or 50 μm.
- Step 6: According to the slice information, a laser beam that travels via deflection mirrors melts the part cross-section. Until the part is finished, the cycle of powder spreading on the SLM platform (Step 5), laser melting (Step 6), and platform lowering is repeated (Mayer et al., 2016). After printing, the

part is removed from the SLM platform and, if required, post-processed as the final step in the process.

2.1 Pre-processing of the STL File

To SLM-manufacture the part, the 3D model should be saved in "*.STL" format.

2.1.1 Part Orientation

The following function can be used for geometry preparation: measuring details, repairing defects (connecting unmatched edges, closing holes, aligning triangular normal, resolving overlapping areas), editing geometry, adjusting mesh, and checking the wall thickness.

The orientation affects the economic production of parts as well as the surface quality of the parts (AutoFab, 2018). To align the part within RDesign software, the following steps are possible:

- Select the surface, which must be oriented toward the build platform.
- Rotate the part into the desired position.
- Define the distance between the part and the build platform (usually 2–3 mm).

The orientation of the part has a huge influence on the stability of the process. A recommendation is to have a small cross-section because otherwise the so-called curling effect will occur, and the part could be destroyed due to thermal stresses (Schwarz et al., 2016). A practical example is shown in Figure 1, where different orientations on the SLM plate were tested.

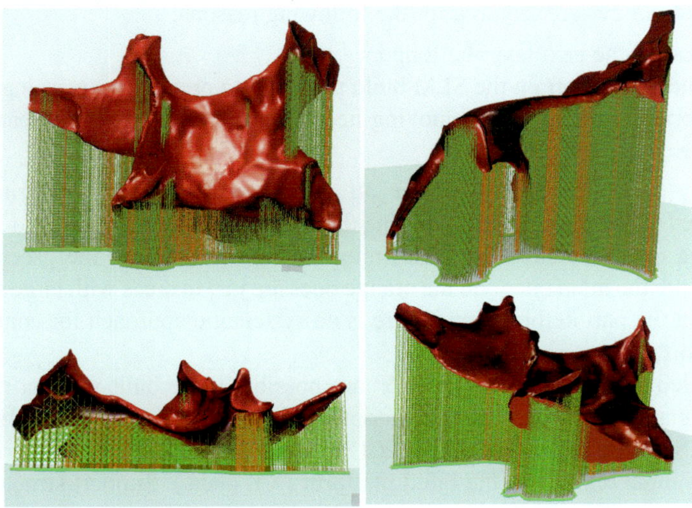

Fig. 1 Different part orientations on SLM plate

On the SLM platform, each part can be positioned and rotated. By orientating the part in different directions, the surface roughness is significantly different.

The influence of build-up orientation is presented and illustrated in important publications, with respect to the roughness of surface (Schwarz et al., 2016) and highlighting the influence onto the ultimate tensile strength (Meier and Haberland, 2008). A vertical orientation of models can reduce the surface roughness, but it will increase production time. Typical surface roughness Ra is between 10 µm and 16 µm on top surfaces and between 7 µm and 10 µm on vertical surfaces (build-up angle 90°).

To reduce the possibility of warping and increase part strength in crucial directions, parts are frequently oriented at an angle. On the other hand, this will raise the overall cost, build time, material waste, and the amount of support needed (Hubs 2024). Orienting the model to have fewer slices but a higher volume of support structures is another way to shorten the construction time.

Besides SLM parameters, the build-up orientation also influences the ultimate tensile strength. If the force is applied parallel to the layer structure, a greater ultimate tensile strength value may be expected. The strength characteristics diminish with the obtuseness of the angle between the direction of processing and the direction of load. Pieces that are printed horizontally have a higher tensile strength than material that is made traditionally (Meier and Haberland, 2008; Chlebus et al., 2011).

The part orientation is a critical aspect, and it should be adapted/optimized for each model according to user needs (for example: reduced build time, low surface roughness, or high mechanical resistance in a specified direction).

2.1.2 Support Structures

Support structures are required for the following reasons:

- to sustain the next layer built upon;
- to anchor the part on the SLM build plate and to prevent the warping effect;
- to serve as a heat sink, removing heat from the component and enabling a more controlled cooling rate.

They can be modeled using a lattice pattern. The lattice pattern can be configured in different ways, modifying the unit cell between 0.5 mm and 2.5 mm. This type of lattice support is illustrated in Figure 2, where the green lattice supports are the main ones and the orange ones are just in areas with large sections to prevent the part deformation. There is no systematic approach for configuring the support structures.

Angled surfaces greater than 45° are possible to be built without supports structures according to multiple experimental studies (Cosma et al., 2019), and German Standard VDI 3405 (VDI, 2015). A self-supporting angle describes the feature's angle relative to the build platform (XY plane). Support structures are needed to anchor surfaces.

The lower the angle, the less likely it is to support itself. As the angle decreases, the downward-facing surface becomes rougher, and eventually the part will fail if the angle is reduced too far (Additiva 2020). It is recommended to avoid designing self-supporting features that are less than 45° degrees relative to the horizontal

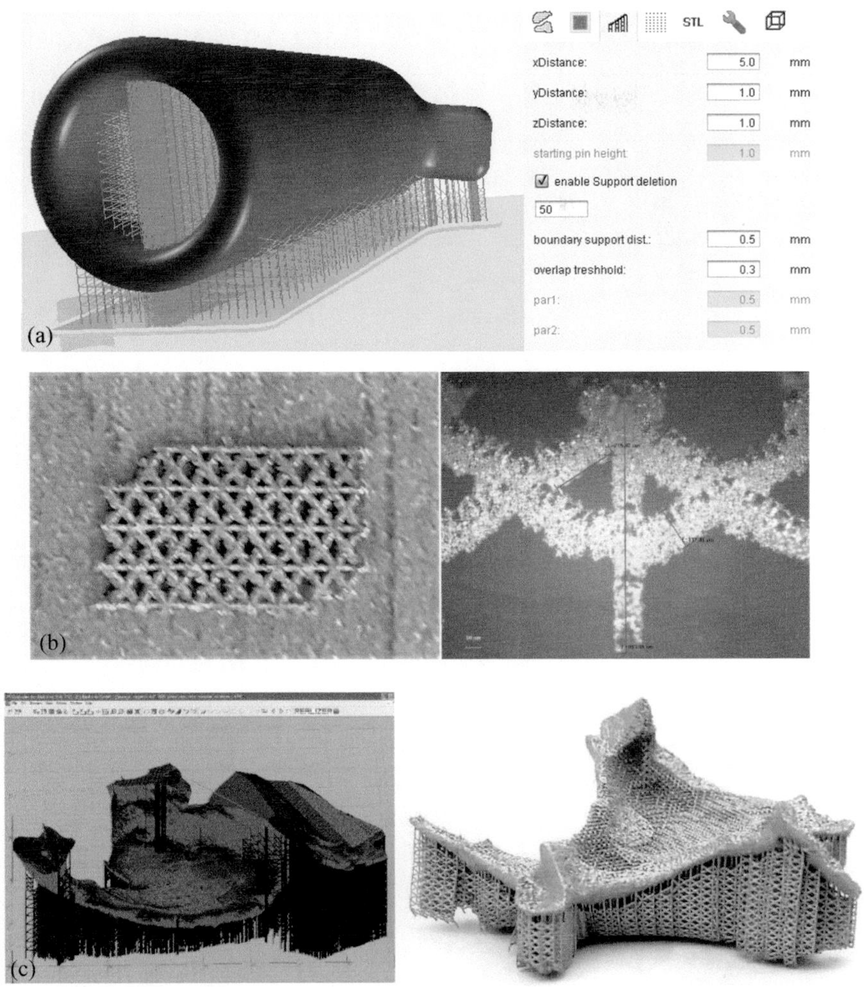

Fig. 2 Support structures type lattice: (a) Modeled in RDesigner software; (b) SLM manufactured; (c) Virtual complex shape anchored with lattice supports (Realizer software) and SLM-manufactured

plane. Changing the gradient of the wall, the surface finishes worse. If the built-up angle is lower than 30° and the surfaces are SLM-manufactured without support, this will compromise their fabrication.

2.2 Set Up the Material File

All the parts should be placed on the virtual platform of the SLM machine at a minimum of 1 mm distance between them without any intercalations using the translation function. The thickness of a layer of powder could be set to typically 50 μm, and all the models should be sliced. By default, RDesigner saves build jobs in its own The REA abbreviation comes from Realizer (name of this SLM

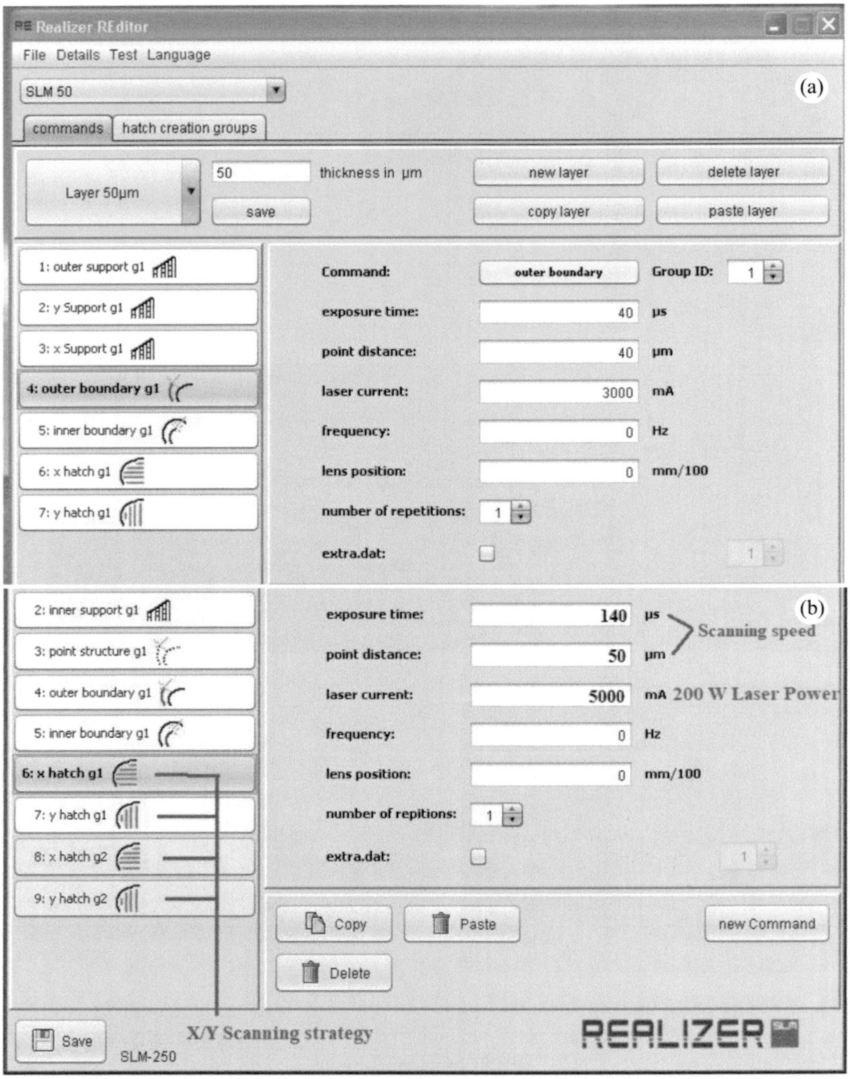

Fig. 3 Configure the SLM process parameters in REditor software: (a) Outer boundary function; (b) Hatch function for scanning in X direction

machine) material file. In addition to the sliced data and generated build paths for laser tracks, this REA file contains all the SLM process parameters (laser power, scan speed, hatch distance, etc.).

To configure the SLM process parameters, REditor software is used. This tool is useful to optimize the SLM parameters, to define the scanning strategy, to organize the laser scanning steps (e.g. first scan the outer boundary, after scanning the hatch area, and in the end the inner boundary), etc. An example is shown in Figure 3, where different functions are included as follows: outer boundary, inner-boundary, and hatch in X and Y directions. For a better understanding of these laser functions used to scan the part, see Figure 3. Besides these functions, the parameters for lattice support structures can be configured in the following functions: X and Y Support, Outer Support, etc.

Each function should be configured, and the main parameters are:

- laser power (10 W–200 W), expressed in mA, where 5000 mA = 200 W;
- scanning speed (100 mm/s to 2000 mm/s), expressed by exposure time and point distance; e.g.: scanning speed 1000 mm/s >> exposure time 40 µs, point distance 40 µm.

Moreover, the laser scanning strategy is another important aspect. Usually, the X/Y strategy is used for SLM printing", which allows scanning in the X direction of "n" layer and in Y direction of the "n + 1" layer (rotating each hatch 90°; details in Figure 4a). Compared to checkerboard or stripe hatch patterns, this hatch pattern has the benefit of potentially contributing to a high build rate. However, compared to the checkerboard hatch pattern, this scanning method may result in an increase in residual stress phenomena; therefore, process parameters need to be carefully considered.

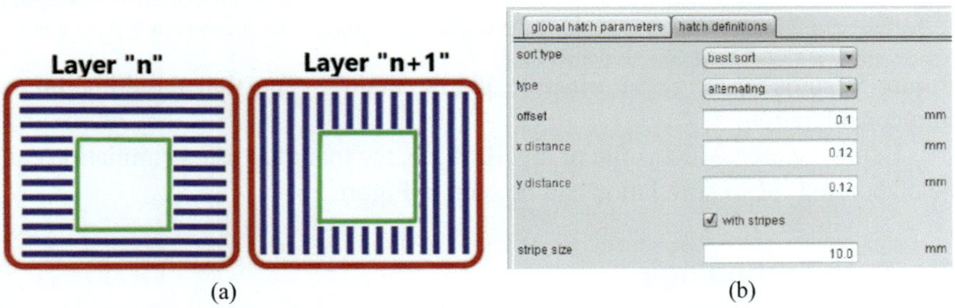

Fig. 4 (a) X/Y alternating laser scanning strategies and laser functions used to scan the part: outer boundary—red, inner boundary—green, hatch area—blue; (b) Hatch distance setup in REditor

Hatch distance represents the scan spacing between two consecutive laser tracks (see Figure 4a, blue tracks). In general, this parameter is set up at approximately the same value as the spot laser diameter (Yap et al., 2015). In this way, a reduced porosity and high mechanical resistance can be expected. If the spot laser diameter is 100 µm, the hatch distance could be 120 µm. An example of how the hatch distance can be configured is illustrated in Figure 4b.

All these SLM parameters should be configured for each powder type and saved. The final REA file of the SLM job will contain slice data and generated build paths for laser tracks, process parameters for each function, scanning strategy, etc. details a typical file used to configure these SLM parameters in REditor.

2.3 *Simulation of the SLM Process*

Because SLM manufacturing is expensive (€50–€80 per hour), simulations are frequently used to forecast how the component will behave during processing.

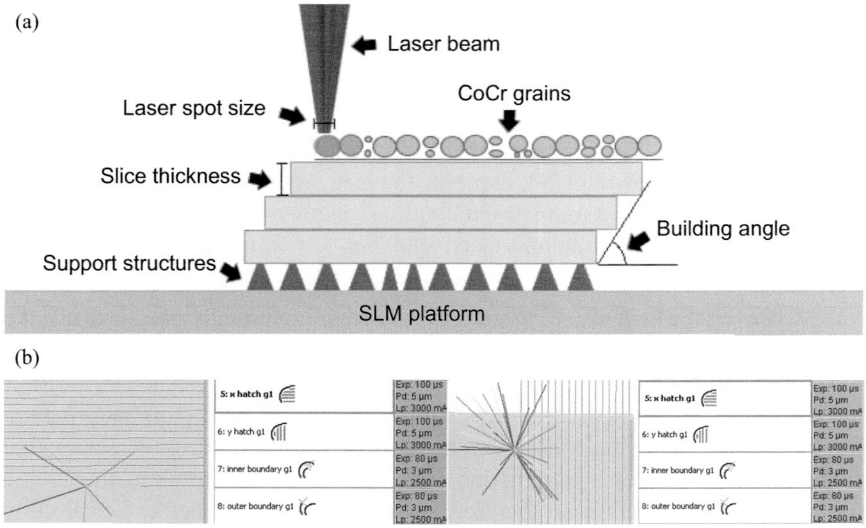

Fig. 5 (a) Schematic illustration showing the SLM principle; (b) SLM simulation in RDesigner software: left image—laser scans in X direction; right image—laser scans in Y direction

Figure 5a sums up the SLM principle and the main aspects that should be taken into consideration.

Using the created REA file of the SLM job, the manufacturing simulation can be elaborated. SLM simulation can be seen in Figure 5b.

3. SLM Technology

Founded by Dr Matthias Fockele, the Realizer company was one of the leading producers of SLM machines. The SLM technology resulted from a German research project developed and patented in 1995 at the Fraunhofer Institute for Laser Technology (ILT, patent DE 19649865). Realizer manufactures four models (SLM 50, 100, 250, and 300) with different printing chambers. In 1999, the company delivered the world's first SLM machine for metals at the Research Centre in Karlsruhe (Realizer).

3.1 Realizer 250 Equipment, Working Principle and Available Powders

In 2017, DMG Mori announced the acquisition of the majority Realizer stake (50.1%). The two companies presented the new LASERTEC 30 SLM metal 3D printer, a new machine for powder bed applications based on Realizer's SLM300i model. Now, DMG is following in the footsteps of other manufacturing giants such as GE and Trumpf, which have invested heavily in metal 3D printing. GE acquired Arcam and Concept Laser and has access to metal Additive Manufacturing (AM)

(EBM and SLM technology). Trumpf, just like DMG Mori, has access to SLM technology. Incidentally, all three groups currently record similar global revenues in the order of $4 billion yearly (Sher 2017).

The SLM technology can produce a functional tool in a short time and in a direct mode. To complete such a task, an SLM Realizer 250 machine (Realizer GmbH, Borchen, Germany) is available at the Technical University of Cluj-Napoca (TUCN) lab. This SLM machine costs €0.5 million. Table 3 presents the main characteristics of the Realizer 250 machine, which is adding layer-by-layer to fabricate metal parts and rapid tools. The Realizer 250 system is equipped with a 200 W solid fiber laser (type Nd:YAG) and emits continuous light in the infrared spectrum with a wavelength of 1064 nm. Other technical specifications are as follows: 100 mm/s to 1500 mm/s scanning speed, 50 μm common layer thickness, inert gas argon or nitrogen depending on power type, heated platform up to 250°C, less than 1% porosity in fabricated parts, always required supports, ± 0.10 mm dimensional accuracy, and typical Ra surface roughness achieved between 6 μm and 15 μm.

Table 3 SLM: Realizer 250 machine

Main Technical Specifications	*System Parameters*
Build chamber in mm (x/y/z)	248 × 248 × 250
Laser power	200/400 W, YLR-Faser Laser
Build speed rate	20 ccm/h
Layer thickness	20 μm – 75 μm
Min Scan line/Wall thickness	150 μm
Operational beam focus variable	60 μm – 100 μm
Scan speed	20 m/s
Inert gas consumption in operation	Argon, 1.5 l/min

Source: Technical data handbook, of the Realizer 250 machine.

The powders that this machine can operate with include stainless steel, tool steel, nickel-based alloy (Inconel 625, 718), Co-Cr, pure titanium (Ti), or titanium alloy (Ti6Al4V). Controlling the SLM process requires knowledge about the size and shape of the particles. Almost all spherical forms are required, and grains range in diameter from 10 μm to 80 μm (Figure 6). The d90 number is a typical way to represent the particle size distribution in percentiles. Based on the results of laser diffraction, this d-value shows that 90% of the particles are finer than this diameter. The common SLM powder was found to have a d90 value between 30 μm and 55 μm. Furthermore, a hall flowmeter funnel with a 2.5 mm aperture may be used to assess the particle flowability, and the flow rate (s/50 g) can be estimated. The hall flow rate is expressed as the time required for a 50 g powder sample to be discharged by gravitational force through the flowmeter funnel (ASTM B213). The typical flow rate is between 17 s and 34 s per 50 g of powder. SLM powders are produced by the gas atomization method.

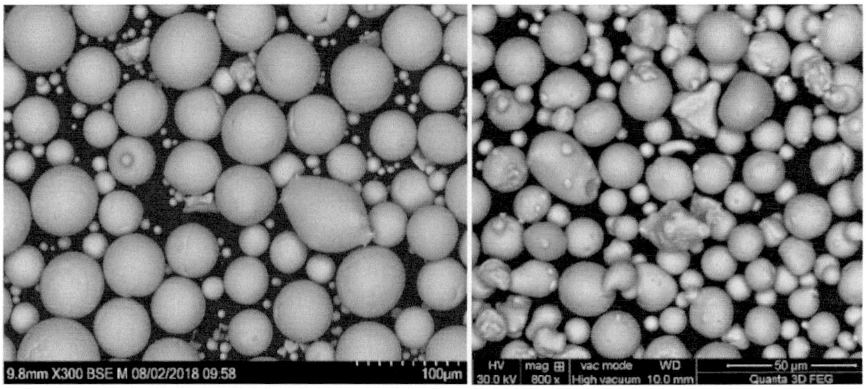

(a) (b)

Fig. 6 SLM powders: (a) Pure Titanium; (b) Stainless steel 316L

The most representative powder makers are:

- Arcast (www.arcastinc.com)
- Carpenter Additive (www.carpenteradditive.com)
- CNPC Powder (www.cnpcpowder.com)
- Constellium (www.constellium.com)
- Fehrmann Alloys (www.alloys.tech)
- GE Additive (www.ge.com/additive/fasterpath)
- Höganäs (www.hoganas.com)
- KBM Advanced Materials (www.kbmadvanced.com)
- Kennametal (www.kennametal.com)
- Material Technology Innovations (www.mt-innov.com)
- Metalpine (www.metalpine.at)
- Oerlikon Additive (www.oerlikon.com/am)
- Osaka Titanium Technology (www.osaka-ti.co.jp)
- Linde (www.linde-amt.com/atomization/additive-manufacturing)
- SAFINA (www.safina.cz)
- Sandvik Osprey (www.materials.sandvik)
- Sino-Euro Materials Technologies of Xi'an Co. (https://en.c-semt.com)
- TANIOBIS (www.taniobis.com)
- Tekna (www.tekna.com)
- Ultra Fine Specialty Products (www.ultrafinepowder.com)
- United States Metal Powders (www.usmetalpowders.com)
- VDM Metals (www.vdm-metals.com)

The powder material prices are subject to weekly changes. Therefore, the following prices are just indicative:

- Stainless Steel 316L (1.4404): 60–90 €/kg
- Tool Steel H13 (1.2344): 50–80 €/kg

- Tool Steel 1.2709 (Maraging 300): 90–130 €/kg
- Titanium Alloy TiAl6V4: 410–520 €/kg
- Titanium Alloy TiAl6Nb7: 530–680 €/kg
- Cobalt-Chromium Alloy (CoCrWMo): 220–340 €/kg
- Aluminum Alloy AlSi10Mg or Alloy AlSi12: 60–95 €/kg
- Inconel 625: 120–170 €/kg
- Inconel 718: 165–190 €/kg

These are the most used powders in SLM

3.2 Technological Parameters

The basic steps of the SLM process are:

- Configure the SLM parameters and create the REA file.
- The building chamber is first filled with inert gas and then heated to the optimal print temperature.
- Under the layer-by-layer principle, the part is fabricated until the whole building is complete.
- After printing, the build needs to cool down, and then the loose powder is extracted.

The main process parameters that can be configured are layer thickness, laser power, scanning strategy, hatch space, and scanning speed. There are SLM machines with high laser power (i.e., 600 W–1000 W), but for most of the SLM equipment, the laser power could be adjusted in the range of 100 W–400 W. Modern SLM machines use two to four lasers at the same time (each laser beam in a different area of the layer) to increase productivity. The scanning speed could be adjusted in the range of 350 mm/s to 2000 mm/s, according to the metallurgical properties of each powder material, which require a specific energy density, to get the suitable melt pool. The layer thickness is an important parameter and could be set between 20 μm and 100 μm. Before starting a job, the SLM platform is preheated at 250°C and maintained constant until the fabrication is finished. The oxygen level is limited to a maximum of 0.5% inside the building chamber, filled with neutral gas (i.e., argon).

Options like voxel size, number of considered neighbors, and minimum and maximum applied laser power could be defined. If chess patterns are used, they are divided into sub-heat patterns. After heat analyses, the heat hatch groups must be calculated. A material file is generated automatically (generation of heat groups). The heat calculations can automatically adapt all the laser parameters for reduced surface roughness using advanced simulations of mass distribution (options available in RDesigner software).

A new layer of powder is perfectly leveled using a special wiper as a soft recoater. The wiper is a composite material reinforced with carbon fibers (Ø 7.5 μm). Sometimes, the metal could be distorted or protruded out of the powder bed, with few small peaks. That is why the recoater should be flexible to avoid stopping the process. Otherwise, the collision with metal peaks could damage the powder bed, or sometimes the process might be blocked if a hard recoater is used, made from rubber or a stainless-steel blade. The SLM process is more stable at TUCN since a recoater (wiper) made of carbon fibers is used to spread the powder on the SLM plate.

To prepare the Realizer machine, the following routine steps should be taken (MCP Realizer 2008):

- Mount the platform with a substrate plate.
- The image to be scanned must be clear and sharp.
- Bring the platform to the right Z-height. This means that the substrate plate must be a little bit lower (0.5 mm) than the build chamber base. This height can be controlled with a spatula.
- Clean the cover of the f-theta-lens with a soft and clean paper towel.
- Close the build chamber and start flooding routine by pressing "flood on" in the Realizer software/motors page.
- Turn the heating on, if desired. A temperature of 240°C is all right.
- When the machine has reached an oxygen value lower than 1%, turn off the flooding routine by pressing "flood off" in the Realizer software/motors page.
- Turn on the gas pump by pressing "gas pump on" and start the flooding routine ("flood on") again till you have reached an oxygen value of 0.2%.
- If 0.2% or lower is reached, turn off flooding routine ("flood off") and press "Gas On".
- Now you must level the platform by driving the recoater to the front of the base plate, drop a little bit of powder by putting a value of 15.000 in the command box of loader 2 in the Realizer software/motors page. Afterwards drive the recoater back to the end of the building chamber.
- Raise platform by 0.1 mm and drive the recoater to the front and to the back.
- Repeat the above step till you can see the complete platform through the powder bed.
- Press "Set Zero" for the elevator in the Realizer software/motors page to determine the actual position of the platform as zero position (Figure 7).
- Open the data you want to build.
- Start the job by pressing the yellow "GO" in the Realizer software.
- Then put in the slice you want to start from. First layer to be built is usually "0", and press "OK".
- Now the machine starts the build.
- During the fabrication, the oxygen concentration should be lower than 1% in a nitrogen/argon protection gas.

The correlation between the laser power and the adequate scanning speed is very important to be optimized for each type of powder material. For the 316L powder, a processability map is illustrated and explained in previous publications (Cosma et al., 2020). For each powder, the SLM process parameters should be optimized to have stable manufacturing and without defects in the manufactured parts. Otherwise, some frequent SLM defects are cracks, shrinkage, severe warping, vaporization, and distortion. These defects may cause a premature failure

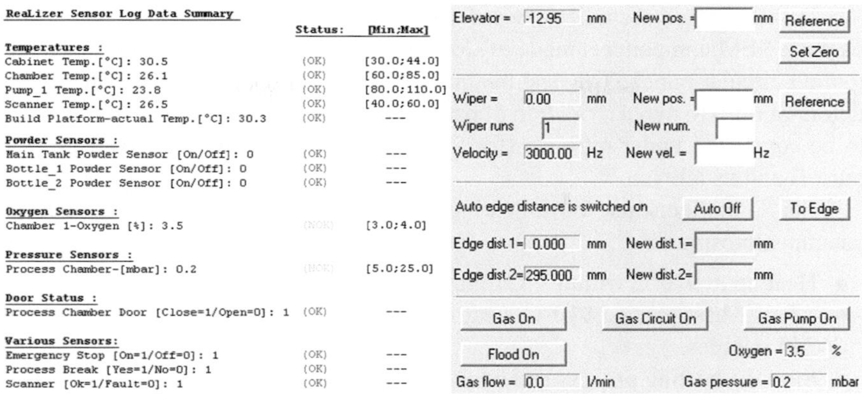

Fig. 7 Machine preparation, Realizer SLM 250

in manufacturing. Based on the temperature gradient mechanism, once a melted pool is heated up rapidly by the laser beam irradiation, the pre-solidified material (underneath the melted layer) tends to expand quickly, but it is kept by the cold (rigid) area of the SLM part. Such heating cycles are repeated multiple times. The processability could be evaluated for each type of powder to point out the range of parameter values when the SLM process is stable and a good SLM part could be printed (Cosma et al., 2020).

4. Post-processing the SLM Parts

Post-processing is essential to obtain good SLM parts, with finished functional surfaces and lower internal stresses.

4.1 Heat Treatment for Stress Releasing

A stress relief heat treatment is recommended because usually the SLM parts do contain residual stress. Before cutting the support structures from the base platform, the SLM parts are introduced in a special oven for a specific stress relief treatment. In the TUCN lab, a Nabertherm furnace is used for thermal treatment (type N 31/H with controller P 470 and protective gas box for heat treatments). The furnace allows specific heat treatment cycles for each material.

A practical example of stress relief treatment is listed here. This heat treatment is applied to Co-Cr parts, and it can be conducted by following the steps presented here (Cosma et al., 2021):

- Heat up to 450°C within 60 minutes and hold for 45 minutes.
- Heat up to 880°C within 45 minutes and hold for 60 minutes.
- Stop the heating process.
- The temperature is cooled down within 30–60 minutes to 300°C.
- After the temperature decreases to 350°C, stop argon/nitrogen supply.
- Open the furnace door after the temperature has decreased to 300°C (or lower) and remove the SLM parts for further processing.

Based on internal studies, it was found that this heat treatment is efficient even on large SLM components made of Co-Cr such as orthopedic prostheses or rapid tooling circular punch. The heat treatment limits significantly residual stress and deformations. To avoid oxidation of parts, stress-relief treatment is recommended to be conducted under nitrogen (for Co-Cr and steel) or argon atmosphere (for pure Ti and its alloys).

For Ti components, the following stress relief treatment can be applied in vacuum atmosphere:

- Heat up to 910°C within 7°C/min.
- After it has reached 910°C, heat up to 940°C within 5°C/min and hold for 240 minutes.
- Stop the heating process and cool down at a rate of 1°C/min to 760°C.
- Then in the argon cool down at a rate of 1°C/min.
- After the temperature decreases to 300°C, stop the argon supply.
- Open the furnace door after the temperature has decreased to 300°C (or lower) and remove the SLM parts for further processing.

After this heat treatment, Ti components can be exposed to a Hot Isostatic Pressing (HIP) treatment with the following parameters: 920°C and 1000 bar for 2 hours.

4.2 Supports Removal and Finishing the Parts

After the stress relief heat treatment, the SLM parts are removed from the machine plate and the support structures are cut using a metal bandsaw. To finish and clean the SLM surfaces, a common method is to sandblast with aluminum oxide (approximately 110–150 μm) at 2–3 bars for 5–10 minutes. The recommended distance between the part and the nozzle is 50 mm, under 45° angle.

To finish the surface where the support anchors the part, initially the peaks were removed manually. Then, the surface is cleaned with carbide/diamond burs suitable for Co-Cr, steel, or Ti alloys and by carborundum polishing. A geometrical inspection of complex facial implants post-processed is illustrated in Figure 8. The SLM implant was compared with its CAD design, and the geometrical accuracy was ±0.30 mm. Current state-of-the-art equipment and processes achieve accuracy of about ±0.20 mm plus 0.10 mm for every 100 mm thereafter.

Certain applications require even a smoother surface, and CNC machining could be used for better finishing (chemical polishing could be a solution, as well).

One of the major advantages of SLM is the reusability of the un-melted powder. After an SLM job is finished, the un-melted powder is collected with a brush and recycled for future fabrication. To recycle the un-used powder, a vibratory sieve shaker is operated (i.e., type Retsch AS 200). There are different sieving procedures, vibratory sieving (vertical movement of the powder) or horizontal sieving. Due to this combined motion, the particles are compared with the aperture sieve mesh, which is commonly 45 μm or 63 μm. It is important to determine the ratio of the particle size to the sieve openings, but there are other factors as well that influence

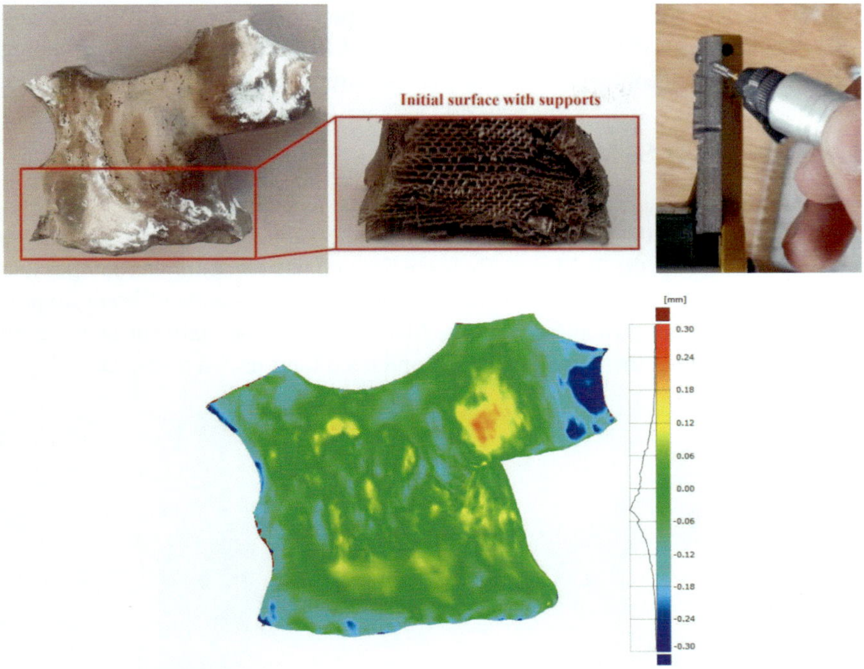

Fig. 8 Initial surface with support structures, after manual finishing, and map with dimensional deviations limited to ±0.3 mm

the probability of a particle skipping through the sieve mesh. The orientation of the particle is also important, and the number of encounters between the particle and the entries of the mesh. The Retsch sieve has the following dimensions Ø200 × 50 mm. After fulfilling the sieve with powder, the amplitude is set up at 80% and the time at 15 minutes. One cycle is enough to recycle the powder, and it is deposited in a collection tray. Generally, before starting an SLM job, 5% of virgin powder should be added to the recycled feedstock. In this topic of powder recyclability, extended studies developed on Inconel 718 and Ti6Al4V reported the following results (LPW Technology, 2015; Strondl et al., 2015):

- disappearance of smaller particles and no change in particle morphology;
- better flowability;
- no change in composition or in processed microstructure;
- reduced changes in mechanical behavior; comparing the ultimate tensile strength obtained from virgin powder and after 25 times recycled ones, it was limited reduced by 2%. The value of elongation at break was similar;
- increased oxygen content;
- decrease in ductility and impact toughness (possibly connected to oxygen pickup);
- slight porridge increase (possibly connected to oxygen pickup).

The details of the results are presented as open source by the LPW company after repeated use of Inconel 718 powder in SLM builds, their research being focused on oxygen content and ultimate tensile strength (LPW Technology, 2015).

5. Design for SLM

One important disadvantage of the SLM process, as compared to selective laser sintering (SLS), is the difficulty of removing the support structures, which are strongly melted onto the SLM part. There are limitations of the complex cavities that could be printed by SLM. Figure 9 presents some SLM parts with different support structures which are difficult to remove. Also, the outer boundary surfaces where the supports contact them are very rough and need long-drawn post-processing.

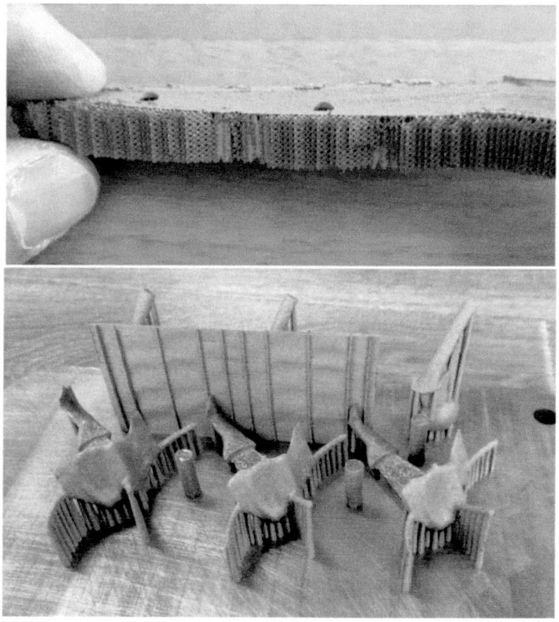

Fig. 9 SLM parts with different support structures which are difficult to remove

By optimizing the design of a part, support structures can be reduced or even avoided. A practical example is presented in previous publications (Cosma et al., 2019). The design for AM rules could be applied to generate special features of the rapid tooling part, depending on the profile (external walls), building angle, and internal features (channels or holes). Examples of optimizing the part design to avoid support structures are presented in previous publications where the redesigned parts are illustrated, in comparison with similar conventional design, for different types of features such as external walls, channels, or holes (Cosma et al., 2019).

6. Applications and Case Studies

Many industrial applications and case studies have been developed at TUCN, within the research projects and PhD theses undertaken by the Additive Manufacturing and Rapid Product Development Research Center–(AMaRaP—https://amarap.utcluj.ro/index.html).

6.1 Case Study: Mold Manufacturing for Injection of Plastic Parts

Case Overview

Beneficiary: Dedienne Romania, location Făgăraş (Brasov), part of Dedienne Multiplasturgy Group from France (www.dedienne.com), a Romanian-based manufacturer of automotive plastics components specialized in injection molding processes.

Objective: Design and SLM manufacture a mold with cooling channels to increase production. The mold will allow to fabricate an automotive plastic part used to control the computer board of the Renault Scenic EG car.

Need: In the field of plastic injection, the time of cooling is often between 60% and 90% of the total time necessary for a production cycle, and new concepts are needed to develop and test.

Difficulties: Conventional casting molds work during many cycles of thermal shocks, which often cause thermal fatigue. Lattice structures could be implemented within some areas of the mold to increase structural resistance to thermal fatigue. The mold should be designed to have adequate compliance for thermal expansion.

Designing software used: SolidWorks, Creo Parametric, Rdesigner, and Autodesk MeshMixer.

Material used: Stainless steel powder (316 L), with 50 μm spherical grains, provided by MCP HEK Tooling Realizer (Germany).

Equipment used: Realizer SLM250 (Germany) providing the SLM technology and containing a Nd:YAG (neodymium-doped yttrium aluminum garnet) solid-state laser type with 200 W maximum laser beam power.

Benefits of SLM manufacturing:

- Special cooling channels could be designed to provide a higher heat transfer rate and a smaller variation of the tool temperatures.
- The cooling channels developed will reduce the cycle time of the plastic injection process, increasing the production rate.
- The active elements of mold will have a reduced weight (15% lighter).

Application details: The functional role of the button is to command, control, and set up the dashboard of a Renault Scenic EG automobile (Figure 10). The button is positioned on the front mask computer board. The active elements of mold will be used to test the injection process of various materials.

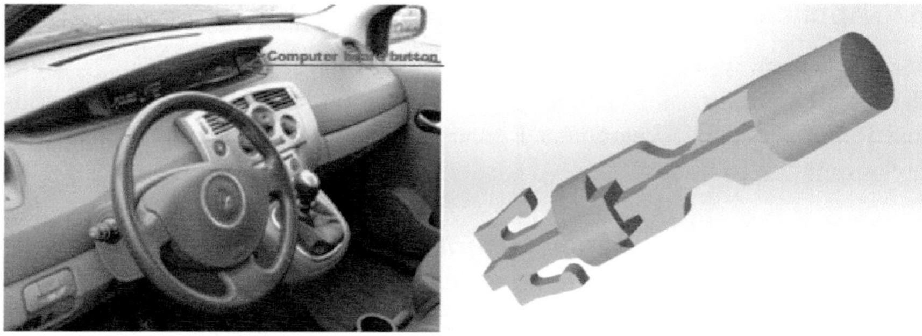

(a) (b)

Fig. 10 (a) Renault Scenic car; (b) Computer board button which requires the new SLM mold

The main activities undertaken during this case study are presented in the following:

- Determination of the *optimum technological parameters required for the injection* of plastic parts such as injection point, injection pressure and the filling time, thickness analysis, volumetric shrinkage, and clamping force.
- *Redesign for additive manufacturing* (SLM process) the active elements of mold. Figure 11 presents the CAD model of mold with integrated cooling channels (a top view). It was designed with 12 main cooling channels with 38 mm length and 1.5 mm width.
- *Design and incorporate the lattice structures* into mold cooling channels. The lattice unit used represents a body-centered cubic structure and it was designed in RDesigner software with an edge length of 0.55 mm (Figure 11). This lattice structure was incorporated into all the mold cooling. The new concept of mold is 15% lighter than conventional counter one.
- *SLM preparation* of new mold. The main steps are detailed earlier in Section 3. Various build-up orientations were simulated. Horizontal orientation is preferred to reduce the production time, but the convex surface of active element will have a pronounced "stairs effect" (normal to high roughness, Figure 12).
- *SLM simulation* of semi-molds fabrication having lattice structures integrated into cooling channels. Different scanning functions can be seen as follows: outer boundary, inner boundary, Hatch in X and Y direction, X and Y Support, outer support. For each of them was configured a value for laser power (in mA, where 5000 mA = 200 W) and scanning speed. The scanning speed is expressed by exposure time (Et) and point distance (Pd). The scanning strategy was alternating X/Y. To limit the impact of internal tension, the semi-molds were fabricated directly on the SLM plate without support structures. An estimation of the manufacturing costs could be done using the RCalculator tool. Using the REA file with all the SLM parameters and parts slices, this app calculates a cost of 985€ for semi-molds, where 80 €/h is the processing fee, and the production time is around 11.3 hours.

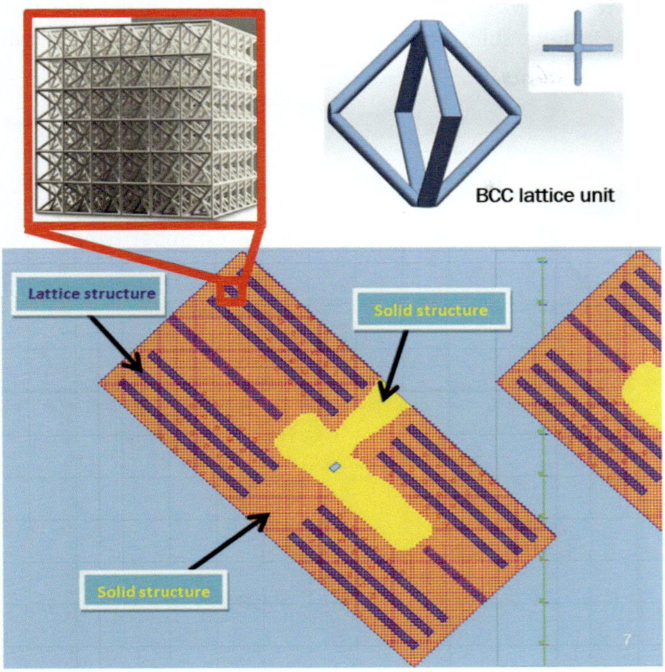

Fig. 11 Virtual model of lattice structure (macro-porous scaffold), active element of mold with solid areas (yellow color) and lattice structures into cooling channels (mauve color)

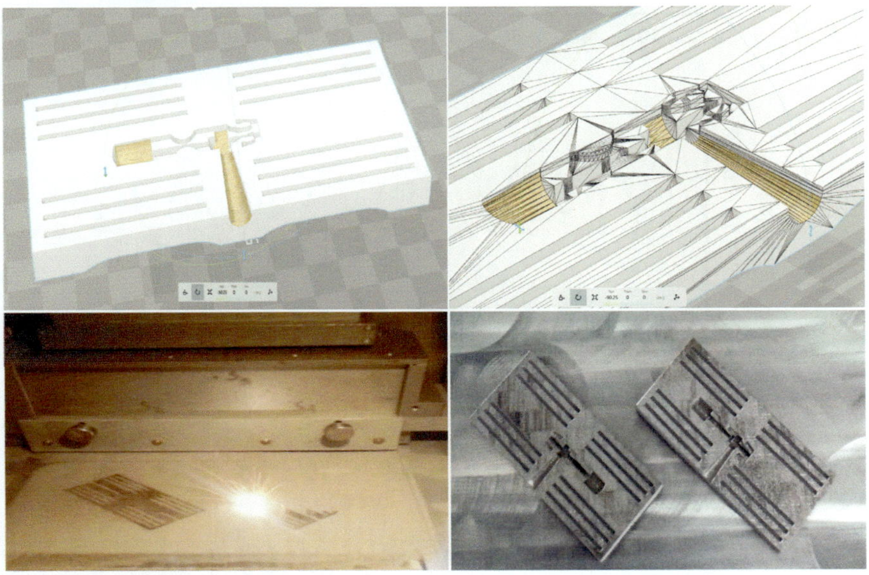

Fig. 12 SLM manufacturing of molds, with horizontal built orientation (minimum time)

- *SLM manufacturing of semi-molds* was developed in optimal conditions without any unfavorable issues or defects (Figure 14). The powder used

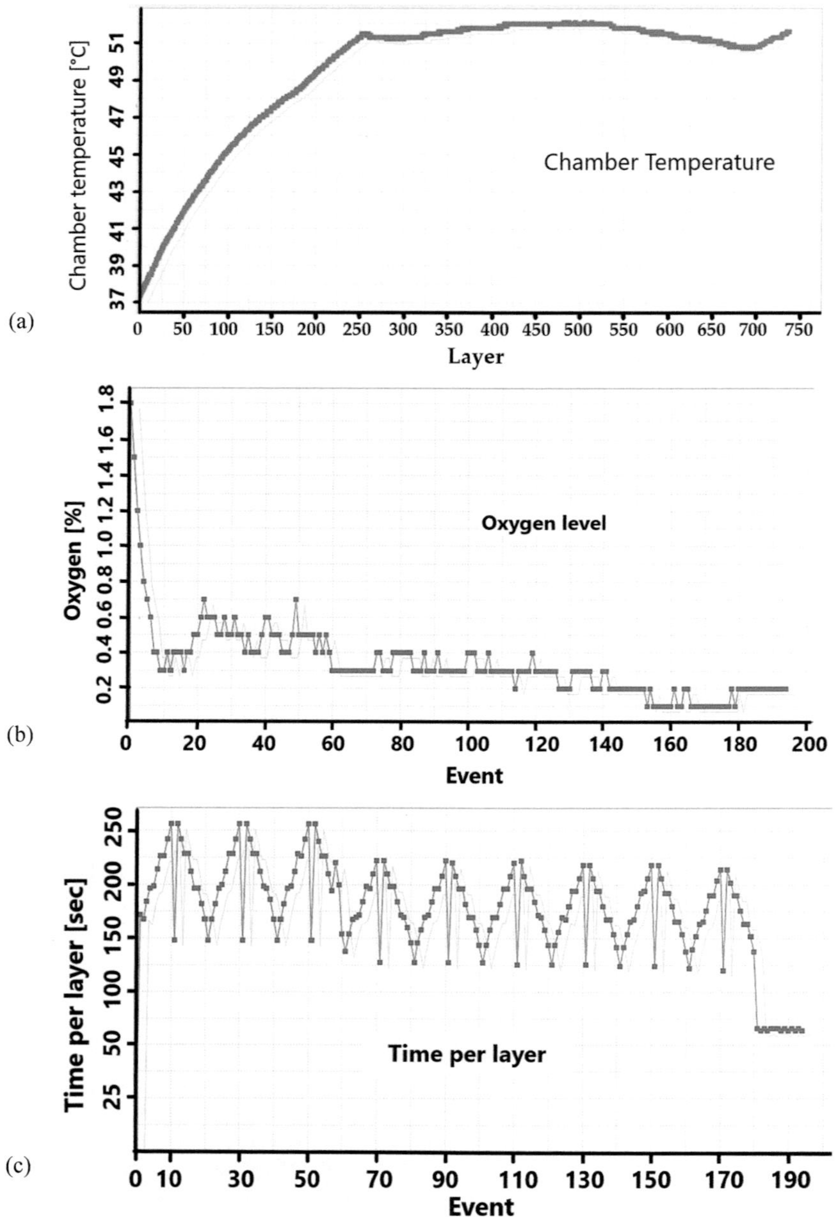

Fig. 13 Diagrams generated during the SLM fabrication by machine sensors: (a) Chamber working temperature; (b) Oxygen content from working chamber; (c) Time per layer expressed in seconds

was stainless steel 316L with 50 μm spherical grains, provided by MCP HEK Tooling (Germany). The production time was 12h because the lattice structures require more time for scanning of each strut slice.

Figure 13 presents some diagrams generated during the SLM fabrication of semi-molds. The working chamber recorded a temperature up to 52°C. After the SLM machine scanned the first 10 layers, the oxygen content decreased from 1.8% to 0.3% and was main constant at approximately 0.4% in an argon atmosphere. In general, the time needed to scan a powder layer varied between 200 s and 300 s being influenced by its area.

- *Post-processing* of SLM-manufactured semi-molds. Initially, the lattice structures were ultrasonic cleaned and then the semi-molds were subject to stress relief treatment at 540°C. The active zone was sandblasted with glass and the contact surfaces were processed using a CNC milling machine (Figure 14).

Fig. 14 Active elements of the SLM molds, with lattice structures in cooling channels (SEM analysis)

The morphological aspects are also illustrated in Figure 14, which presents the lattice struts using a scanning electron microscope (SEM). The stair effect could be seen on this type of lattice structure, where there is some balling phenomenon, unmolten or half-attached particles on struts. The important idea is to avoid the need for support; that is why the lattice struts have an angle of 45°, which is the minimum value, when the support is not necessary.

- *Injection molding tests* were performed at Dedienne Romania. The injection molding process parameters were optimized for using this ABS material. The main injection parameters configured were melting temperature of ABS (270°C) and injection pressure (10 bar). An injection cycle takes 12 s/piece, and the ABS part has a Ra roughness around 3.8 μm (Figure 15).

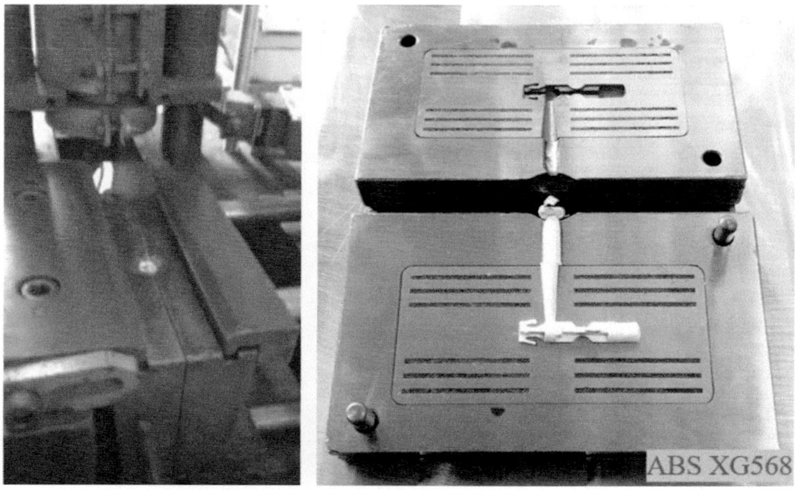

Fig. 15 Injection test using the SLM mold and ABS plastic

The presented case study demonstrates that the design solution is feasible and the SLM process is capable of manufacturing cooling channels into molds having lattice structures. Compared with conventional molds CNC machined from bulk raw material, the actual molds have a reduced weight by 15%. Integrating these lattice structures into the cooling channels improves heat transfer and reduces the cooling cycle time by 14%, increasing the production rate of plastic parts.

6.2 SLM Applications Developed by TUCN

Figure 16 presents a few rapid tooling and rapid prototype applications developed within the AMaRaP Research Centre at TUCN, using the Realizer SLM 250 machine.

7. Research Perspectives and Future Trends in SLM

Combining the design for AM methods and the SLM process could be a promising approach to obtain optimized shapes for industrial tools. The manufacture of

Fig. 16 SLM applications developed at TUCN, using (a) Pure Ti; (b) Co-Cr; (c) Ti-6Al-7Nb; (d) Stainless steel 316L

tooling for injection molding is a key business sector that has a significant impact on overall costs and lead times in the product development cycle (Irrinki et al., 2016). Molding cycle times account for 35% of the part cost, and innovative mold designs and materials using SLM manufacturing appear to offer the promise for further impacting the cost-per-part produced by injection molding.

SLM technology offers the opportunity to make rapid cutting tools with lower weight and higher durability. Guhring GmbH recently developed complex tools made from H13 tool steel. It is an end-milling cutter with a better design, which reduces the weight (60%), decreases the production time (65%), and reduces the cost (75%), for low-volume production tooling. This rapid tool will allow the Guhring team to open new revenue streams and better service to their customers, including different coolant pathways. Such an end-milling cutter with 60% reduced weight obtained with large hatch distance parameter is open source illustrated (www.guhring.com).

Cutting tool manufacturer Mapal leveraged SLM technology to produce smaller versions of its through-coolant insert drills where the coolant pathway has a triangular cross-section. The spiral coolant pathway runs parallel to the flute, which improves the core stability and improves cooling effectiveness. This SLM drill with triangular passageways increases the quantity of coolant flow by 30%.

Tool-manufacturing Horn is also using SLM to fabricate special tools and tool holders made of aluminum (AlSi10Mg) and stainless steel (1.4404).

Sandvik Coromant recently developed a lightweight mill-cutter using topological optimization method and SLM manufacturing. Such lighter tools work with increased productivity, smaller vibration, and are more secure during machining with long overhangs.

Published articles present novel concepts of lightweight milling cutter integrating lattice structures into its body based on finite element analysis (Hanžl et al., 2019). The virtual model of cutter has been created to predict the strain and deformation of the lattice struts. Finite Element Analyses (FEA) simulations were elaborated according to static loads and demonstrate that this new concept of lightweight cutter meets the strength requirements, though its stiffness does not reach the fully filled version.

Arno Tools developed new toolholders for stationery cutting tools. These toolholders are used in turning and offer curving channels to direct coolants to where it can best aid the cutting edge (Zelinski, 2019).

A 3D-printed frame for Ducati Panigale V4 was presented at the 2024 Acronym of International Motorcycle and Accessories Exhibition. Motorcycle Show (Italy), being designed by Angelelli Automobili and fabricated by WEAREAM Company, it is the name. A topological optimization approach was used to design a seamless, monolithic structure (Davide, 2024) SLM technology will move soon from the research laboratories to industrial production and will increase the flexibility for low volume production of customized products. The advantages of the SLM process include reduced manufacturing time for prototypes and complex components, plus material and energy savings. Due to the high resistance of

SLM parts (attributed to the fine crystalline structure created), they will generate benefits in terms of total production cost, making this emerging technology more adopted by companies. AM technology is already a key component of the Industry 4.0 strategy. Around 84% of companies from Europe are willing to invest in the digitalization of manufacturing technologies to transform them into Smart Factories. Smart Factories are developed to meet the expectations of companies, with respect to higher productivity, better efficiency, cost reductions, higher profits, more customized products for better customer satisfaction, and shorter times to market.

Acknowledgement

This work was supported by a European grant from the Erasmus Plus Programme of the European Union, project number 2019-1-RO01-KA203-VET-063486 within KA2 action.

References

3DHubs. 2022. Automotive 3D Printing Applications. Available at www.3dhubs.com/knowledge-base/automotive-3d-printing-applications, (Accessed on 5 May 2021).

Additiva. 2020. Design for Additive Manufacturing. Additiva Company. Modena, Italy. Available at www.additivalab.com, (Accessed on 8 November 2021).

Aniwaa. 2019. Best Metal 3D Printers. Available at https://www.aniwaa.com/buyers-guide/3d-printers/best-metal-3d-printer/, (Accessed on 10 November 2022).

AutoFab. 2018. AutoFab User Manual. Marcam Engineering GmbH, Germany.

Cherry, J.A., Davies, H.M., Mehmood, S., Lavery, N.P., Brown, S.G.R., Sienz, J. 2015. Investigation into the Effect of Process Parameters on Microstructural and Physical Properties of 316L Stainless Steel Parts by Selective Laser Melting. *The International Journal of Advanced Manufacturing Technology* 76: 869–879.

Chlebus, E., Kuźnicka, B., Kurzynowski, T. and Dybala, B. 2011. Microstructure and Mechanical Behaviour of Ti-6Al-7Nb Alloy Produced by Selective Laser Melting. *Materials Characterization* 62(5): 488–495.

Cosma, C., Kessler, J., Gebhardt, A., Campbell, I. and Balc, N. 2020. Improving the Mechanical Strength of Dental Applications and Lattice Structures SLM Processed. *Materials* 13(4): 905.

Cosma, C., Moldovan, C., Campbell, I., Cosma, A.C. and Balc, N. 2018. Theoretical Analysis and Practical Case Studies of Powder-based Additive Manufacturing. *Acta Technica Napocensis* 61(3): 396–378.

Cosma, C., Moldovan, M., Simion, M. and Balc, N. 2021. Impact of Laser Parameters on Additively Manufactured Cobalt-chromium Restorations. *The Journal of Prosthetic Dentistry* 128(3): 421–429.

Cosma, C., Zaharia, B., Berce, P. and Balc, N. 2019. Design for Additive Manufacturing to Produce Complex Metal Parts. *Romanian Journal of Technical Sciences. Applied Mechanics* 64(3): 231–241.

Davide, S. 2024. Angelelli Automobili Presents 3D Printed Ducati Motorcycle Frame. Automotive AM. VoxelMatters Ltd. Available at https://www.voxelmatters.com/angelelli-automobili-presents-3d-printed-ducati-motorcycle-frame/, (Accessed on 30 October 2024).

Fateri, M., Gebhardt, A., Thuemmler, S. and Thurn, L. 2014. Experimental Investigation on Selective Laser Melting of Glass. *Physics Procedia* 56: 357–364.

Hanžl, P., Zetek, M., Rulc, V., Purš, H. and Zetková, I. 2019. Finite Element Analysis of a Lightweight Milling Cutter for Metal Additive Manufacturing. *Manufacturing Technology* 19(5): 753–758.

Hubs. 2024. Additive technologies compared, Metal 3D printing, https://www.hubs.com/guides/metal-3d-printing/, accessed on 9 January 2024.

Irrinki, H., Barmore, B., Kate, K. and Are, S. 2016. Material Selection for the Production of Injection Moulding Tooling by Additive Manufacturing. *Metal AM* 2(2): 77–89. Available at https://www.metal-am.com/articles/material-selection-for-the-3d-printing-of-injection-moulding-tooling/, (Accessed on 20 October 2024.).

Jaya, B. 2019. SLM Application to Produce Tools; Polishing? No, Only Milling. Vogel Communications Group. Available at www.etmm-online.com/slm-application-to-produce-tools-polishing-no-only-milling-a-866376/, (Accessed on 30 October 2024).

LPW Technology. 2015. Case Study 05: Powder Degradation. LPW Technology Ltd, UK.

Majumdar, T., Bazin, T., Massahud Carvalho Ribeiro, E., Frith, J.E. and Birbilis, N. 2019. Understanding the Effects of PBF Process Parameter Interplay on Ti-6Al-4V Surface Properties. *PLoS One* 14(8): e0221198.

Mayer, D., Stoffregen, H., Heuss, O., Thiel, J., Abele, E. and Melz, T. 2016. Additive Manufacturing of Active Struts for Piezoelectric Shunt Damping. *Journal of Intelligent Material Systems and Structures* 27: 743–754.

MCP Realizer. 2008. Operating Manual SLM—REALIZER 250. MCP HEK Tooling GmbH. Germany.

Meier, H. and Haberland, C. 2008. Experimental Studies on Selective Laser Melting of Metallic Parts. *Materials Science & Engineering Technology* 39(9): 665–670..

Schwarz, A., Kessler, J. and Gebhardt, A. 2016. Design and Construction in 3D Printing. Summer School, Aachen University of Applied Sciences, Germany.

Sher, D. 2017. DMG Mori Acquires a Majority Stake in SLM 3D Printer Manufacturer Realizer. Available at https://www.3dprintingmedia.network/dmg-mori-acquires-majority-stake-slm-3d-printer-manufacturer-realizer/, (Accessed on 20 May 2024).

Strondl, A., Lyckfeldt, O., Brodin, H., Ackelid, U. 2015. Characterization and Control of Powder Properties for Additive Manufacturing. *JOM* 67: 549–554.

VDI 3405. 2015. Additive Manufacturing Processes, Rapid Manufacturing: Design Rules for Part Production Using Laser Sintering and Laser Beam Melting.

Wang, Z., Xiao, Z., Tse, Y., Huang, C. and Zhang, W. 2019. Optimization of Processing Parameters and Establishment of a Relationship between Microstructure and Mechanical Properties of SLM Titanium Alloy. *Optics and Laser Technology* 112: 159–167.

Yap, C.Y., Chua, C.K., Dong, Z.L., Liu, Z.H., Zhang, D.Q., Loh, L.E., Sing, S.L. 2015. Review of Selective Laser Melting: Materials and Applications. *Applied Physics Reviews* 2: 041101.

Zelinski, P. 2019. AM is Advancing into Tools for Machining: Examples from EMO. Additive Manufacturing, Gardner Business Media USA. Available at https://www.additivemanufacturing.media/articles/am-is-advancing-into-tools-for-machining-examples-from-emo, (Accessed on 10 February 2023).

8 Numerical Simulation of Polytetrafluoroethylene Orthogonal Cutting Process

Felicia Stan[1] and Catalin Fetecau[2*]

1. Introduction

Numerical modeling and simulation of manufacturing processes is extensively used due to its ability to predict and provide relevant information in different stages of manufacturing (design, validation, or control), thus improving productivity by reducing the number of trial-and-error experiments. Considerable efforts have been devoted over the last years to the development of computational methods and simulation strategies for predicting fundamental variables and industrial relevant outcomes in the field of metal machining (Miguélez et al., 2013; Wang et al., 2013; Ezilarasan et al., 2014; Chen et al., 2013; Melkote et al., 2022; Mir et al., 2022). As a result, finite element (FE) simulation has become probably the only instrument in predicting both mechanical and thermal behavior of both material and tool in metal cutting processes. FE simulation provides useful information for optimizing the cutting parameters, cutting forces, or other machining performances, like tool wear, tool-life, and surface integrity (Maranhão and Davim, 2010; Miguélez et al., 2013; Wang et al., 2013; Ezilarasan et al., 2014; Chen et al., 2013; Melkote et al., 2022; Mir et al., 2022). In addition, the FE simulation can be used to study the chip formation mechanism as it provides information on stress, plastic strain, strain rate, and temperature in both workpiece and tool (Atlati et al., 2011; Miguélez et al., 2013; Wang et al., 2013; Ezilarasan et al., 2014; Chen et al., 2013; Melkote et al., 2022) that can hardly be obtained by experimentation. However, simulation of the machining processes is a challenging task mainly due to the geometric complexity of the tool-chip-workpiece system. The correct selection of the constitutive model in conditions similar to the cutting process is often regarded as a critical step (Ozel and Zeren, 2004; Dixit et al., 2011; Arrazola et al., 2013), especially for newly developed materials, including polymers and polymer-based composites. A critical role is also played by the friction at the tool-chip interface which influences not only the value of the cutting forces but also the distribution of the shear plane angle and the cutting temperature on the tool rake face (Ozel, 2006; Ozel and Arrazola, 2010). Consequently, without successful constitutive and

[1,2] *Professor*, Center of Excellence Polymer Processing, Dunarea de Jos University of Galati.
E-mail: felicia.stan@ugal.ro[1]

* Corresponding author: catalin.fetecau@ugal.ro

friction models, expensive and rather slow experiments will continue to dominate the development of the machining processes (Arrazola et al., 2013).

Polymers are used widely in many key industrial applications that require precise machining, and numerical prediction of machining performance (e.g., cutting force, surface roughness, temperature, and shear angle) is essential for manufacturers in ensuring high quality products. Despite the significant progress in the modeling of metal machining processes, in literature, studies on the machining simulation of polymers are scarce and mainly focused on polymer-based composites (Rao et al., 2008; Dandekar and Shin, 2008; Mkaddem and Mansori, 2009; Davim, 2009; Dandekar and Shin, 2012). Moreover, the experimental analysis still dominates the process development in the field of polymer machining, investigating fundamental aspects such as cutting forces (Davim and Reis, 2004; Davim and Mata, 2007; Mata et al., 2009; Mkaddem and Mansori, 2009; Dandekar and Shin, 2012; Fetecau and Stan, 2012; Stan et al., 2013; Ni et al., 2023; Cui et al., 2022), surface roughness (Davim and Mata, 2005; Palanikumar et al., 2006; Palanikumar, 2008; Petropoulo et al., 2008; Cui et al., 2022; Fetecau and Stan, 2012), temperature, and wear (Palanikumar and Davim, 2009). However, other fundamental process variables such as strain rate and shear angle cannot be directly measured using experimental testing procedures. The modeling of the polymer machining process is limited largely due to the difficulties related to the identification of material constitutive and friction/contact models. This is because polymers exhibit a rich variety of material behavior originating from their particular behavior that is strongly dependent on time, temperature, and pressure.

Aiming at assessing the challenges in numerical simulation of polymer cutting process, the numerical simulation of polytetrafluoroethylene (PTFE) orthogonal cutting process is presented in this study, and the effect of the main cutting parameters (tool rake angle, feed rate, and friction coefficient) on the machining characteristics such as cutting forces, temperature at tool-chip interface, plastic strain, and shear angle is discussed and analyzed.

2. Theoretical Aspect of Modeling of Cutting Process

2.1 Johnson–Cook Constitutive Model

To carry out numerical simulations of the cutting process, a constitutive model representing the material flow stress with respect to strain, strain rate, and temperature at different machining conditions must be provided. For cutting speeds typically used in machining, the average strain rates developed in the shear zone (10^3 s^{-1} to 10^5 s^{-1} or even higher (Grzesik, 2008) are much higher than the strain rates normally developed in compression and tension tests (10^{-3} s^{-1} to 10^{-1} s^{-1}).

In the attempt to describe the plastic and visco-plastic material behavior during machining simulation, several empirical models have been proposed: Johnson-Cook, Zerilli-Armstrong, Bodner-Partom, Khan-Huang, and modified Bai-Wierzbicki (extended MBW model) material models (Johnson and Cook, 1983;

Liang and Khan, 1999; Milani et al., 2009; Notta-Cuvier et al., 2013; Buchkremer et al., 2014). Among them, the Johnson-Cook (J-C) model is one of the most widely used for modeling and simulation of industrial applications, particularly for modeling high strain rate deformation of metals in adiabatic transient dynamic analysis. Because the empirical models rely on determining material constants by fitting the model to experimental stress-strain responses at different temperatures and strain rates, an important advantage in using the J-C model over other models is the small number of constants to be determined. Another advantage of using the J-C model is that numerous studies have already been conducted to identify the material constants for different materials (Johnson and Cook, 1983; Liang and Khan, 1999; Milani et al., 2009; Notta-Cuvier et al. 2013; Buchkremer et al., 2014) and thus this model has been implemented in almost all commercial simulation software.

The constitutive model proposed by Johnson and Cook describes the flow stress of a material as a product of strain, strain rate, and temperature, individually determined by the following equation (Johnson and Cook, 1983):

$$\sigma = \left(A + B\varepsilon_p^n\right) \cdot \left(1 + C\,\ln\left(\frac{\dot{\varepsilon}_p}{\dot{\varepsilon}_0}\right)\right)\left(1 - T^{*m}\right) \tag{1}$$

where the nondimensional temperature T^* is given by

$$T^* = \frac{T - T_R}{T_{\text{melt}} - T_R}, \quad \text{if} \quad T_R \leq T \leq T_{\text{melt}} \tag{2}$$

in which σ is the effective stress, ε_p is the effective plastic strain, $\dot{\varepsilon}_p$ is the strain rate, $\dot{\varepsilon}_0$ is the reference strain (s^{-1}), n is the hardening coefficient, m is the thermal softening coefficient, A is the yield strength (MPa), B is the hardening modulus (MPa), C is the strain rate sensitivity coefficient (viscosity), T_R is the reference temperature at or below which the response is not temperature dependent (°C), T_{melt} is the melting temperature of the work material (°C), and T is the current temperature of the work material (°C).

2.2 Friction at the Tool-Chip Interface

Friction along the tool-chip interface is a complex and intriguing phenomenon, playing a significant role in the mechanics of cutting processes because of its direct effect on chip formation, cutting temperature, cutting forces, tool wear, etc. (Ozel, 2006; Rech, 2006; Vaz et al., 2007; Filice et al., 2007; Zemzemi et al., 2009; Rech et al., 2009; Ozel and Arrazola, 2010; Molinari et al., 2011; Jamshidi and Nategh, 2013; Ezilarasan et al., 2014; Storchak, 2024). On the other hand, friction along the tool-chip interface is influenced by cutting parameters, such as cutting speed, feed rate, and rake angle (Rech, 2006; Vaz et al., 2007; Filice et al., 2007; Zemzemi et al., 2009; Rech et al., 2009; Molinari et al., 2011; Jamshidi and Nategh, 2013; Storchak, 2024). Therefore, it is necessary to understand the frictional behavior along the tool face and around the tool edge in order to accurately predict the cutting forces, strain rate, temperature, etc.

The friction mechanism of polymers is different from that of metals since both time-dependent and time-independent deformations, as well as the low surface energy, are very important in polymer tribological systems (Lafaye et al., 2006; Samyn and Schoukens, 2008; Quaglini et al., 2009; Quaglini and Dubini, 2011). The apparent friction coefficient can be decomposed into two components, adhesion and ploughing (Quaglini et al., 2009; Quaglini and Dubini, 2011). Since the friction of polymers is closely related to their viscoelastic behavior (Quaglini et al., 2009; LeMaitre, 2001), the ploughing coefficient may be decomposed into a term due to viscoelasticity and another due to plasticity. However, at low and medium pressure levels, as is the case of polymer machining, the ploughing of a polymer on smooth metal surfaces can be neglected, and friction can be represented in terms of the adhesion mechanism only (Quaglini et al., 2009; Lafaye et al., 2006).

As indicated by several publications, generally, in a polymer-metal tribological system, two mechanisms contribute to the friction (Lafaye et al., 2006; Samyn and Schoukens, 2008; Quaglini et al., 2009; Quaglini and Dubini, 2011; LeMaitre, 2001): shearing of the junctions formed by the adhesion of asperities and the dissipation of mechanical energy due to deformation of asperities. The contribution of the two mechanisms to the overall friction force depends on the normal load or contact pressure, sliding velocity, and chemical, mechanical, and geometrical properties of both surfaces (Samyn and Schoukens, 2008; LeMaitre, 2001). Due to the flexibility and orientation of the polymer molecular chains, for most polymers, at a moderate velocity, friction increases with sliding velocity until a maximum value is reached, followed by a decrease of the friction coefficient (Samyn and Schoukens, 2008; Quaglini et al., 2009; Quaglini and Dubini, 2011). However, for stiff polymers an opposite behavior was reported by Quaglini et al. (2009).

Although it is generally accepted that a constant friction coefficient cannot entirely describe the friction at the tool-chip interface during machining, and yet it is almost impossible to have detailed information about the local pressure and temperature along the tool-chip interface (Ozel, 2006; Filice et al., 2007; Ozel and Arrazola, 2010; Arrazola et al., 2013; Molinari et al., 2011), the friction models are generally defined as: (i) Coulomb-type models, or (ii) friction models depending on sliding velocity (Ozel, 2006; Ozel and Arrazola, 2010; Molinari et al., 2011; Dixit et al., 2011, Arrazola et al., 2013; Ezilarasan et al., 2014), in which friction is defined as a constant, a function of time, or a function of interface pressure (Ezilarasan et al., 2014).

The simplest friction model incorporated in commercially available software is the Amontons-Coulomb law model (also referred to as the Coulomb friction model) (Ernst and Merchant, 1941; Ozel, 2006; Grzesik, 2008)

$$\tau = \mu \sigma_n \tag{3}$$

where τ is the frictional shear stress, σ_n is the normal stress, and μ is the so-called apparent friction coefficient.

The apparent friction coefficient is generally computed from the overall force components along the tool-chip interface as follows (Ernst and Merchant, 1941; Grzesik, 2008; Milani et al., 2009):

$$\mu = \frac{F}{N} = \frac{F_C \sin\gamma + F_T \cos\gamma}{F_C \cos\gamma - F_T \sin\gamma} \tag{4}$$

where F is the friction force on the tool-chip interface, N is the normal force acting on the tool-chip interface, F_T is the thrust force, F_C is the cutting force, and γ is the tool rake angle, as shown in Figure 1.

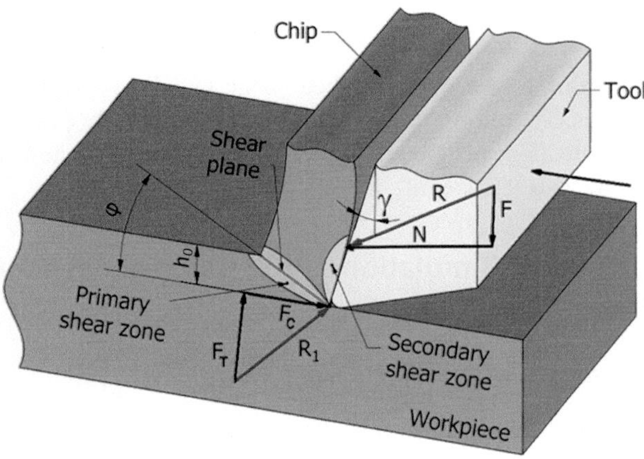

Fig. 1 Forces acting on the tool-chip interface during the orthogonal cutting process

2.3 Shear Angle

The shear angle is an important parameter with impact on all fundamental variables including cutting forces and tool-chip interface temperature. Several approaches are commonly used to calculate the shear angle, such as (i) the Merchant model, (ii) the chip compression ratio, and (iii) direct measuring from the strain and strain rate plots (Ernst and Merchant, 1941; Grzesik, 2008; Storchak et al., 2022a).

The most common method for the estimation of shear plane angle is based on the Merchant model (Ernst and Merchant, 1941). The simplifications associated with the Merchant model might result in an unrealistic estimation of the shear plane angle, but in the literature this model is still extensively used to estimate the shear angle (Ernst and Merchant, 1941; Grzesik, 2008; Storchak et al., 2022a, 2022b). According to the Merchant model (Ernst and Merchant, 1941), the shear angle can be expressed as follows (Figure 1):

$$\varphi = \frac{\pi}{4} - \frac{\beta}{2} + \frac{\gamma}{2} \tag{5}$$

where γ is the rake angle and β is the associated friction angle at the tool-chip interface defined by (Ernst and Merchant, 1941; Grzesik, 2008)

$$\beta = \text{atan}(COF) = \text{atan}\left(\frac{F}{N}\right) \tag{6}$$

in which F and N are the friction and normal forces, respectively, on the rake face computed based on the cutting forces, as in Eq. (4).

Another method by which the shear angle can be determined involves the chip compression ratio, k, defined as the ratio of the chip thickness, h, to undeformed chip thickness, h_0 (Figure 1). The shear angle can be expressed in terms of the chip compression ratio as (Storchak et al., 2022a; Storchak, 2024)

$$\varphi = \text{atan}\left(\frac{\cos\gamma}{k - \sin\gamma}\right) \tag{7}$$

During machining, the primary shear zone (PSZ) is characterized by the highest strain rates. Therefore, the shear angle can be directly evaluated from the strain rate and equivalent plastic strain plots, i.e., the shear plane is drawn to the middle of the PSZ (Storchak et al., 2022a; Bil et al., 2004).

3. Finite Element Simulation of the Orthogonal Cutting Process

3.1 Two-dimensional Finite Element Model

The FE AdvantEdge^{TM} software (Third Wave Systems, USA), which is an explicit dynamic, thermo-mechanically coupled FE software using a Lagrangian updated formulation, was used to simulate the orthogonal cutting process (Figure 2a). The material models implemented in AdvantEdge^{TM} account for elastic-plastic strains, strain rates, and thermal softening, which generally occur during machining operations.

Figure 2b shows the input parameters necessary for simulating the cutting process. In the simulation of the cutting process using AdvantEdge™ software, several assumptions are generally considered, such as:

(i) the tool is assumed stationary, and the workpiece moves along the cutting direction with a constant cutting speed;
(ii) the workpiece height is at least 5 times the feed rate, and the workpiece length is at least equal to its height;
(iii) chip formation occurs by the continuous plastic flow of the material around the tool.

To ensure steady state conditions during machining simulation, the workpiece length was set to 10 mm while the length of the cut was set to 5 mm, which is more than 15 times the maximum uncut chip thickness. The plane strain conditions are satisfied since the width of the cut (4 mm) is greater than the uncut chip thickness (which for the orthogonal cutting is equal to the feed rate), and the tool cutting-edge length is much larger than its width.

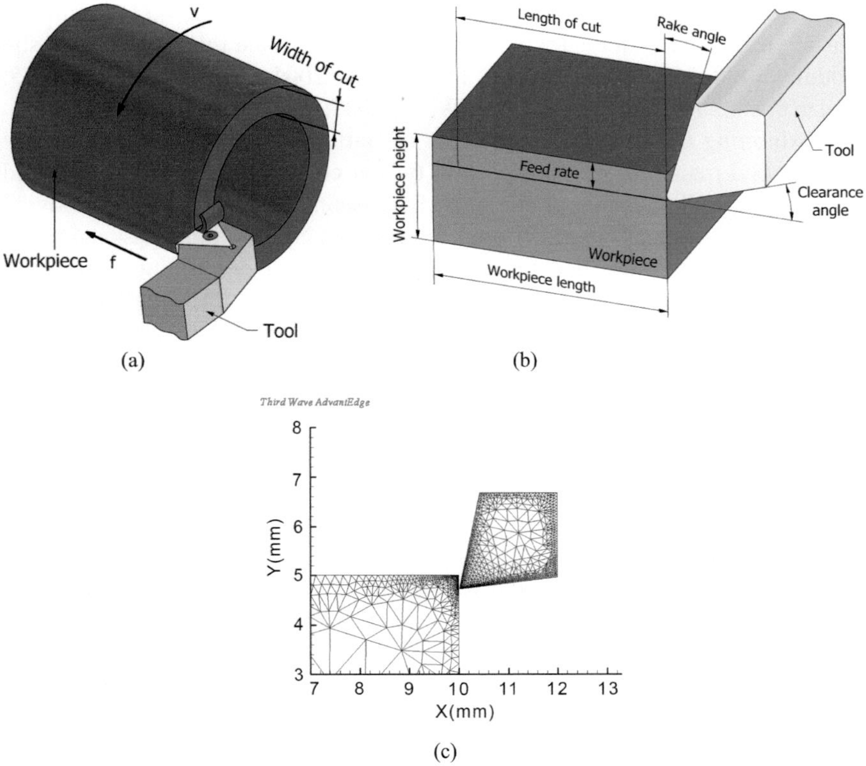

Fig. 2 (a) Schematic representation of the orthogonal cutting process; (b) Finite element model; (c) Initial mesh for the workpiece-tool system in machining simulation (triangle elements with 6 nodes)

Figure 2c shows the initial FE mesh for the workpiece-tool system. The workpiece was modeled with plane strain six-node quadratic triangular elements, and the initial mesh consists of at least 60,000 elements. It should be noted that the initial FE model was identical for all simulation tests taken into consideration, and adaptive remeshing was used to accommodate chip formation and alleviate the excessive deformation of the workpiece-chip mesh (AdvantEdge^{TM}, Third Wave Systems).

3.2 Material Properties

The polymer investigated in this study is polytetrafluoroethylene (PTFE). PTFE is widely used in key industrial applications that require good resistance to chemicals and stability at high temperatures. PTFE is a semi-crystalline polymer with several remarkable properties, including high melting temperature (up to 345°C), high molecular weight, and melt viscosity (10^{10} to 2×10^{11} poise at 380°C), high thermal stability, low dielectric constant (~ 2), very high resistivity (> 10^{18} Ωm), uniform plastic yielding behavior, strong strain rate sensitivity, and relatively low elastic modulus (400–750 MPa) (Rae and Dattelbaum, 2004; Rae and Brown, 2005; Jordan et al., 2007; Ehrenstein, 2001; Willis, 2007; PTFE technical data sheets).

Exhibiting the smallest coefficients of static and dynamic friction of any known solid materials, i.e., 0.04–0.1 at typical contact conditions (Samyn and Schoukens, 2008; Quaglini et al., 2009; Ehrenstein, 2001), PTFE is extensively used in industrial and consumer applications, namely, bearings, gaskets, pump interiors, washers, rings and seals, and running under dry sliding conditions. However, while showing excellent antifriction properties, PTFE suffers from a high wear rate (~ 10^{-7} cm/cm) that greatly limits its industrial applications (Willis, 2007; Samyn and Schoukens, 2008; Quaglini et al., 2009).

The manufacturing of PTFE components includes techniques below the melting temperature, such as compression molding and sintering, while techniques typically used for thermoplastics, like screw extrusion or injection molding, are hindered due to its high melt viscosity. The melting of semi-crystalline polymers with very high molecular weight, as is the case of PTFE, leads to the systems with high concentration of entanglements persisting in the amorphous phase and, consequently, drastically reduced deformability (Rae and Dattelbaum, 2004; Rae and Brown, 2005; Jordan et al., 2007). For this reason, machining processes are considered the most common and versatile of all manufacturing processes (Stan et al., 2013; Cui et al., 2022; Ni et al., 2023).

However, in the field of machining, there is a general lack of modeling and simulation results for PTFE polymer due to limited material information (e.g., deformation hardening, thermal softening, and high strain rate). In the attempt to describe the complex behavior of the PTFE polymer during deformation, several constitutive models have been developed, such as the Johnson-Cook model (Khoddamzadeha et al., 2009; Sonne and Hattel, 2013), the Zerilli-Armstrong model with a thermal activation plastic component combined with a Maxwell-Weichert linear viscoelastic (Zerilli and Armstrong, 2002, 2007), the inelastic model (Kletschkowski et al., 2002a), and the endochronic visco-plastic models (Kletschkowski et al., 2002b). Given the ductile nature of the PTFE and because of the increasing hardening of this polymer with the strain rate, several publications have reported the use of the J-C constitutive model for modeling PTFE and PTFE-based composites under different conditions (Khan and Zhang, 2001; Bergstrom and Hilbert, 2005; Sonne and Hattel, 2013; Herbold et al., 2007; Cai et al., 2008; Raftenberg et al., 2008).

In this study, the J-C parameters derived by Sonne and Hattel (2013) and reported in Table 1 were used to represent the deformation behavior of the PTFE under conditions relevant to the machining process. The additional thermo-mechanical properties used in the simulation are given in Table 2.

Table 1 J-C model parameters for the PTFE (Sonne and Hattel, 2013)

Model Parameter	*Value*
A (MPa)	9.2
B (MPa)	10.5
C (–)	0.041
n (–)	0.36
m (–)	0.4
T_R (°C)	23
$\dot{\varepsilon}_0$ (1/s)	1
T_{melt} (°C)	331

For machining simulations, the properties reported in Table 1 and Table 2 were implemented into the AdvantEdge™ simulation software. The tool material was selected from the standard tool material library of the software (AdvantEdge™, Third Wave Systems), i.e., polycrystalline diamond tool.

Table 2 Thermo-mechanical properties of the PTFE used for simulation

Property	*Value*
Elastic modulus (MPa)	625
Yield stress (MPa)[1,2]	9.2
Density (kg/m^3)[2, 3]	2200
Thermal conductivity (W/m °C)[2]	0.25
Heat capacity (J/kg °C)[2]	1300
Poisson's ratio[3]	0.46
Melting temperature (°C)[1]	331

[1] Values from experimental tests
[2] Rae and Dattelbaum (2004); Rae and Brown (2005); Jordan et al. (2007).
[3] PTFE technical data sheets

It should be noted that the melting temperature of the PTFE in Table 1 and Table 2 was verified by performing Differential Scanning Calorimetry (DSC) analysis at a heating/cooling scan of 10°C/min. Tensile tests were also carried out at different strain rates and temperatures to verify the value for the yield strength and Young's modulus reported by the Sonne and Hattel (2013).

3.3 Simulation Experiments

To study the effect of cutting parameters on fundamental variables such as strain rate, temperature at the tool-chip interface, shear angle, and cutting forces, the FE simulations of the orthogonal cutting process were carried out for different combinations of cutting parameters. Table 3 summarizes the factors (feed rate, rake angle, and friction coefficient) and levels used in the simulation experiments. All other input parameters were kept constant: 4 mm depth of cut, 7° clearance angle, 0.02 mm edge radius, and 155 m/min cutting speed. The minimum relative tool sharpness, i.e., the ratio between the uncut thickness and the tool edge radius, being higher than 3 ensures that the orthogonal cutting conditions are satisfied during simulation.

Table 3 Process parameters and their levels

Run	*Cutting Speed (m/min)*	*Rake Angle* γ *(°)*	*Feed Rate* f *(mm/rev)*	*Friction Coefficient* μ_0 *(–)*
1	155	−6	0.08	0.4
2	155	−6	0.14	0.3
3	155	−6	0.28	0.3
4	155	0	0.08	0.4
5	155	0	0.14	0.3
6	155	0	0.28	0.3
7	155	6	0.08	0.4
8	155	6	0.14	0.3
9	155	6	0.28	0.3

It should be noted that previous experimental cutting tests have shown that the effect of the cutting speed (in the range of 60–155 m/min) on the cutting forces is not statistically significant (Stan et al., 2013). Therefore, in this study, the effect of the cutting speed was not included in the simulation.

The friction between the cutting tool and the PTFE workpiece was implemented through the constant Coulomb friction model, as described in the second section. In the AdvantEdge™ software, within the Coulomb friction model, the friction stress on the rake face is calculated from the normal stress acting on the same surface. To investigate the role of friction in the numerical simulation, the (global apparent) coefficient of friction for the Coulomb model was considered as an input variable. Since the PTFE display nonstick properties, the decomposition of the chip-tool interface into a sticking region and a sliding/slipping region was not considered in this study.

To determine the friction coefficient, the experimental values of the cutting forces measured at 155 m/min were used (Stan et al., 2013). It should be noted that the statistical analysis of the friction coefficient has shown that the effect of the tool rake angle is not statistically significant (p–value = 0.425 > 0.05), and thus only the effect of feed rate on the friction coefficient was considered, as shown in Table 3. In addition, to further investigate the effect of friction on the cutting forces, additional simulation tests were carried out using a constant (artificial) friction coefficient of 0.1.

4. Results and Discussion

4.1 Chip Formation, Plastic Strain, and Temperature

Figure 3 shows the mesh for the tool-chip-workpiece system at the final stage of the machining simulation with 155 m/min, for different values of the tool rake angle and feed rate. The simulation indicates that continuous chips are formed during machining simulation of PTFE, and the curvature of the chip decreases with decreasing feed rate and increases with increasing rake angle from negative to positive values. With increasing feed rate from 0.08 mm/rev to 0.28 mm/rev, two types of chips were predicted: curl chip at 0.08 mm/rev and severe deformed chip at 0.28 mm/rev. It should be noted that the chip is formed by continuously remeshing the workpiece since no separation criteria is defined, and chip is assumed to be due to plastic flow (AdvantEdge™, Marusich, 2001).

During machining of PTFE, continuous chips were generated without fracture in the form of a coil or slightly curved form, regardless of the machining parameters, as shown in Figure 4, and these results are in good agreement with other experimental studies (Ni et al., 2023; Cui et al., 2022). Therefore, it can be concluded that the simulation model is computationally efficient for predicting chip formation during machining simulation of PTFE.

Figure 5 and Figure 6 show the plastic strain and temperature contour plots in machining simulation of PTFE with a cutting speed of 155 m/min, for different feed rates and tool rake angles. These figures were taken at the end of the 5 mm length of cut where the steady state conditions were fully developed. The use of negative rake angles increases the tool-chip contact, as shown in Figure 5. The chip has to be more deformed, requiring more work to be done on it and more pressure on the tool face, leading to the increase of cutting forces and heat generated in the shear band. The plastic deformation of the chip material is nearly uniform with no shear instability during machining simulation with both negative and positive rake angles. The continuous chip formation is caused by the long-range homogeneous plastic flow of the chip materials.

The outside of the chip is obviously more affected by the plastic strain when compared to the inside zone, as shown in Figure 5. The thickness of the chip layer affected by the plastic strain increases with increasing feed rate and decreases with increasing rake angle from negative to positive values. The higher the feed rate, the wider the zone affected by the plastic strain. The plastic shear strain decreases as the rake angle increases from −6° to 6°, as shown in Figure 5. During the machining simulation of the PTFE polymer, the average shear strain reached the order of 3–4 mm/mm, close to the tool tip and in the outer layer of the chip. As the feed rate increases from 0.08 mm/rev to 0.28 mm/rev, a larger strain to an extent of 5.4 mm/mm was generated.

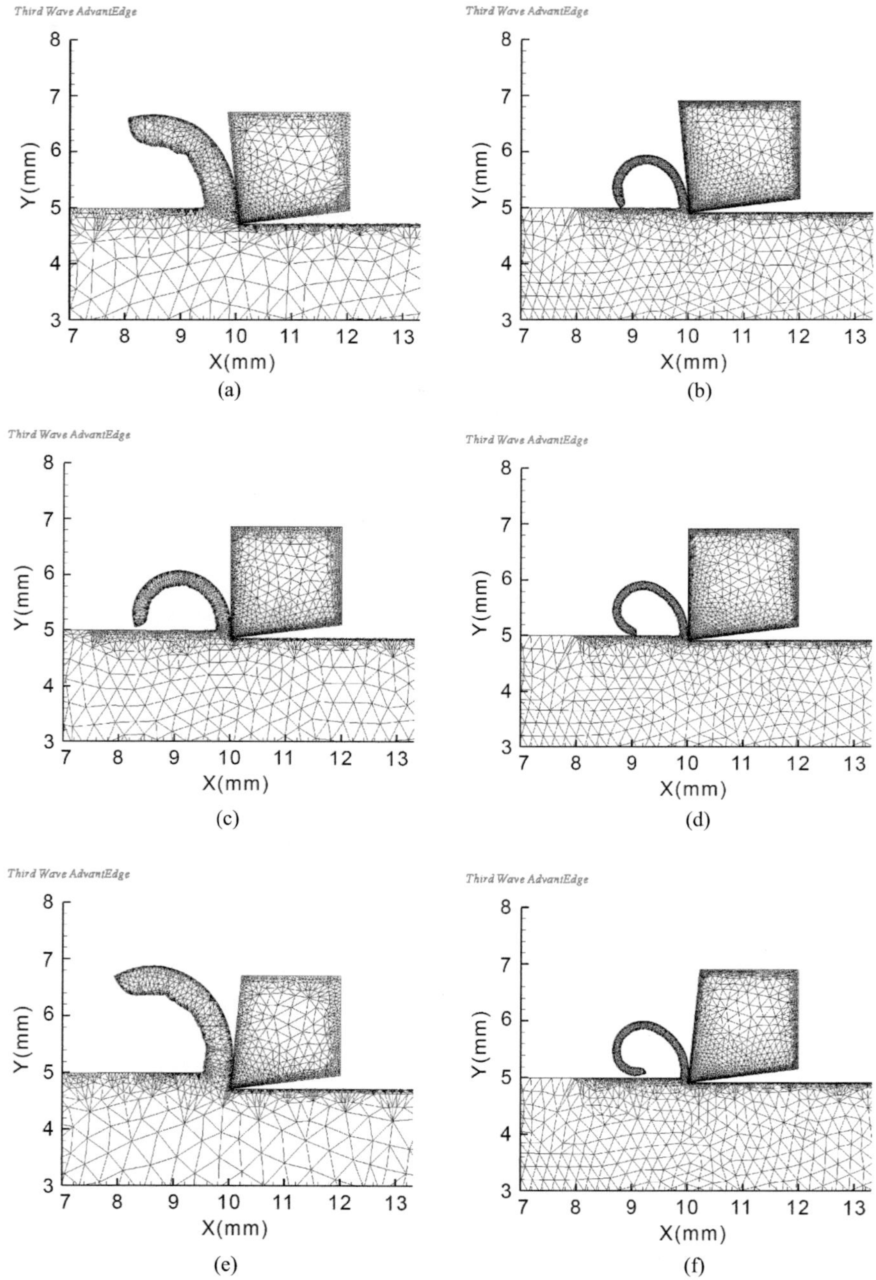

Fig. 3 Mesh at the end of length of cut in machining simulation of PTFE: (a) $\gamma = -6°, f = 0.28$ mm/rev; (b) $\gamma = -6°, f = 0.08$ mm/rev; (c) $\gamma = 0°, f = 0.28$ mm/rev; (d) $\gamma = 0°, f = 0.08$ mm/rev; (e) $\gamma = 6°, f = 0.28$ mm/rev; (f) $\gamma = 6°, f = 0.08$ mm/rev

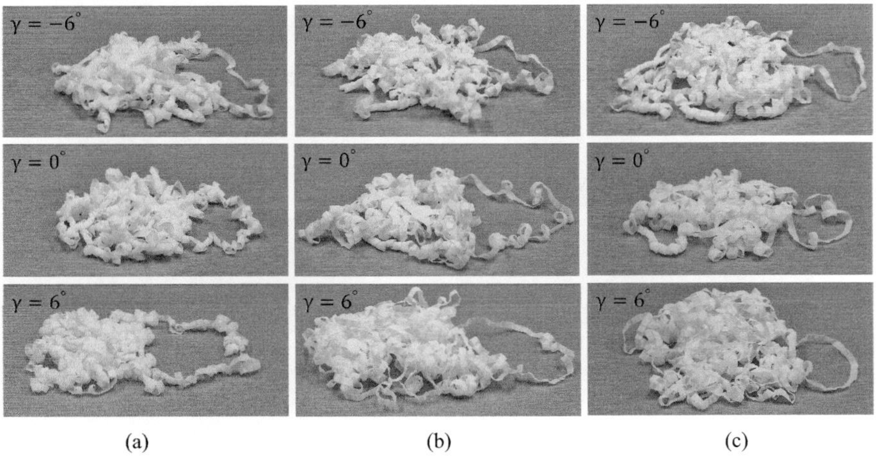

Fig. 4 Chip formation during orthogonal machining of PTFE with 155 m/min: (a) 0.08 mm/rev; (b) 0.14 mm/rev; (c) 0.28 mm/rev

A modest increase in temperature was observed during the machining simulation of the PTFE polymer. The temperature increases from ambient temperature to a maximum value of 32°C to 38°C, depending on the cutting conditions (Figure 6). With the increase of the feed rate and the decrease of the rake angle, the tool-chip contact length also increases, resulting in increasing heat energy and temperature. The highest cutting temperature of about 38°C occurred at a combination of 155 m/min cutting speed, 0.28 mm/rev feed rate, and −6° rake angle, close to the tool tip. The predicted cutting temperature is in good agreement with the experimentally measured temperature reported in the literature for machining of PTFE (Ni et al., 2023; Cui et al., 2022).

As shown in Figure 6, most of the predicted temperature values are in the temperature range in which the crystal is triclinic and the amorphous material is above the γ and below the β transitions, and below the glass transition temperature for PTFE (Rae and Dattelbaum, 2004; Rae and Brown, 2005; Jordan et al., 2007; Brown et al., 2006). Generally, the heat during orthogonal cutting is generated by friction at the tool-chip interface (influencing the temperature rise in the second shear zone) and by internal heating (energy dissipation) caused by the plastic deformation of the material in the primary shear zone (heat generation during chip formation) (Puerta Velasquez, 2007; Bouvard et al., 2011). Due to the extremely low thermal conductivity of PTFE (0.25–0.35 W/m-K), a small percentage of heat is transferred through conduction to the uncut work material ahead of the cutting tool, while the highest percentage of heat remains in the removed chip, producing a local increase of temperature in a narrow zone where high strain occurs (Yamaguchi, 1990; Pascoe and Tabor, 1995). On the other hand, during deformation, PTFE stores much more of the work done as structural change (~30%) than do typical metals (<10%) as indicated by several publications (Rae and Dattelbaum, 2004; Rae and Brown, 2005; Brown et al., 2006).

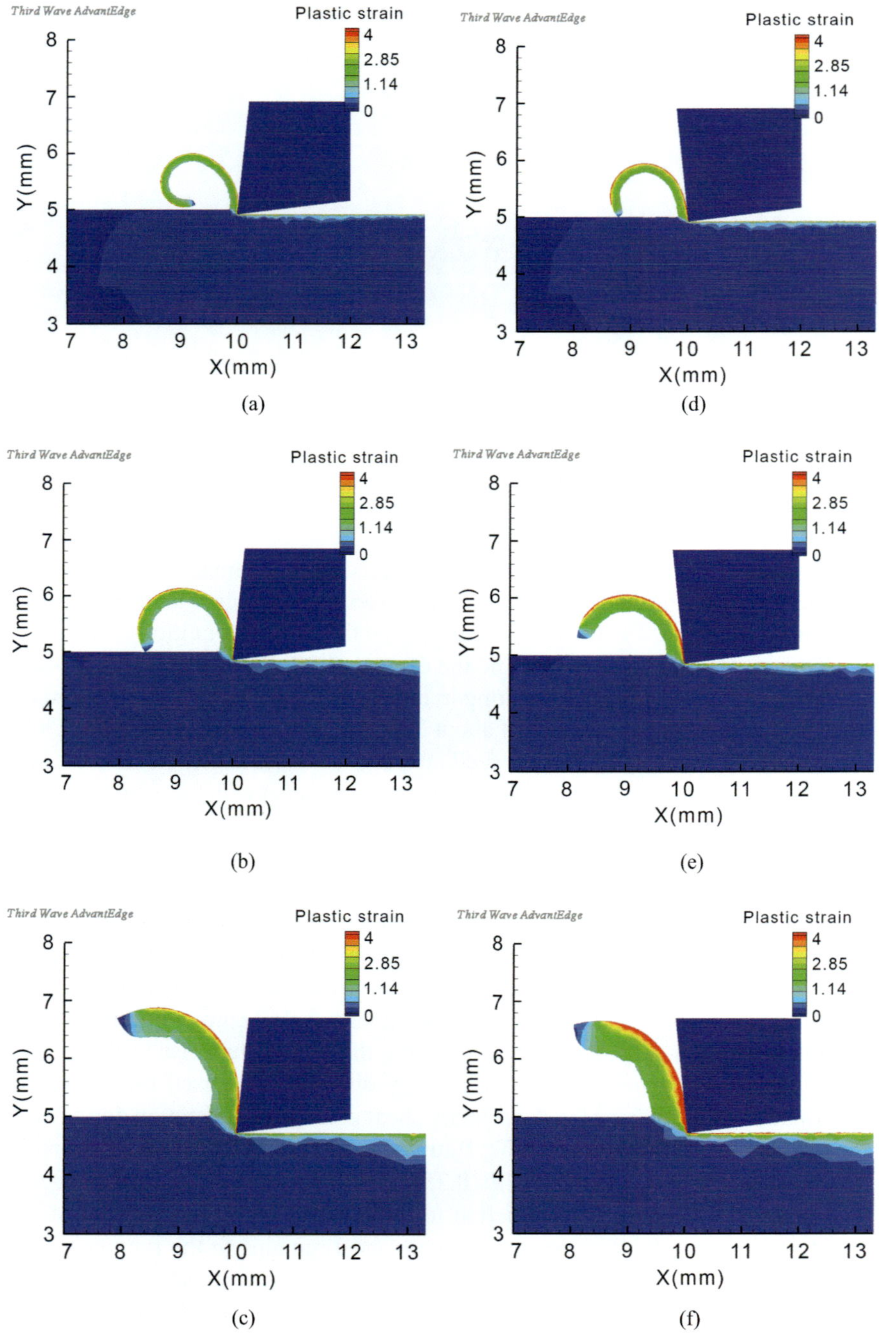

Fig. 5 Distribution of equivalent plastic strain in machining simulation of PTFE: (a) $\gamma = 6°$, $f = 0.08$ mm/rev; (b) $\gamma = 6°$, $f = 0.14$ mm/rev; (c) $\gamma = 6°$, $f = 0.28$ mm/rev; (d) $\gamma = -6°$, $f = 0.08$ mm/rev; (e) $\gamma = -6°$, $f = 0.14$ mm/rev; (f) $\gamma = -6°$, $f = 0.28$ mm/rev

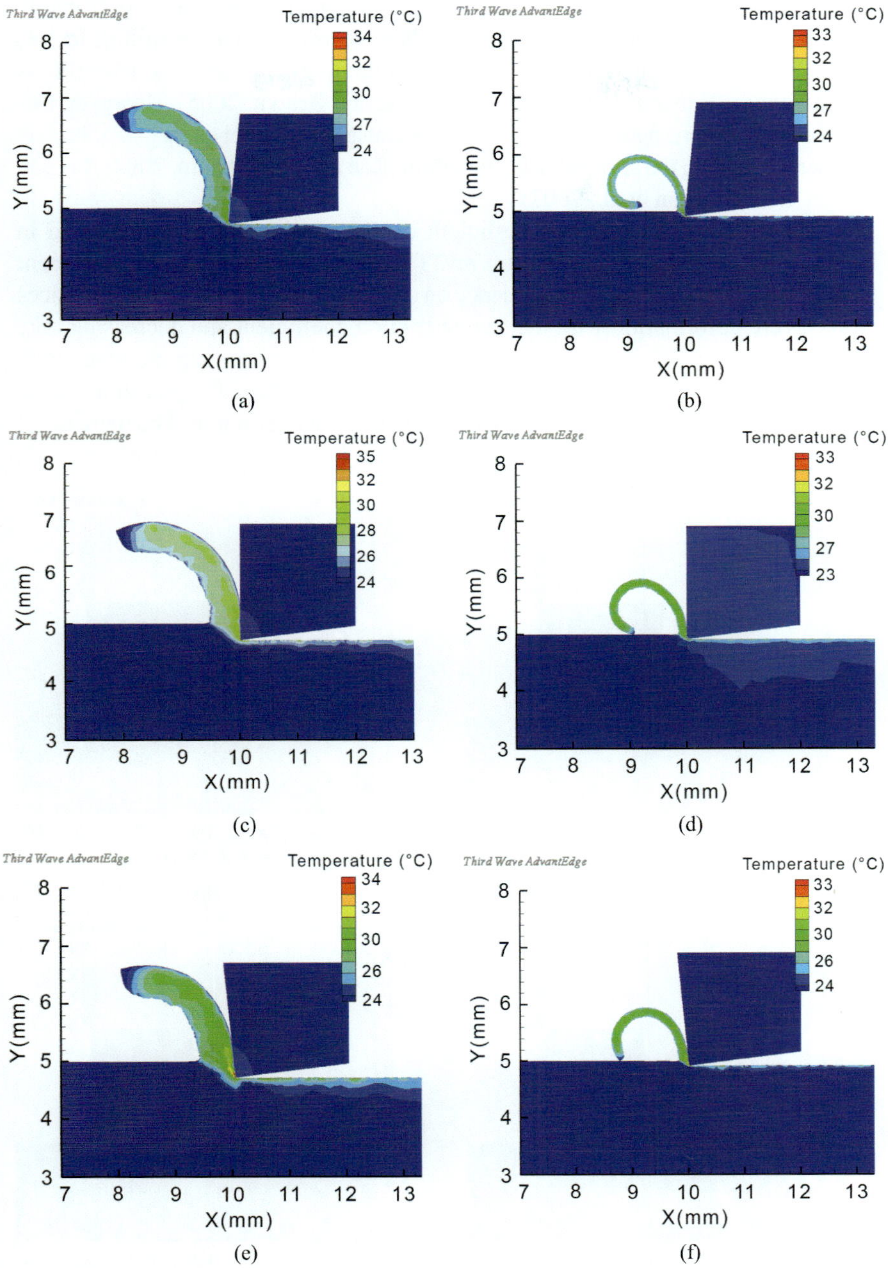

Fig. 6 Contour plot of temperature in machining simulation of PTFE: (a) $\gamma = 6°$, $f = 0.28$ mm/rev; (b) $\gamma = 6°$, $f = 0.08$ mm/rev; (c) $\gamma = 0°$, $f = 0.28$ mm/rev; (d) $\gamma = 0°$, $f = 0.08$ mm/rev; (e) $\gamma = -6°$, $f = 0.28$ mm/rev; (f) $\gamma = -6°$, $f = 0.08$ mm/rev

Comparing with the metals, in the case of polymers, this energy is probably used in crystallographic and amorphous chain rearrangements resulting in less sample heating, and therefore the thermal softening effect due to self-heating is not so intense (Rae and Dattelbaum, 2004; Rae and Brown, 2005). However, this strain-induced rearrangement is temporary because the internal energy of polymers is expended during the spring-back/relaxation (Rae and Dattelbaum, 2004; Rae and Brown, 2005; Jordan et al., 2007).

The effect of the friction coefficient on the plastic strain is illustrated in Figure 7, for a feed rate of 0.28 mm/rev. The increase of the friction coefficient from 0.3 to 0.4 has a significant influence on chip deformation, the chip experiences more severe deformation with increasing friction coefficient and decreasing rake angle from 6° to −6°. The equivalent plastic strain along the chip increases with increasing friction coefficient, as shown in Figure 7, while the curvature of the chip (chip bending) decreases with decreasing friction coefficient. The increase of

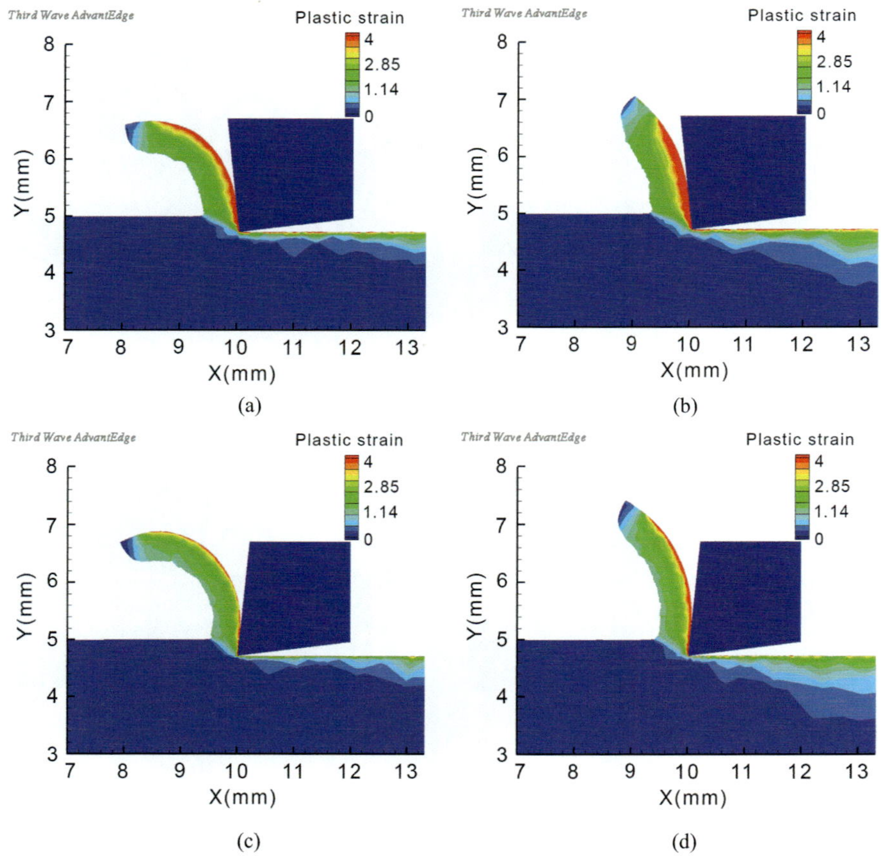

Fig. 7 Effect of friction on the plastic strain during machining simulation of PTFE with 0.28 mm/rev feed rate: (a) $\mu_0 = 0.3$, $\gamma = -6°$; (b) $\mu_0 = 0.4$, $\gamma = -6°$; (c) $\mu_0 = 0.3$, $\gamma = 6°$; (d) $\mu_0 = 0.4$, $\gamma = 6°$

the rake angle from negative to positive values decreases the thickness of the chip layer affected by the plastic strain, and increases the chip length. However, chip thickness remains uniform and continuous along its length.

Figure 8 shows the effect of friction coefficient on the temperature at the tool-chip interface in machining simulation of PTFE with −6° tool rake angle and 0.28 mm/rev. The temperature in the secondary deformation zone (SDZ) increases with increasing friction coefficient from 0.3 to 0.4 due to the heat generated through friction at the tool-chip interface. The low thermal coefficient of PTFE hinders heat dissipation and causes the increase of chip temperature. Therefore, at higher friction coefficient, the temperature rise is mainly caused by friction at the tool-chip interface. For a −6° rake angle, temperatures as high as 38°C are obtained in a narrow layer of the chip close to the tool and below 24°C in most part of the workpiece.

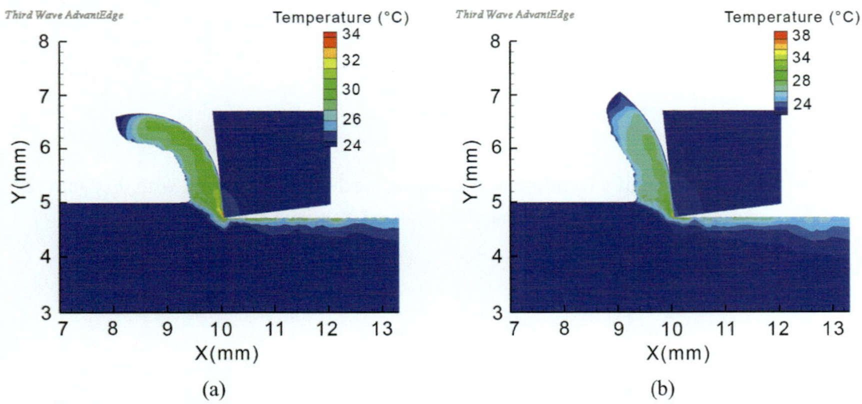

Fig. 8 Effect of friction on the temperature during machining simulation of PTFE at 0.28 mm/rev: (a) $\mu_0 = 0.3$; (b) $\mu_0 = 0.4$

4.2 Prediction of Cutting Forces

Figure 9 shows an example of the variation of cutting forces for the time interval of machining simulation of PTFE with 0.28 mm/rev feed rate and different tool rake angles. It can be seen that when machining with negative tool rake angle (Figure 9a), the cutting force displays some amplitude fluctuation; however, as the tool advances into the workpiece, the chip formation is fully established and cutting becomes steady with a reasonable constant load, regardless of the tool rake angle.

Figure 10 shows the variation of the numerically predicted cutting force components, i.e., cutting force and thrust force, with respect to the feed rate for different values of the tool rake angle, during machining simulation with 155 m/min. The values are the average cutting forces over 5 mm length of cut. For a given value of the tool rake angle, the simulated cutting and thrust forces are increasing considerably with the feed rate, while for a fixed value of the feed rate,

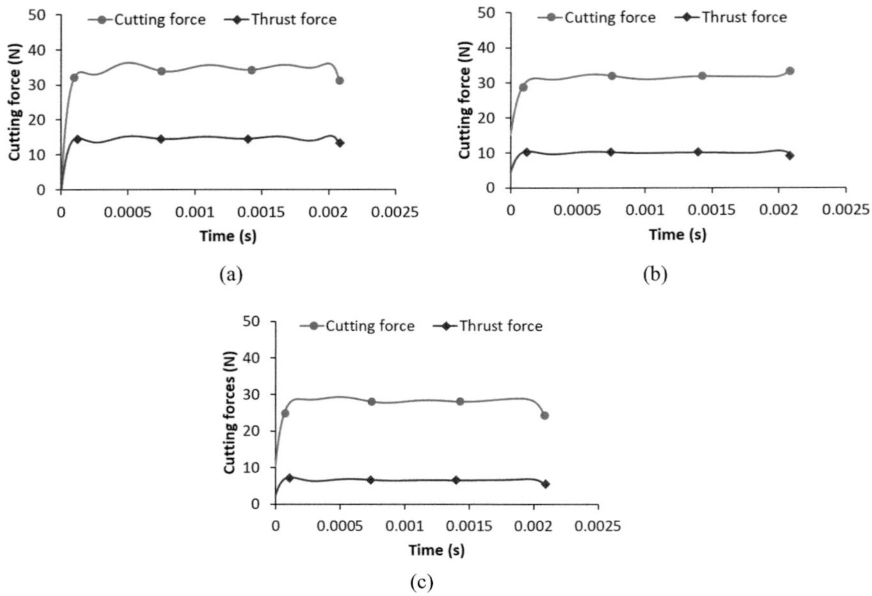

Fig. 9 Time variation of the cutting forces during machining simulation with 0.28 mm/rev feed rate and (a) -6° rake angle; (b) 0° rake angle; (c) 6° rake angle

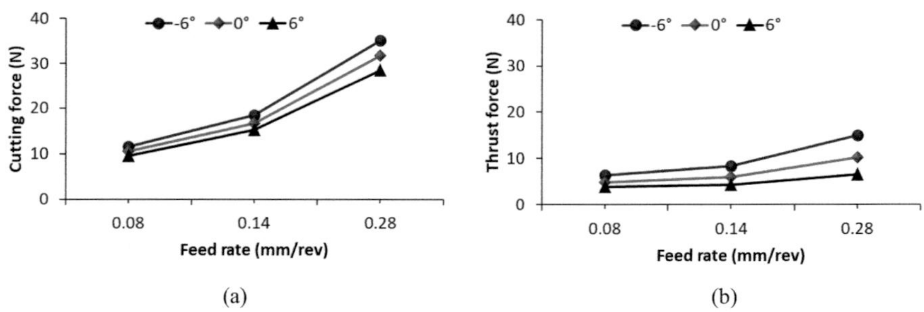

Fig. 10 Dependence of simulated cutting force (a) and thrust force (b) on the feed rate at different tool rake angles in machining simulation of PTFE with 155 m/min

the cutting and thrust forces are decreasing with increasing tool rake angle from negative to positive. Note that in the simulation experiments, the friction model, e.g., the Coulomb friction coefficient, was calibrated based on the experimental cutting tests, as shown in Table 3. As shown in Figure 10, the main cutting force varies between 11.5 N and 35 N, while the thrust force varies between 6.5 N and 15 N, depending on the cutting conditions. The lowest values of the cutting force components were observed at a tool rake angle of 6° because at this angle the lowest values of plastic deformation of machined PTFE are obtained in comparison with a tool rake angle of –6°.

To assess the capacity of the J-C model in predicting cutting forces, in Figure 11, the simulated cutting forces as a function of feed rate and tool rake angle were compared with the corresponding experimental results reported by Stan et al. (2013). It should be noted that cutting and thrust forces were measured by means of a dynamometer for the same cutting conditions presented in Table 3.

The relative error used for evaluating the accuracy (deviation) of the numerically predicted results was defined as

$$\Delta = \frac{(F_{EXP} - F_{FEM})}{F_{FEM}} \times 100(\%) \tag{8}$$

where F_{FEM} is the simulated value and F_{EXP} is the experimental value.

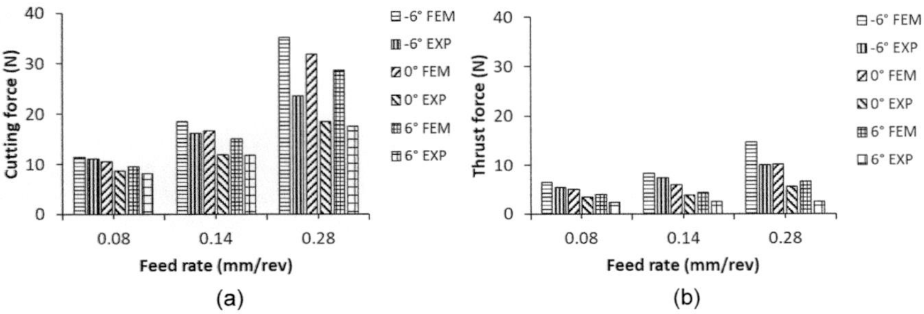

Fig. 11 Comparison between simulated and experimental cutting force (a) and thrust force (b)

The simulated cutting force and thrust force have the same trend to increase with increasing feed rate and decrease with increasing tool rake angle from –6° to 6° as experimentally measured cutting forces. However, only few simulation results fall within the experimental values. The cutting forces are generally best predicted when machining simulation with low feed rates. At a feed rate of 0.08 mm/rev, there is a good agreement between the numerical and the experimental cutting force results, with a maximum deviation between experimental and simulation cutting force that does not exceed 15%, which can be considered satisfactory, given the assumptions used in describing the material behavior with the J-C model. The largest deviations between the experimental and simulated cutting forces were observed at 0.28 mm/rev, and were about 40%.

The simulated thrust force is generally over predicted by the FE analysis for all examined tool rake angles and feed rates. However, for the lowest feed rate of 0.08 mm/rev, the deviation between the simulated and experimental thrust force ranges from 12% to 16%, which is considered acceptable.

The level of the cutting and thrust forces is the result of the action of both the material response within the primary shear zone (where the chip is formed) and the tool-chip interface. Thus, the differences between the simulations and the experiments are possibly induced by the difference of workpiece material used in determining the parameters for the J-C constitutive model and by the material behavior, as represented by the J-C constitutive model. On the other hand, the real contact phenomena in the cutting zones are significantly more complicated than the Coulomb friction model. More than 50% difference in the results of experimental and FE simulation studies has been found when using constant friction coefficient (Arrazola et al., 2008; Mir et al., 2022).

To further investigate the effect of friction on the numerically predicted cutting forces, simulation tests were carried out with a constant friction coefficient of 0.1. Figure 12 indicates a better match between the simulated values of the cutting forces and the corresponding experimental values, with deviations generally below 20%, and the largest deviation not exceeding 30%.

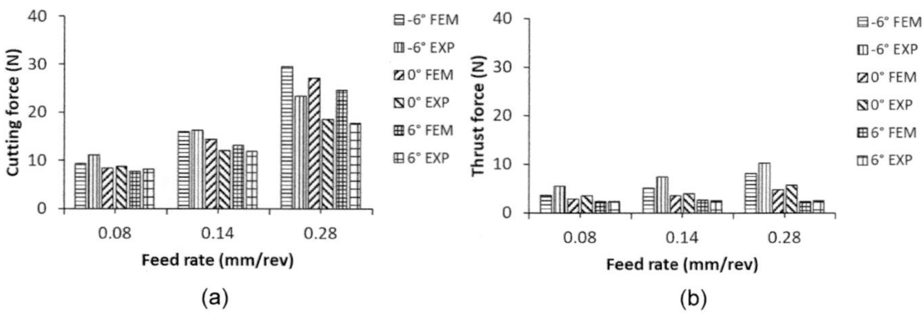

Fig. 12 Comparison between predicted and experimental cutting force (a) and thrust force (b) in machining simulation of PTFE with a friction coefficient of $\mu_0 = 0.1$

4.3 Prediction of Shear Angle

The dependence of the shear angle determined by the Merchant model using the simulated and experimentally measured cutting forces on the feed rate for different tool rake angle values is shown in Figure 13. The shear angle extracted from the strain rate plots (PSZ) is also shown in Figure 13. As can be seen, the simulated Merchant shear angle values are consistent with the corresponding experimental data, with a maximum relative error that generally does not exceed 10%. Moreover, the shear angle extracted from the PSZ based on the shear rate plots shows a good agreement with the calculations based on the Merchant model using both experimental and simulated cutting forces. The deviations of the PSZ shear angles from the Merchant model are not more than 6%. Statistical analysis of the shear plane angle indicates that the difference between the three methods is not statistically significant (p – value = 0.44).

In order to analyze the effect of the feed rate and rake angle on the numerically predicted Merchant shear angle, the main effect plot was generated in Figure 14. As shown in this figure, the shear angle increases monotonically with increasing tool rake angle and feed rate. The result of increased shear angle with increasing tool rake angle is a thinner chip, as can be seen in Figure 3. The shear angle increases with increasing feed rate since an increase in the feed rate results in an increase in the chip cross-section removed in the unit of time.

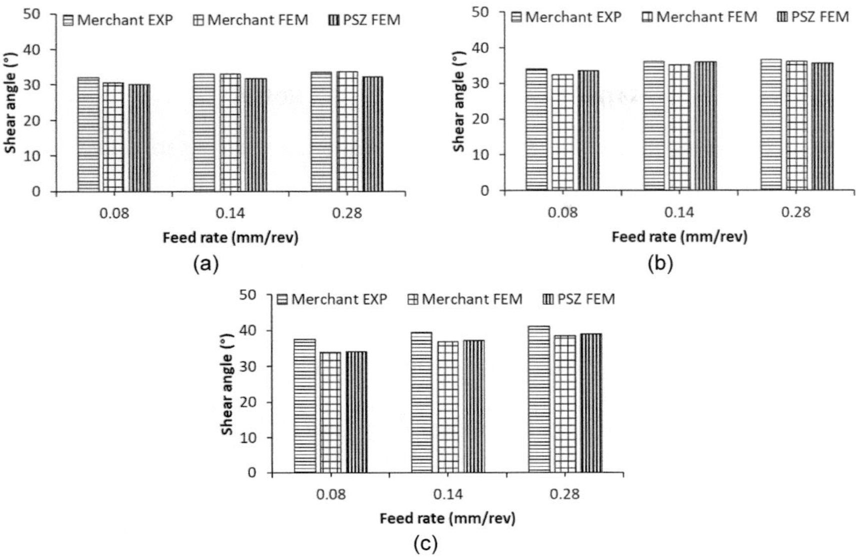

Fig. 13 Comparison of experimental and simulated shear angle for different feed rates and rake angles: (a) –6° tool rake angle; (b) 0° tool rake angle; and (c) 6° tool rake angle

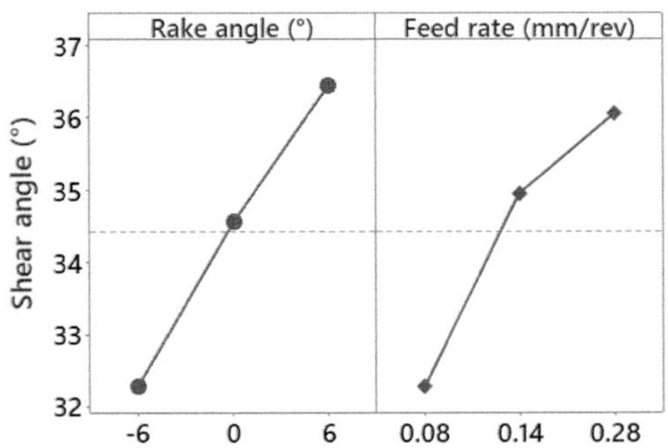

Fig. 14 Main effect plots for the simulated shear angle based on the Merchant model

The analysis of variance (ANOVA) for the simulated Merchant shear angle (Table 4) indicates that both tool rake angle and feed rate have a significant effect on the shear angle (p – value < 0.05). The tool rake angle has the most significant contribution of 53% followed by feed rate which contributes by 46%.

Table 4 ANOVA for simulated Merchant shear angle

Source	*DF*	*Seq SS*	*Adj SS*	*Adj MS*	*F-value*	*p-value*
Rake angle (°)	2	26.07	26.07	13.04	105.95	0.00
Feed rate (mm/rev)	2	22.90	22.90	11.45	93.07	0.00
Error	4	0.49	0.49	0.12		
Total	8	49.47				

Considering the data presented in Figure 14 and Table 4, a model was derived for the shear angle using the regression analysis,

$$\phi = 31.594 + 0.347 \times \gamma + 17.00 \times f \tag{9}$$

that can be used to predict the shear angle as an alternative to experimental values.

4.4 Prediction of Friction Coefficient

Generally, the effects of sliding friction and sticking contact are merged into the friction coefficient (Quaglini et al., 2009; Quaglini and Dubini, 2011). However, for the work material under consideration, the sticking between the chip and the tool is not activated; therefore, the friction coefficient is equal to the apparent friction coefficient.

The apparent coefficient of friction (*COF*) was calculated separately for each combination of cutting parameters in Table 3, using the numerically predicted cutting forces, as shown in Eq. (4). A comparison of the simulated *COF* with the corresponding experimental values μ_0 (input parameter in the FE simulation) is shown in Figure 15, for different values of the feed rate and tool rake angle.

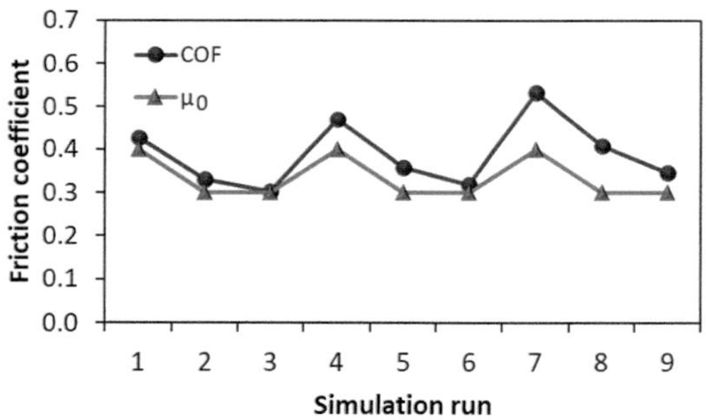

Fig. 15 Comparison between the experimental and simulated apparent friction coefficient

The patterns of friction in simulations and experiments are almost similar. The largest deviations (about 25%) were observed at tool rake angle of 6° and feed rate of 0.08 and 0.14 mm/rev (Experiment run 7 and 8, respectively). For other cutting conditions, the deviation does not exceed 10%, providing a fairly good agreement. Statistical analysis of the friction coefficient indicated that the difference between the simulated *COF* and input friction coefficient μ_0 is not statistically significant ($T-\text{value}=-1.41, p-\text{value}=0.117$). It seems, therefore, logical that other factors should be considered in addition to the constant friction coefficient to explain the differences between the simulated and experimental cutting forces.

5. Concluding Remarks and Future Prospects

Numerical simulation of the polytetrafluoroethylene orthogonal cutting process was carried out in the commercial software AdvantEdge™ using the J-C material model and Coulomb friction model. Chip formation, temperature at the tool-chip interface, cutting forces, and shear angle, along with their trend as a function of the two most important cutting parameters (tool rake angle and feed rate) and the friction coefficient were analyzed to evaluate the cutting performances. The predictive capability was assessed by comparing the numerically predicted cutting forces, shear angle, and friction coefficient with experimental results available from our previous research. The main conclusions of this study are as follows:

- Overall, the J-C model and Coulomb friction model have demonstrated the ability to simulate the mechanical behavior of PTFE during orthogonal cutting process, continuous chip formation, shear strain, and temperature at the chip-tool interface.
- The J-C constitutive model with constant friction coefficient was able to reasonably predict the cutting forces during machining simulation of PTFE at small feed rates (e.g., 0.08 m/rev and 0.14 m/rev) and failed to predict the cutting forces at high feed rates (e.g., 0.28 mm/rev). However, the cutting and thrust forces determined by numerical simulation retain the tendency to increase with increasing feed rate and decrease with tool rake angle as the experimental cutting forces. Increasing the rake angle from negative (e.g., −6°) to positive (e.g., 6°) values results in thinner chip and a reduction of the cutting forces.
- The friction coefficient plays a critical role in the machining simulation of PTFE even though this polymer has a very low friction coefficient. The gap between the simulated and experimental cutting forces can be reduced by the appropriate selection of the coefficient of friction used in the simulation.
- Despite various simplifying assumptions, the FE simulation provides a good estimation of the friction coefficient and shear angle, as highlighted by the values determined from the experimental cutting forces.

The discrepancy between the simulation and experimental cutting force results is possibly induced by the fact that no time-dependent behavior of PTFE was considered in the constitutive model, no damage/failure was included in the model, and the friction coefficient was considered constant even though large deformations take place along the shear plane angle and chip-tool interface.

The current work could be considered as an attempt to simulate the machining of the PTFE polymer using the J-C model and Coulomb friction model and to numerically address the basic cutting physics governing the simulation of PTFE machining. However, a more detailed analysis using the complete stress–strain behavior of the PTFE polymer, including Design of Experiment sensitivity analysis, is required. Also, the dependence of the friction coefficient on contact conditions (sliding velocity, contact pressure, and contact temperature) and different constitutive models must be considered during modeling and simulation of the PTFE cutting process.

References

AdvantEdge™ v5.4. 2010. *User's Manual*. Minneapolis: Third Wave Systems.

Arrazola, P.J., Ozel, T., Umbrello, D., Davies, M. and Jawahir, I.S. 2013. Recent Advances in Modelling of Metal Machining Processes. *CIRP Annals Manufacturing Technology* 62: 695–718.

Arrazola, P.J., Ugarte, D. and Domínguez, X. 2008. A New Approach for the Friction Identification during Machining through the Use of Finite Element Modeling. *International Journal of Machine Tools and Manufacture* 48(2): 173–183.

Atlati, S., Haddag, B., Nouari, M. and Zenasni, M. 2011. Analysis of a New Segmentation Intensity Ratio "SIR" to Characterize the Chip Segmentation Process in Machining Ductile Metals. *International Journal of Machine Tools and Manufacture* 51: 687–700.

Bergstrom, J.S. and Hilbert, J.L.B. 2005. A Constitutive Model for Predicting the Large Deformation Thermomechanical Behavior of Fluoropolymers. *Mechanics of Materials* 37(8): 899–913.

Bil, H., Kılıc, S.E. and Tekkaya, A.E. 2004. A Comparison of Orthogonal Cutting Data from Experiments with Three Different Finite Element Models. *International Journal of Machine Tools and Manufacture* 44: 933–944.

Bouvard, J.L., Ward, D., Marin, E.B., Bammann, D. and Horstemeyer, M.F. 2011. Development of an Internal State Variable Model to Describe the Mechanical Behavior of Amorphous Polymer and Its Application to Impact Testing. In *Time Dependent Constitutive Behavior and Fracture/Failure Processes*, Vol. 3. Conference Proceedings of the Society for Experimental Mechanics Series, 75–76. New York: Springer.

Brown, E.N., Rae, P.J., Orler, B.E., Gray, G.T. and Dattelbaum, D.M. 2006. The Effect of Crystallinity on the Fracture of Polytetrafluoroethylene (PTFE). *Materials Science And Engineering: C* 26(8): 1338–1343.

Buchkremer, S., Wub, B., Lung, D., Münstermann, S., Klocke, F. and Bleck, V. 2014. FE-simulation of Machining Processes with a New Material Model. *Journal of Materials Processing Technology* 214: 599–611.

Cai, J., Nesterenko, V.F., Vecchio, K.S., Herbold, E.B., Benson, D.J., Jiang, F., Addiss, J.W., Walley, S.M. and Proud, W.G. 2008. The Influence of Metallic Particle Size on the Mechanical Properties of PTFE-Al-W Powder Composites. *Applied Physics Letters* 92: 1–3.

Chen, G., Ren, C., Zhang, P., Cui, K. and Li, Y. 2013. Measurement and Finite Element Simulation of Micro-cutting Temperatures of Tool Tip and Work-piece. *International Journal Machine Tools and Manufacture* 75: 16–26.

Cui, Z., Ni, J., He, L., Guan, L., Han, L. and Sun, J. 2022. Investigation of Chip Formation, Cutting Force and Surface Roughness during Orthogonal Cutting of Polytetrafluoroethylene. *Journal of Manufacturing Processes* 77: 485–494.

Dandekar, C.R. and Shin, Y.C. 2008. Multiphase Finite Element Modeling of Machining Unidirectional Composites: Prediction of Debonding and Fiber Damage. *Journal of Manufacturing Science and Engineering* 130(5): 051016.

Dandekar, C.R. and Shin, Y.C., 2012. Modeling of Machining of Composite Materials: A Review. *International Journal of Machine Tools and Manufacture* 57: 102–121.

Davim, J.P., ed. 2009. *Machining Composites Materials*. Wiley-ISTE.

Davim, J.P. and Mata, F. 2005. Optimisation of Surface Roughness on Turning Fibre-reinforced Plastics with Diamond Cutting Tools. *The International Journal of Advanced Manufacturing Technology* 26: 319–323.

Davim, J.P. and Mata, F. 2007. New Machinability Study of Glass Fibre Reinforced Plastics Using Polycrystalline Diamond and Cemented Carbide (K15) Tools. *Materials & Design* 2: 1050–1054.

Davim, J.P. and Reis, P. 2004. Machinability Study on Composite (Polyetheretherketone Reinforced with 30% of Glass Fiber-PEEK GF 30) Using Polycrystalline Diamond (PCD) and Cemented Carbide (K20) Tools. *International Journal of Advanced Manufacturing Technology* 23: 412–418.

Dixit, U.S., Joshi, S.N. and Davim, J.P. 2011. Incorporation of Material Behavior in Modeling of Metal Forming and Machining Processes: A Review. *Materials & Design* 32: 3655–3670.

Ehrenstein, G. 2001. *Polymeric Materials: Structure, Properties, Applications*. Munich: Hanser.

Ernst, H. and Merchant, M.E. 1941. Chip Formation, Friction and High Quality Machined Surfaces. *Surface Treatment of Metals ASM* 29: 299–378.

Ezilarasan, C., Senthil Kumar, V.S. and Velayudham, A. 2014. Theoretical Predictions and Experimental Validations on Machining the Nimonic C-263 Super Alloy. *Simulation Modelling Practice and Theory* 40: 192–207.

Fetecau, C. and Stan, F. 2012. Study of Cutting Force and Surface Roughness in the Turning of Polytetrafluoroethylene Composites with a Polycrystalline Diamond Tool. *Measurement* 45(6): 1367–1379.

Filice, L., Micari, F., Rizzuti, S. and Umbrello, D. 2007. A Critical Analysis on the Friction Modelling in Orthogonal Machining. *International Journal of Machine Tools and Manufacture* 47(3–9): 709–714.

Grzesik, W. 2008. *Advanced Machining Processes of Metallic Materials: Theory, Modelling and Application*. Elsevier Science.

Herbold, E.B., Cai, J., Benson, D.J. and Nesterenko, V.F. 2007. Simulation of Particle Size Effect on Dynamic Properties and Fracture of PTFE-W-Al Composites. *AIP Conference Proceedings* 955: 785–788.

Jamshidi, H. and Nategh, M.J. 2013. Theoretical and Experimental Investigation of the Frictional Behavior of the Tool–chip Interface in Ultrasonic-vibration Assisted Turning. *International Journal of Machine Tools and Manufacture* 65: 1–7.

Johnson, G.R. and Cook, W.H. 1983. A Constitutive Model and Data for Metals Subjected to Large Strains, High Strain Rates and High Temperatures. *Proceedings 7th International Symposium on Ballistics*, 541–547.

Jordan, J.L., Siviour, C.R., Foley, J.R. and Brown, E.N. 2007. Compressive Properties of Extruded Polytetrafluoroethylene. *Polymer* 48(14): 4184–4195.

Khan, A. and Zhang, H. 2001. Finite Deformation of a Polymer: Experiments and Modeling. *International Journal of Plasticity* 17: 1167–1188.

Khoddamzadeha, A., Liu, R. and Wu, X. 2009. Novel Polytetrafluoroethylene (PTFE) Composites with Newly Developed Tribaloy Alloy Additive for Sliding Bearings. *Wear* 266: 646–657.

Kletschkowski, T., Schomburg, U. and Bertram, A. 2002a. An Inelastic Material Model for Filled Polytetrafluoroethylene. *Archive of Applied Mechanics* 72: 293–299.

Kletschkowski, T., Schomburg, U. and Bertram, A. 2002b. Endochronic Viscoplastic Material Models for Filled PTFE. *Mechanics of Materials* 34(12): 795–808.

Lafaye, S., Gauthier, C. and Schirrer, R. 2006. Analysis of the Apparent Friction of Polymeric Surfaces. *Journal of Materials Science* 41(19): 6441–6452.

LeMaitre, J., ed. 2001. *Handbook of Materials Behavior Models, Nonlinear, 1*. Academic Press.

Liang, R. and Khan, A.S. 1999. A Critical Review of Experimental Results and Constitutive Models for BCC and FCC Metals over a Wide Range of Strain Rates and Temperatures. *International Journal of Plasticity* 15: 963–980.

Maranhão, C. and Davim, J.P. 2010. Finite Element Modelling of Machining of AISI 316 Steel: Numerical Simulation and Experimental Validation. *Simulation Modelling Practice and Theory* 18: 139–156.

Marusich, D.T. 2001. Effects of Friction and Cutting Speed on Cutting Force. Proceedings of ASME Congress 2001, November 11–16, New York, NY.

Mata, F., Gaitonde, V.N., Karnik, S.R. and Davim, J.P. 2009. Influence of Cutting Conditions on Machinability Aspects of PEEK, PEEK CF 30 and PEEK GF 30 Composites Using PCD Tools. *Journal of Materials Processing Technology* 209: 1980–1987.

Melkote, S., Liang, S.Y., Ozel, T., Jawahir, I.S., Stephenson, D.A. and Wang, B. 2022. A Review of Advances in Modeling of Conventional Machining Processes: From Merchant to the Present. *Journal of Manufacturing Science and Engineering* 144: 110801.

Miguélez, M.H., Soldani, X. and Molinari, A. 2013. Analysis of Adiabatic Shear Banding in Orthogonal Cutting of Ti Alloy. *International Journal of Mechanical Sciences* 75: 212–222.

Milani, A.S., Dabboussi, W., Nemes, J.A. and Abeyaratne, R.C. 2009. An Improved Multi-objective Identification of Johnson–Cook Material Parameters. *International Journal of Impact Engineering* 36(2): 294–302.

Mir, A., Luo, X., Llavori, I., Roy, A., Zlatanovic, D.L., Joshi, S.N. and Goel, S. 2022. Challenges and Issues in Continuum Modelling of Tribology, Wear, Cutting and Other Processes Involving High-strain Rate Plastic Deformation of Metals. *Journal of the Mechanical Behavior of Biomedical Materials* 30: 105185.

Mkaddem, A. and Mansori, M.E. 2009. Finite Element Analysis when Machining UGF-reinforced PMCs Plates: Chip Formation, Crack Propagation and Induced-damage. *Materials & Design* 30: 3295–3302.

Molinari, A., Cheriguene, R. and Miguelez, H. 2011. Numerical and Analytical Modeling of Orthogonal Cutting: The Link between Local Variables and Global Contact Characteristics. *International Journal of Mechanical Sciences* 53(3): 183–206.

Ni, J., Yuan, Y., Cui, Z., Lou B. and He L. 2023. Investigation of Machinability in Turning of PTFE based on MRR and Cutting Energy. *Journal of Manufacturing Processes* 85: 122–131.

Notta-Cuvier, D., Langrand, B., Markiewicz, E., Lauro, F. and Portemont, G. 2013. Identification of Johnson–Cook's Viscoplastic Model Parameters Using the Virtual Fields Method: Application to Titanium Alloy Ti6Al4V. *Strain* 49: 22–45.

Ozel, T. 2006. The Influence of Friction Models on Finite Element Simulations of Machining. *International Journal of Machine Tools and Manufacture* 46: 518–530.

Ozel, T. and Arrazola, P. 2010. Investigations on the Effects of Friction Modeling in Finite Element Simulation of Machining. *International Journal of Mechanical Sciences* 52: 31–42.

Ozel, T. and Zeren, E. 2004. Determination of Work Material Flow Stress and Friction for FEA of Machining Using Orthogonal Cutting Tests. *Journal of Materials Processing Technology* 153–154: 1019–1025.

Palanikumar, K. 2008. Application of Taguchi and Response Surface Methodologies for Surface Roughness in Machining Glass Fiber Reinforced Plastics by PCD Tooling. *The International Journal of Advanced Manufacturing Technology* 36: 19–27.

Palanikumar, K. and Davim, J.P. 2009. Assessment of Some Factors Influencing Tool Wear on the Machining of Glass Fibre-reinforced Plastics by Coated Cemented Carbide Tools. *Journal of Materials Processing Technology* 209: 511–519.

Palanikumar, K., Karunamoorthy, L. and Manoharan, N. 2006. Mathematical Model to Predict the Surface Roughness on the Machining of Glass Fiber Reinforced Polymer Composites. *Journal of Reinforced Plastics and Composites* 25: 407–419.

Pascoe, M.W. and Tabor, D. 1995. The Friction and Deformation of Polymers. *Proceedings of the Royal Society of London. Series A, Mathematical and Physical Sciences* 235: 210–224.

Petropoulo, G., Mata, F. and Davim, J.P. 2008. Statistical Study of Surface Roughness in Turning of PEEK Composites. *Materials & Design* 29: 218–223.

PTFE technical data sheets. Available at http://www.ceproinv.ro. 2024.

Puerta Velasquez, J.D., Bolle, B., Chevrier, P., Geandier, G. and Tidu, A. 2007. Metallurgical Study on Chips Obtained by High Speed Machining of a Ti-6 wt.%Al-4 wt.%V Alloy. *Materials Science and Engineering: A* 452–453: 469–474.

Quaglini, V. and Dubini, P. 2011. Friction of Polymers Sliding on Smooth Surfaces. *Advances in Tribology*: 178943.

Quaglini, V., Dubini, P., Ferroni, D. and Poggi, C. 2009. Influence of Counterface Roughness on Friction Properties of Engineering Plastics for Bearing Applications. *Materials & Design* 30(5): 1650–1658.

Rae, P.J. and Brown, E.N. 2005. The Properties of Poly(tetrafluoroethylene) (PTFE) in Tension. *Polymer* 46(19): 8128–8140.

Rae, P.J. and Dattelbaum, D.M. 2004. The Properties of Poly(tetrafluoroethylene) (PTFE) in Compression. *Polymer* 45: 7615–7625.

Raftenberg, M.N., Mock, W. and Kirby, G.C. 2008. Modeling the Impact Deformation of Rods of a Pressed PTFE/Al Composite Mixture. *International Journal of Impact Engineering* 35(12): 1735–1744.

Rao, G.V.G., Mahajan, P. and Bhatnagar, N. 2008. Three-dimensional Macro-mechanical Finite Element Model for Machining of Unidirectional-fiber Reinforced Polymer Composites. *Materials Science and Engineering: A* 498: 142–149.

Rech, J. 2006. A Multi-view Approach to the Tribological Characterization of Cutting Tool Coatings for Steels in High Velocity Dry Turning. *International Journal of Machining and Machinability of Materials* 1: 27–44.

Rech, J., Claudin, C. and D'Eramo, E. 2009. Identification of a Friction Model—Application to the Context of Dry Cutting of an AISI1045 Annealed Steel with aTiN-coated Carbide Tool. *Tribology International* 42: 738–744.

Samyn, P. and Schoukens, G. 2008. Experimental Extrapolation Model for Friction and Wear of Polymers on Different Testing Scales. *International Journal of Mechanical Sciences* 50: 1390–1403.

Sonne, M.R. and Hattel, J.H. 2013. Modeling the Constitutive and Frictional Behavior of PTFE Flexible Stamps for Nanoimprint Lithography. *Microelectronic Engineering* 106: 1–8.

Stan, F., Vlad, D. and Fetecau, C. 2013. Statistical Cutting Force Model for Orthogonal Cutting of Polytetrafluoroethylene (PTFE) Composites. In *ASME 2013 International Manufacturing Science and Engineering Conference*, June 10–14.

Storchak, M. 2024. Determination of Chip Compression Ratio for the Orthogonal Cutting Process. *Journal of Manufacturing and Materials Processing* 8: 190.

Storchak, M., Möhring, H.C. and Stehle, T. 2022b. Improving the Friction Model for the Simulation of Cutting Processes. *Tribology International* 167: 107376.

Storchak, M., Stehle, T. and Möhring, H.C. 2022a. Determination of the Shear Angle in the Orthogonal Cutting Process. *Journal of Manufacturing and Materials Processing* 6: 132.

Vaz, M., Owen, D.R.J., Kalhori, V., Lundblad, M. and Lindgren, L.E. 2007. Modelling and Simulation of Machining Processes. *Archives of Computational Methods in Engineering* 14: 173–204.

Wang, B., Liu, Z. and Yang, Q. 2013. Investigations of Yield Stress, Fracture Toughness and Energy Distribution in High Speed Orthogonal Cutting. *International Journal of Machine Tools and Manufacture* 73: 1–8.

Willis, O.R. 2007. Characterizing Fluoropolymeric Materials for Microelectronics and MEMS Packaging. Dissertation, State University of New York at Binghamton.

Yamaguchi, Y. 1990. *Tribology of Plastic Materials*. Amsterdam: Elsevier.

Zemzemi, F., Rech, J., Salem, W.B., Dogui, A. and Kapsa, P. 2009. Identification of a Friction Model at Tool/Chip/Workpiece Interfaces in Dry Machining of AISI4142 Treated Steels. *Journal of Materials Processing Technology* 209: 3978–3990.

Zerilli, F.J. and Armstrong, R.W. 2002. Thermal Activation Constitutive Model for Polymers Applied to Polytetrafluoroethylene. *AIP Conference Proceedings* 620: 657–660.

Zerilli, F.J. and Armstrong, R.W. 2007. A Constitutive Equation for the Dynamic Deformation Behavior of Polymers. *Journal of Materials Science* 42: 4562–4574.

9 A CAD Framework for 3D Design and Optimization of Solid End Mill Geometry

Nikolaos Tapoglou[1*] and Anastasios Tzotzis[2]

1. Introduction

In today's advanced manufacturing landscape, digital tools play a crucial role in supporting production environments at every stage of the process. These tools are integral to design, production scheduling, operational planning, and machining process development. A key factor in the effectiveness of any digital tool is the accuracy and fidelity of the information it utilizes, particularly in machining operations.

High-value components often require multiple precise machining operations to meet stringent geometric and tolerance requirements. For prismatic parts, roughing operations typically employ inserted cutters due to their lower cost, while finishing operations rely on solid carbide end mills to achieve the desired surface quality and geometric accuracy. Solid carbide tools are preferred in finishing operations because they deliver superior surface quality and ensure geometric conformity. The intricate geometry of these cutting tools plays a critical role in determining machining performance, making it essential to carefully optimize various tool parameters for improved production efficiency.

In end milling operations, the success of any digital simulation tool hinges on the precise representation of the end mill's geometry, encompassing all its detailed features. This representation must also be adaptable, allowing for customization at any point to optimize the machining process.

The use of digital tools in machining research has been an ever-increasing field of development (Altintas et al., 2014; Bergs et al., 2023). In machining, the primary fields investigated include the determination of the final part quality (Tosello et al., 2023), the investigation of in-process parameters, such as cutting forces, stresses, and temperatures (Nikolaos and Aristomenis, 2012), and the prediction of surface quality and residual stresses (Siebrecht et al., 2015; Tzotzis et al., 2023). Research has focused on all aspects of machining, ranging from simulation models for milling and turning operations (Kyratsis et al., 2021; Tapoglou et al, 2023) to more

[1] *Assistant Professor*, Department of Industrial Engineering and Management, International Hellenic University.

[2] *Postdoctoral Researcher*, Department of Product and Systems Design Engineering, University of Western Macedonia.
E-mail: a.tzotzis@uowm.gr[2]

* Corresponding author: ntapoglou@ihu.gr

complex kinematics processes like gear hobbing and power skiving (Tapoglou, 2019).

Research in the field of cutting tool design and optimization started with digital models with a view to understanding the mechanics and dynamics of the cutting process. The main means of creating these initial models was using discritised geometries and kinematics to predict cutting forces and process stability (Altintas and Lee 1996; Engin and Altintas, 2001). More recently, Sanz-Calle et al. (2024) investigated the optimal helix angle of end mills for reducing chatter occurrences.

In the field of cutting tool design, ISO/TS 13399 (ISO/TS 13399-303 2016) has set the basis for the standardization of the data representation and exchange of cutting tool data. Relevant to the research presented, ISO/TS 13399-303 focuses on the data required for solid end mills and the required planes and coordinate systems representation.

Li et al. focused on the development of a method for designing and optimizing solid carbide end mills using three-dimensional finite element simulation. The study focused on optimizing key geometrical parameters like helix angle, rake angle, and the number of flutes to minimize cutting forces and temperatures during the milling of titanium alloys Ti-6Al-4V (Li et al., 2014).

Lin et al. developed a design system for end mills based on Open CASCADE. The system was based on the mathematical equations of the geometry of the tool (Lin et al., 2023). Similarly, Wang et al. (2024) developed a mathematical model for the generation of the chisel edge on ball end mills. Jiang et al. (2024) presented an optimized calculation method for determining the grinding wheel profile used in helical flute forming grinding.

This research introduced a novel computer-aided design (CAD)-based framework that enables designers to create and optimize solid end mills with greater precision. Tool parameters such as diameter, rake angle, helix angle, and the number of teeth serve as input variables, allowing the designer to influence and refine the overall tool design for improved machining efficiency.

The remainder of the chapter is structured as follows; the second section presents the key details of the end mill grinding process; the third section outlines the architecture of the developed simulation platform while the fourth section outlines the resulting geometries acquired through the simulation platform with an investigation of key parameters on the geometry. The fifth section presents the concluding remarks and future work.

2. End Mill Manufacturing

End mills are manufactured through a series of steps with both material and geometric considerations in each step. The first step in the manufacturing process is the generation of the tool blank, that is produced through sintering of the tool blank. The blank end mill goes through a number of grinding operations next to ensure all the geometric features are generated on the tool. The main grinding steps include the grinding of the flute geometry, grinding the clearance angle, and forming the end faces of the tool with grinding of the end face and the gashes of the tool. After all the grinding steps are completed, the tool undergoes edge preparation processes before undergoing a coating process to increase the wear performance characteristics of the end mill. Having in mind the end application of each tool, the blank material composition, all the geometric characteristics of the end mill and the existence of coating and its composition must be considered, when designing a tool.

It is apparent that one of the most critical steps in the end mill manufacturing process is the grinding steps in which the majority of the geometrical features of the tool are generated. Grinding of all the geometric features includes complex kinematics as both the grinding wheel and the end mill are positioned and rotated in the 3D space. Moreover, the grinding wheel itself has a geometry that is not plain cylindrical and has features like angled surfaces, curved edges, chamfers, and edge roundings that are required to produce certain geometrical features on the end mill.

3. End Mill Manufacturing Simulation

Understanding all the parameters of end mill manufacturing process can enable the accurate simulation of the generation process and lead to high fidelity simulation results. Using a novel simulation environment, the generation of solid 3D geometries of end mills can be achieved. The simulation software, developed and embedded in a CAD environment, can digitally simulate the manufacturing process, taking into consideration all the process variables of the actual process. The kinematics of the end mill grinding process are presented in Figure 1.

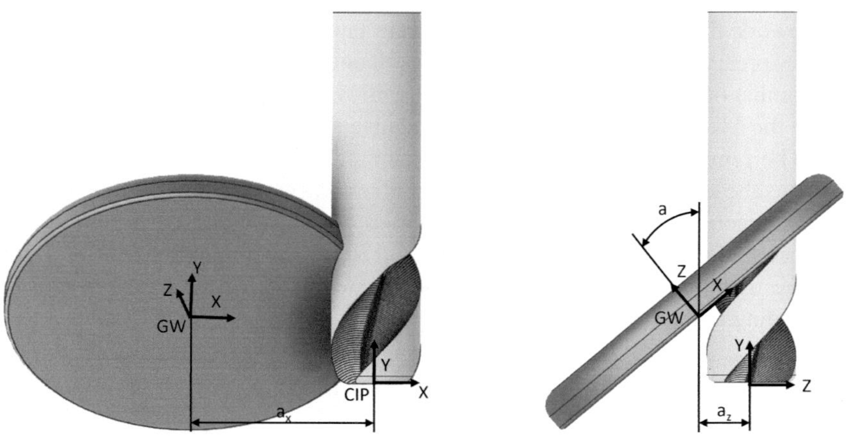

Fig. 1 End mill flute grinding process

The process uses two rotations along with one translational movement to generate the flute and the clearance face geometries. The grinding wheel is positioned relative to the end mill and rotates around its axis (GW_Z) to generate the required cutting motion of the grinding process. The generating motions, critical for the generation of the final geometry, are the rotation of the end mill and the translation of the grinding wheel along the axis of the tool (CIP_Z). These last two motions need to be synchronised as their relation determines the load on the tool and, more importantly, the helix angle of the end mill. The generation of the end face and the gashes of the tools is more straightforward as it only involves the rotation of the grinding wheel and a translation of the wheel along one axis. The developed simulation platform uses the accuracy of a CAD kernel to accurately represent all the aforementioned kinematics. The simulation process is summarized in Figure 2.

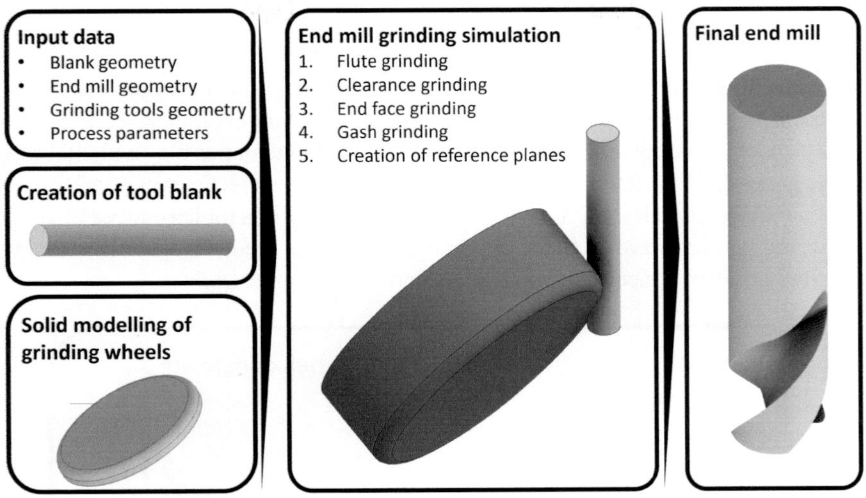

Fig. 2 Simulation framework flowchart

The first step of the process is the entry of all the necessary parameters for the simulation of the end mill manufacturing process. The data are split into four categories, namely, blank and final end mill geometry, grinding tool geometry, and process parameters. The blank end mill geometry includes parameters such as the diameter and length of the blank as well as any taper or corner radius of the blank. The grinding tool geometry is one of the most critical aspects of the process. The profile of the grinding wheel is delivered into the simulation code by means of a series of 2D sketches. This allows for the full customization of the end mill geometry. The geometry of the profile incorporates the diameter of the wheel. Finally, the process parameters include the grinding wheel posture (axial and radial offsets and wheel tilt angle), the designed helix angle, and number of teeth of the end mill as well as values for the clearance angles and the gash depth.

After all the required data are entered on the model, the calculation phase can be initiated. The simulation algorithm automatically generates the solid geometry of the blank end mill based on the user's data. Next, the solid geometry of the grinding wheel is generated through the user data. The tool is then assembled, in the appropriate posture, with the end mill blank to simulate the cutting process. The virtual grinding of the end mill requires the assembly process to be repeated several times until the final flute length is reached. This process is followed by all grinding steps following the required kinematics of the step. At the end of the process, the reference planes and coordinate systems, based on ISO/TS 13399-303, are created in order to make the design standard compliant. The final geometry is saved as a native file on the CAD environment as well as in ISO 10303 format for use in computer-aided manufacturing (CAM) and computer-aided engineering (CAE) software.

4. Simulation Results

The developed simulation algorithm was used in order to produce a range of geometries focusing primarily on the profile of the flute geometry. As it can be easily understood, the tool geometry and the applied kinematics play a pivotal role in the generated flute geometry. This chapter outlines the effect parameters have on the final flute geometry. Figure 3 presents the geometry of a tool produced through the developed framework. In the geometry one can see the faces generated due to the flute grinding process, the clearance, and gash geometries.

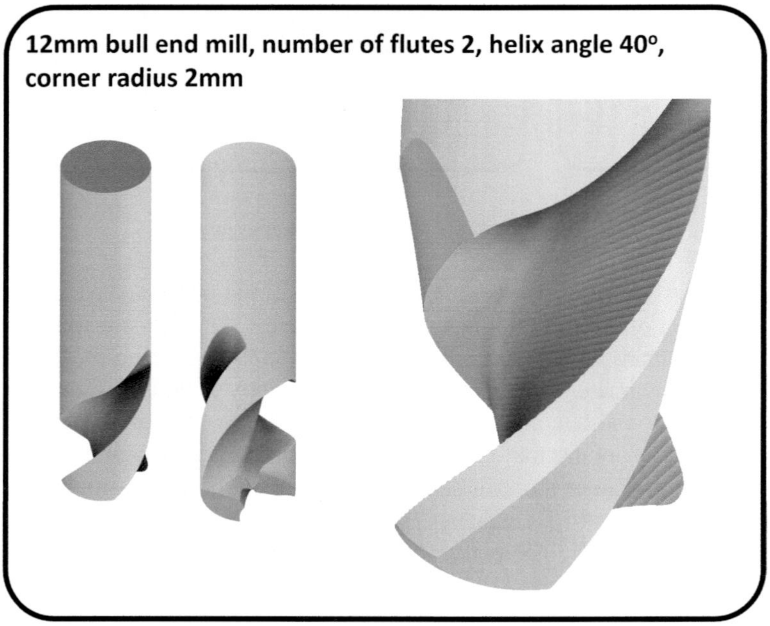

Fig. 3 Simulation platform results

Looking at Figure 4, the effect of the alteration of helix angle, the diameter of the grinding wheel, and the corner radius of the wheel can be seen. Out of the investigated parameters, as expected, the helix angle has the strongest effect on the end mill geometry. The corner radius of the grinding wheel has a strong effect on the flute internal geometry, which in turn influences the flow characteristics of the chip. The increase in the wheel corner radius leads to a more uniform curvature of the flute geometry with a constant change of curvature through the flute. Lower values of the corner radius allow for a sharp change of curvature which can lead to a modest chip-breaking action.

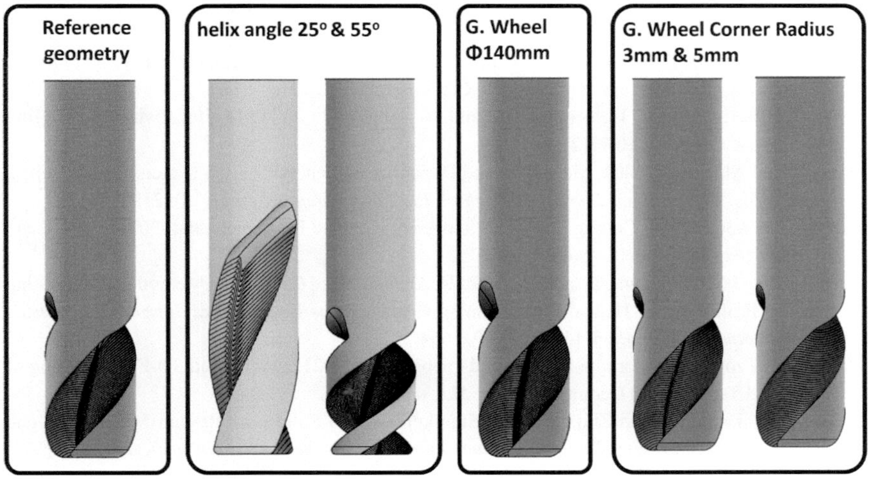

Fig. 4 Effect of grinding parameters on end mill geometry

Conclusions

In this work, a novel CAD-based framework for the design and optimization of solid carbide end mill geometry has been presented. The framework offers a highly adaptable and precise tool for designers, allowing them to model, simulate, and optimize critical parameters such as tool diameter, helix angle, rake angle, and flute profile.

One of the key contributions of this research is the ability to simulate the complete grinding process, incorporating intricate geometric features such as the flute and clearance faces, end face geometries, and gashes. The simulation process utilizes ISO/TS 13399-303 standards to ensure compatibility with existing CAM and CAE systems, making the generated geometries immediately usable in broader manufacturing workflows.

Future work will focus on expanding the capabilities of the framework, including the integration of more complex cutting tool geometries and further refining the simulation accuracy. Additionally, further validation of the framework against physical trials will be carried out to ensure it meets industry requirements and can be applied to a broader range of machining processes.

Acknowledgements

The work presented in this chapter is partially supported by the Eureka/Eurostars funded research project "SCORE—Sustainable maChining Of aeRospace comoditiEs" (project no: 2823).

References

Altintas, Y. and Lee, P. 1996. A General Mechanics and Dynamics Model for Helical End Mills. *CIRP Annals* 45(1): 59–64.

Altintas, Y., Kersting, P., Biermann, D., Budak, E., Denkena, B. and Lazoglu, I. 2014. Virtual Process Systems for Part Machining Operations. *CIRP Annals* 63(2): 585–605.

Bergs, T., Biermann, D., Erkorkmaz, K. and M'Saoubi, R. 2023. Digital Twins for Cutting Processes. *CIRP Annals* 72(2): 541–567.

Engin, S. and Altintas, Y. 2001. Mechanics and Dynamics of General Milling Cutters. Part I: Helical End Mills. *International Journal of Machine Tools and Manufacture* 41(15): 2195–2212.

ISO/TS 13399-303. 2016. Cutting Tool Data Representation and Exchange—Part 303: Creation and Exchange of 3D Models—Solid End Mills.

Jiang, L., Yang, Z., Li, Y., Ding, G. and He, X. 2024. An Optimized Calculation Method of the Grinding Wheel Profile for the Helical Flute Forming Grinding. *The International Journal of Advanced Manufacturing Technology* 132(3): 1649–1664.

Kyratsis, P., Tzotzis, A., Markopoulos, A. and Tapoglou, N. 2021. CAD-based 3D-FE Modelling of AISI-D3 Turning with Ceramic Tooling. *Machines* 9(1): 4.

Li, A., Zhao, J., Pei, Z. and Zhu, N. 2014. Simulation-based Solid Carbide End Mill Design and Geometry Optimization. *The International Journal of Advanced Manufacturing Technology* 71: 1889–1900.

Lin, Z., Yue, C., Liu, X., Liu, X., Yu, Z., Hu, D. and Zhang, Q. 2023. Development and Application of Parametric Design System for Solid End Mills based on Open CASCADE. *The International Journal of Advanced Manufacturing Technology* 127(9): 4659–4676.

Nikolaos, T. and Aristomenis, A. 2012. CAD-based Calculation of Cutting Force Components in Gear Hobbing. *Journal of Manufacturing Science and Engineering* 134(3): pp. 031009-1–031009-8.

Sanz-Calle, M., Iglesias, A., de Lacalle, L.L., Dombovari, Z. and Munoa, J. 2024. Optimal Milling Cutter Helix Selection for Period Doubling Chatter Suppression. *International Journal of Machine Tools and Manufacture* 202: 104211.

Siebrecht, T., Kersting, P., Biermann, D., Odendahl, S. and Bergmann, J. 2015. Modeling of Surface Location Errors in a Multi-scale Milling Simulation System Using a Tool Model based on Triangle Meshes. *Procedia CIRP* 37: 188–192.

Tapoglou, N. 2019. Calculation of Non-deformed Chip and Gear Geometry in Power Skiving Using a CAD-based Simulation. *The International Journal of Advanced Manufacturing Technology* 100(5): 1779–1785.

Tapoglou, N., Efstathiou, C., Tzotzis, A. and Kyratsis, P. 2023. Study of the Topography of Face Milled Surfaces Using CAD-based Simulation. In *Computational Design and Digital Manufacturing*, 159–166. Cham: Springer International Publishing.

Tosello, G., Bissacco, G., Cao, J. and Axinte, D. 2023. Modeling and Simulation of Surface Generation in Manufacturing. *CIRP Annals* 72(2): 753–779.

Tzotzis, A., Tsagaris, A., Tapoglou, N., and Kyratsis, P. 2023. High-precision CAD-based Simulation for Turning Considering Tool Microgeometry. *International Journal of Mechatronics and Manufacturing Systems* 16(1): 83–95.

Wang, H., Liu, X., Yue, C., Jiang, Z., Zheng, G., and Zhao, K. 2024. An Accurate Mathematical Model for Ball End Mills with Chisel Edge. *The International Journal of Advanced Manufacturing Technology* 131(12): 5729–5743.

10 | Technicity between Prediction and Autonomy

Ceci n'est pas une chaise (This is not a chair)

Valerio Perna[1*]

1. Introduction

The last 10 years have seen an evergrowing interest in the implementation of so-called intelligent tools in design production and research. Neural networks, generative adversarial networks (GANs), text-to-image, etc., have become everyday buzzwords in dedicated design literature and part of pedagogy processes, where students are engaged for a proper and conscious use of their potentialities and characteristics. Furthermore, contemporary design processes aim to manage a complex level of information concerning material, styles, forms, and prototyping in a time-effective manner, where the mental space between the idea and its concretization becomes day-by-day shorter and dictated by market rules. Indeed, incorporating artificial intelligence (AI) in product design processes (PDP) can promote a faster shift toward the potentiality of user-based customization, considering that customers today prefer tailor-made solutions to their tastes rather than rationalized and homogenous outcomes (Ananto et al., 2021; Cao et al., 2023).

Besides the burgeoning number of images populating online platforms and individual portfolios, the rise of artificial intelligence generated content (AIGC) marked an important milestone in the possibility of transcending the physical barriers in production limitations of designers, opening the way for a series of unlimited potential generations to be quickly explored and discerned, with the inner desire to guarantee quicker times of reaction in the creative process, enhance the design phase effectiveness, and significantly reduce the time-to-market toward competitive advantage (Zuoxu et al., 2022).

From a methodological point of view, the diffusion of these novel tools aspires to improve collaborative processes among professionals and influence a new open-source-based philosophy rooted in the possibility of creating your own dataset model to be shared and enlarged within a specific community. Today, according to Liu et al. (2021), 78% of the people engaged in the implementation of AI technologies in their workflow have the main goal of optimizing the convey of the design concept, delivering more coherent results in a shorter amount of time, and revealing design information daily with a more nuanced approach.

[1] *Lecturer*, Department of Architecture, EPOKA University.

* Corresponding author: vperna@epoka.edu.al

To draft a whole branch of all the different platforms available out there starts being complex, considering the blossoming of new opportunities daily. Just to mention some numbers, according to the portal Everypixel (2024), from 2022 to 2023, more than 15 billion AI-generated images were produced, more than Shutterstock's database of photos, illustration, and vectors ("only" 386 million). Among the most used ones are the chatbot ChatGPT™ that can engage in meaningful conversation with the user in terms of linguistic codes and facilitate the production of texts, essays, briefs, and – recently – also of image-based contents; Stable Diffusion™, where the biggest number of images have been produced in the last two years (almost 13 billion); and Midjourney™ which, thanks to its easy learning curves, can satisfy the needs of both professionals and the broader audience to craft beyond the physical act of making and enhance human–computer interaction (Xi et al., 2023).

Ludwig Wittgenstein, in his *Tractatus Logico-Philosophicus* (1921), stated that "*the limits of my language mean the limits of my world*," marking the importance of the cognitive structure of language as a way to determine the new possibility of being and creating, underlying the importance of linguistic studies as the capacity to assign names to objects and communicate effectively through this process is one of the fundamental traits that distinguishes humans from the rest of the animal kingdom. In this perspective, the combinatory use of language patterns and their direct (style) transfer into visual stimuli leads the user to the generation of novel images, overcoming the notion of form and geometry toward aesthetic and combinatory results nurtured by a post-human intelligence and design approach (Del Campo et al., 2019).

Starting from these premises, the chapter aims to delve into the importance of an inform implementation of AI-based tools in design practices and to analyze and interrogate the possibilities of a pedagogic methodology with the students based on concepts such as chaos, variability, and the post-human. The main objective is to demonstrate how students can benefit from the use of such novel tools as long as the in-class approach is not just dedicated to visuals but could also be enhanced by the reproducibility of the results toward coherency and later optimization for digital prototyping purposes. The philosophical background of this work is rooted in Gilbert Simondon's work and the concept of *technicity* (1958) intended as how humans affect their environment through the use of technology and how such technologies reciprocally produce them in the first place.

2. Chasing the Post-human: Amidst Chaos, Variability, and Technicity

The rise of new technology has always been a profound component of change for the society that developed (Gourdoukis, 2020). Whether it refers to analogue technologies or digital ones, each advancement brings a series of new symbolic and cognitive patterns which, to some extent, are not easily readable as soon as the novel tools are developed and implemented. If this is true, and considering the relatively young appearance of AI tools in design, designers and professionals are still far from the full consciousness on how to complete the exploitation of the cognitive mutation they are submerged in this historical phase and how they affect the environment humans live in. According to Pierre Lévy, it may take a long time—even a generation span—before assimilating and turning into practice the novelties a new technology might bring in. Nevertheless, it is undeniable that the notion of intelligence has become one of the buzzwords of the last two decades, comprising every possible (design) action, theory, praxis, or process (Lévy, 2010). Furthermore, referring to Simondon's idea of technicity, time is one of the fundamental substances through which a technology shifts from its initial use as a mere utensil to a more meaningful and consistent individuality that transduces itself in more than was devised in the first place (Simondon, 1958). The French philosopher's theorization shows how humans do not just create the technology they use but, at the same time, they are also shaped by that in their modes of existence and perceiving the world. Through the phenomenon of "transduction"—the act of creating meaning through relations between disparate individuals—humans create novel information through actions and new affordances, leading to an unexpressed potential of modifying the environment (*umwelt*) they inhabit. When it comes to the topic of AI and its impact on contemporary design practices, the dynamic of transduction has enlarged its realm, referring not just to relationships between biological individuals but rather between two different kinds of intelligences: one embodied into a "skin and bones" body while the other depicted as a tangle of invisible networks simulating the human brain.

It is clear then, how a new agenda for design is shifting day by day toward the notion of intelligence and the implications it generated from a specific set of conditions in which it is accepted that a human and a non-human (machine) could exist and synergy of "human intelligence + artificial intelligence", or "fusion intelligence", is likely to emerge (Boden and Edmonds, 2009) through the interpolation of libraries based on data and knowledge, application, and algorithm (Wang et al., 2021).

This mutation in the notion of intelligence also leads to a paradigmatic shifts, where the human component is to be understood as part of a new worldview. Humans are no longer dominant and expressing the major form of intelligence in the whole living realm but are a singular part of a more extended and interconnected network of relationships, and every form of intelligence has to be considered dynamically interdependent, nurturing in men what can be described as a post-human condition (Pepperell and Yazdani, 1995; Vogiatzaki, 2017).

Considering man as a part of a bigger whole and not in a dominant position concerning species embeds the original aim of digital tools, when they appear on the design scene. Such instruments were seen as pervaded by a liberatory force that would free design procedures from standardized solutions toward customized and non-predictable outcomes where each item would be different considering the complexity of the *a priori* condition through which it was generated (Migayrou, 2003). The main dynamic behind these liberatory forces could be explained from a philosophical point of view considering Simondon's idea of transindivituation, which is the reciprocal transformation of individuals through the use of technology and the enhancement provided by the new communications methods and mental categories that technology injects, stimulating deeper and unexpected systems of relations. With the implementation of AI technologies in contemporary design practices, it is then clear that the encountering of ovel theoretical frameworks and practical applications related to the concepts of innovation, creativity, and imagination—three primary driving forces behind design activities. Indeed, nowadays, the dynamics and complexity of the mind–body–environment continuum make human predictions a somewhat unreliable guide for shaping reality. This is not only because such predictions are inherently short-term but, more importantly, also because they are influenced by pre-existing human stereotypes, entrenched prejudices, and ongoing conflicts. In contrast, imagination—particularly, collective imagination—is proposed as a more equitable alternative for creating reality. Collective imagination, as a function of collective intelligence, is envisioned as a means toward an unpredictable and yet unwritten future. This perspective optimistically suggests the existence of a far more liberated world and it serves as a vehicle for shifting from rational to speculative thinking, though this shift is by no means an invitation to abandon our critical faculties, moral values, or socio-cultural foundations.

Nevertheless, the implementation of new technologies and non-human intelligent tools does not completely stand for the liberation from standardized solutions toward more in-formed individually customized ones. This phenomenon was already highlighted by architect, designer, and researcher Greg Lynn at the edge of the 2000s who, commenting on the outcomes of some students' exploration at Columbia University, stated: "*(t)hey all looked the same. It's the technology. We were figuring out the limitations of the software...*" (Cramer and Guiney, 2000).

In his claim, Lynn subtly wants to state not only the misuse and unfamiliarity of the designers with the new tools but also the peculiar dynamic through which the standardization of forms was not just a casualty but rather a persistent element generated by the communication between human and digital intelligence. In contemporary exploration, where human cognitive and mental abilities are transferred to machines, resulting in an increased level of responsibility being assigned to them, the risk of the burgeoning image-generation coming from AI tools such as Midjourney™, Stable Diffusion™ and Dall-E™ is to point again to a homogeneity of results that denies the original purposes for which they were created. To avoid this issue, this chapter—in its last part—delves into the possibility of proactively acting to overcome such issues and, through the pedagogic experiment in a design studio with the students, to set the tone for a more conscious application and implementation of AI-based tools and to indicate the way they are closely linked to prior cultural productions, drawing from extensive archives of images and verbal texts on which they are trained, where inputs are deconstructed and processed into a rich array of visual patterns.

Before that, it is important to introduce the concept of variability and chaos, while drafting a brief history of two different positions in the conception and implementation of intelligent technologies that still orientate the contemporary debate and the presence of a predominant position on the other subtly preferring dependence rather than autonomy in processes (Varela, 1989).

3. From Dartmouth to Autonomy and Networks: A Brief History of AI and Its Development

In 1956, Dartmouth College hosted the *Dartmouth Summer Research Project on Artificial Intelligence*. The wording AI—artificial intelligence—was enunciated by John McCarthy (1955) in a former preparatory document, and the event could be considered as an extended workshop of several weeks where different pioneers in the field (Allen Newell, Arthur Samuel, Oliver Selfridge, Herbert Simon, etc.) reflected on the assumptions that every aspect of learning, or any other feature of intelligence, can, in principle, be described with such precision that a machine could be designed to simulate it (McCarthy et al., 1955). The discussion approached several cutting-edge fields of technologies such as computers, natural language processing, neural networks, theory of computation, abstraction and creativity, and was grounded in previous research such as the theory of computability and Alan Turing's work (Hodges, 1983; Davies, 2000).

Seminal to this meeting was an article written by Claude Shannon (1950) that analyzed the possibility for a machine to play chess and evaluate its response in terms of performance and moves to be deployed, thus suggesting a first framework for the evaluation of machine behaviours. Furthermore, the features highlighted by Shannon transform the computer from a mere calculator (performing only arithmetical operations) into a symbol manipulator or general-purpose machine (not numerical tasks but more complicated ones) and endowing the computer to take decisions starting from a discrete process made of trial and errors (indeed transforms the machine into a learning entity). Indeed, Shannon recognizes the capacity for a computer to deal with complex problems and offer alternative solutions into a broader range of possibilities, not filtered into an optimal state one but a series of considerable options (Shannon, 1950).

Following these premises, Otto Schmitt's *The Brain as a Different Computer* brought new considerations to the table. As a biologist, Schmitt argues for the need to understand the contradiction between the contemporary perception of the computer and the inner capacities of the biological brain. According to him, a machine should not have the objective of performing absolute calculation but rather imitate the flexibility of reasoning usually shown by living entities, thus developing a kind of loose or "grey logic" rather than a "black-n-white" one based on "true or false" categories (McCulloch et al., 1956). Such a shift in technology production could enhance machines not just to perform tasks but also to understand and manage abstract concepts and/or inappropriate knowledge usually available to living beings in real-life experiences (Table 1).

Table 1 Computers vs Schmitt's idea of intelligence in machine's environment

Computers	*Schmitt's Idea*
Determinacy	Indeterminacy
"Purely digital" computations	"Statistical", "distributed" computations
"Black and white" logic	"Grey" logic
Logical reasoning	Common-sense (knowledge-based) reasoning
"Systematic", fully-informed procedures	"Non-systematic", partially informed procedures

Some of the points highlighted in Schimtt's part of the chart influenced the further development of the notion of intelligence in the way machines process and understand a specific set of data through which they are trained. Considering the latest developments in the fields of AI and text-to-image generators, it is reasonable to suppose that the very *grey logic*, *common sense*, and *statistical* procedures are at the core of these novel tools at disposal today and question from an ontological point of view the limit of their capacities and the extent to which they can be considered simple machines rather than intelligent beings simulating human creativity and complex interrelations (Del Campo and Leach, 2022).

Undoubtedly, the Dartmouth Conference played a seminal role in determining the direction toward which AI and its studies were directed and then developed. Nevertheless, to fully understand possible limitations and discrepancies in the epigenetic process of development of such technologies, it is important to reflect on the work of two important figures that emerged during and after the Second World War, John von Neumann and Norbert Wiener, and the different tendencies they shaped in the field: prediction and autonomy (McCarthy et al., 1955).

After the Second World War, a team of scientists and mathematicians led by John Von Neumann—started developing a predictive numerical methodology for weather conditions. Until that point the methods used were mostly empirical and based on facts rather than having a scientific basis. The first one to claim the possibility of applying numerical models for predictions was Lewis Fry Richardson who, even though showing no concrete results, was considered by Von Neumann as the base study to further develop his experiments. The idea behind the research was that weather conditions could be divided into stable and unstable phenomena, and regarding the first ones, prediction was possible, and where prediction could arrive, control could be implanted to develop a proper model. The first experiments exploited the technological possibility of the time and, if the first attempt needed 24 hours for a single prediction—meaning that the latter and event were almost happening in real time—further ones could be extended for a time span of 40 days after which the systems became unstable. If a part of the scientific community saw these experiments as a disruptive success that would change how weather phenomena could be understood and predicted, another part led by Norbert Wiener showed strong criticism toward such experiments. According to the Canadian physician, we sought to convince the world that weather prediction could be approached as a deterministic problem, suggesting that atmospheric patterns could be fully understood, controlled, and forecasted with precision. This perspective overlooked the inherent complexities and chaotic nature of weather systems, which are influenced by countless dynamic variables and interactions, and was later demonstrated by Lorenz (1963) shortly after Von Neumann's death (Gourdoukis, 2020).

Today, it is well-known that it is almost impossible to predict weather conditions for a period longer than 30 days, while the effectivity in calculation for shorter duration is proven. Wiener's approach could be considered more suitable for the time even though the main interest of this chapter on this debate is not the weather conditions themselves but the conceptual shifts, the two positions marked in the development of contemporary digital technologies and calculation tools. Both scientists were present at the 'Macy Conference on Cibernetics' in 1946, where concepts like biological computations and reasoning systems were presented and discussed. Von Neumann and Wiener, as keynote speakers, both expressed their points of view and perspectives concerning the direction toward which digital technologies should be pushed and developed. The first expressed his ideas of a machine that could calculate anything and that any logical problem presented to it could have a solution, while the other one contested this statement and explained that "*von Neumann machines, faced with a paradox, enter into endless oscillations [...] From this, he developed the concept of reflex and then finalized activity*" (Varela, 1989).

As analyzed by Varela in his book, the starting point of these two positions are innerly different and even more subtle when it comes to computer sciences and the way of developing and using digital technologies. Von Neumann's approach is based on the interest in heterogeneous dynamics looked at from the outside: computations are seen as a set of unchangeable rules confined to a rational process where every kind of question could be solved within pure calculations as the whole reality is 'simplified' as a matter of univocal and unilateral data following a linear path. The perfect metaphor for such a perspective is the 'black box' where the reaction that happens in the middle of a process is never visible while the inputs and the outputs are always at disposal to be analyzed. It is impossible then to understand the complexity of the different interactions and, most importantly, time is flattened in the linear path marked by input and reaction/results, thus making the only controllable variables those that perfectly align with the machine's predefined pattern.

On the other hand, Wiener introduced the concept of variability where the focus of processes is not intended on calculation power and linear dynamics but more on the relationship happening during the elaboration. As reality is a complex entity, we have to seek an interest in parallel processes that can happen at the same time while inputs and outputs are not the centre of the discussion but just some more or less casual outcomes that could happen. Referring again to the black box previously mentioned, Wiener's whole idea is about opening the box and revealing what is happening at the core of the cybernetic interrelations along the process path it follows. Indeed, Von Neumann's approach can be understood as a deterministic top–down focused on the idea of prediction and estrangement from the dynamics happening and can be defined using the words command, heteronomy, and serialism, while Wiener's can be seen as a bottom–up level based on variability and autonomy where the interaction is the key of the whole process and grounded on autonomy and parallelism.

Besides these technical notes, what is important in terms of design and digital tools are the two different outcomes that these positions caused in the further development of machines and the way they have affected processes. For Von Neumann, the computer must be a determinist tool able to eliminate autonomy and variability, confining unpredictability as a dangerous possibility both on a scientific and political level (the Hungarian-American mathematician and physicist was one of the main figures in the project Manhattan and in the development of the atomic bomb). On the other hand, for Wiener, the systems must not be influenced from the outside through the representation of dyamics but rather be understood from the inside where autonomy and systems are based on the casual and unexpected interactions between the different elements; thus evolution and progress are not predetermined and based on rational conditions but something that 'evolves' as in natural sciences from casuality and random combinations of phenotypes as in biological processes (Gould, 2002). In this perspective, the theory of nonlinearity—often connected to concepts like chaos and emergence—suggests that natural systems can undergo sudden, unpredictable shifts from one state to another. This idea challenges the view that all modifications are gradual or linear. Indeed, in complex systems such as weather, ecosystems, or economies, certain thresholds or "tipping points" can lead to rapid transformations that are not easily anticipated by traditional, linear models (Carpo, 2013).

Von Neumann's approach became the foundation for the development of digital computers, where the latter is intended as a rational brain able to perform not more than rational tasks. In his book, Mario Carpo highlights this discrepancy during the roaring years (1990–early 2000s) of digital architecture and design when the groundbreaking presence of computers and advanced digital tools and the "irrational exuberance" of the beginning were quickly crystallized in more standard procedures where the digital turned into an optimization tool to perform practical operations rather than a speculative instrument to challenge the limits of the current design production and reality. Nevertheless, the 15 years have seen the rise of a renewed interest in Wiener's position where the brain and—by transitive property—the computer are seen as a neural network based on the concept of openness and non-determinacy. The impact of AI in design marked a paradigmatic shift in the use and the development of technologies, shifting indeed towards variability and autonomy, interactions and relations within the systems. The black box, as intended by Von Neumann, has been questioned and disarmed to go at the heart of the processes following the until-now peripheral vision of Wiener. Tectonics, style, estrangements, and composition are some of the keywords at the core of the most recent design experimentation driven by AI technologies and their applications. The next part claims for a more in-formed process in design education that could empower the students in the use of such tools under the aegis of variability and consistency, elaborating a proper methodology of communication with the neural networks and the managing of the magmatic chaos contained in the datasets through which such bits of intelligence are trained and developed.

4. This is not a chair: Theories and Methods behind the Praxis

Ceci n'est pas une chaise (This is not a chair) obviously inspired by René Magritte *La trahison des images* or *Ceci n'est pas une pipe* (1929) (Magritte, 1929), is a pedagogy experiment that wants to highlight the difference in tangibility and consistency that the world of reality has with that of signs, while at the same time inviting reflection on the complexity of language in the era of AI technologies and their implementation in the design field. The theoretical background used is also grounded in the field of semiotic studies and linguistics since the platform used is a computation tool—based on variability—that can produce a vast number of original images based on the training they have received through a large dataset of visual-verbal and multimodal documents and items.

Two semiotic mechanisms are used in this framework. The first one is linked with the idea of enunciation (Benveniste, 1970) which is considered "*the very act of producing an utterance and not the text of the utterance that is our object. This act is the act of the speaker who mobilizes language for his purposes [...]*". Even though the analytical in the neural network used is not completely revealed to the user and a part of it may resemble the black box idea perpetrated by Von Neumann, the act of the enunciation can be considered as the basic step to guarantee a sort of "communication" based on language between the human part and the machine operator. Thus, from a semiotic point of view, the most recent text-to-image-based AI platforms can be considered as an enuncial machine considering that the outcome they produce can be an enunciation in the form of a visual (Midjourney^{TM}, Dall-E^{TM}, Stable Diffusion^{TM}, etc.) or textual (ChatGPT^{TM}, Gemini AI^{TM}, etc.) utterance.

The second gimmick is based on the concept of generative semiotics, which differs from Benveniste's approach. While Benveniste focuses on the act of enunciation itself, this perspective emphasizes its cultural and historical components, where pre-existing speech forms are acculumated as patterns. Such patterns, become readily available for reuse, both in verbal or visual form (as in the case of AI)—forms ready to be used again. This concept is thorough when it comes to describing the operating dynamics of the datasets, which are trained on specific existing data which are deconstructed and reprocessed according to the utterance generated from the human counterpart. Generative AIs, in this regard, function as machines deeply embedded in enunciative praxis—the process through which meaning is produced in language and communication. Their output is shaped by the act of "summoning" or drawing from vast datasets of prior human knowledge, and then "actualizing" these inputs into new, unique outputs. This process relies on probabilistic models and algorithms that analyze patterns in large datasets, simulating the way humans might create, interpret, and re-enact ideas (Greimas and Fontanille, 1991).

Given these conditions and having explained to the students the importance of such semiotic concepts prior to the design operations, the decision about the platforms to be used fell on a specific class of AI generative models: the so-called Diffusion Models™. Diffusion Models™ such as Midjourney™ and Dall-e™ (del Campo, 2024) are trained following two progressive operations: the first one where some noise is sequentially added to an image in a process known as "forward diffusion". This step involves gradually corrupting the original image by introducing random noise at each stage, thereby transforming it into a nearly unrecognizable, noisy version. The model then learns to reverse this process through "reverse diffusion", where it predicts and sequentially removes noise from the noisy image, ultimately recovering a coherent representation of the initial input. This reverse process is guided by denoising operations, with each step conditioned on the noise level to iteratively reduce it and reconstruct the image accurately. Such models use a "large language model" (LLM) to understand the composition of language-based commands (prompts) and transform them into embeddings (lists of numbers) that can then be used from the neural network to elaborate the resulting images.

For pedagogy purposes, the design operation was limited to a single object, used every day and that could be denoised from the model database. The final decision fell on a chair that from a normal design class perspective can be seen as a perfect design assignment encapsulating several factors to be addressed such as ergonomics, aesthetics, materials, functionality, and durability. Furthermore, chairs have played a defining role in design history, symbolizing innovation through evolving materials, technologies, and aesthetics. Often viewed as both functional objects and cultural symbols, chairs have consistently reflected shifts in style, engineering, and social values. Iconic designs, like the Thonet No. 14, the Eames Lounge Chair, and the Barcelona Chair, are examples of breakthroughs in material application and form that have repeatedly re-entered the spotlight over time. These designs return to prominence due to their innovation and versatility, blending functionality with aesthetics. Their ability to capture the essence of an era while remaining adaptable allows them to "steal the limelight" time and again, reinforcing the chair as a symbol of enduring design excellence.

Given these conditions, the first step done with the students was to challenge the dataset through which the neural networks were trained and understand which ideal of the chair they stored in their database. This operation comprehends a philosophical layer that relates to the need to avoid prioritizing abstract concepts concerning ongoing relationships and open up to the possibility of creating a non-hierarchical and inclusive ontology of the chair denying the assumptions that what is perfect and unperishable is already known. In a few words, what was tackled was the notion derived from Platonism where the design and transcript relations are imposed according to a transcendental idea that can never be understood and, thus, reproduced or concretely realized (Smith, 2012), and everything is reduced to a vague projection/appearances of something existing somewhere else. Debating on these pure ideas deriving from Platonism opens up the door for a more relational logic not based on the binary of true and false, and relates with the simulacrum proposed by Smith. The concept of the simulacrum challenges traditional ideas of representation by dismissing fixed characteristics or absolute truths. Unlike representations that mirror an original ideal or archetype, the simulacrum operates independently of any ultimate "idea" or standard form. Indeed, Deleuze describes it as a dynamic entity, formed from relationships rather than any intrinsic qualities or predefined models, thereby subverting established systems of classification and evaluation.

For this reason, in our enunciative in the use of AI and for our need to create utterances that could help us in challenging the preconceived idea of a chair that might be stored in the used neural networks, we started interrogating different neural networks with specific prompts that could generate several iterations and outcomes testing the encoded data regarding the subject of our investigation. Furthermore, another gimmick was used in this phase. In many attempts, a "style layer" within the submitted prompt was added and the diffusion model was asked to present the results in Bernd and Hilla Becher's iconic typological photography style to start depicting a taxonomy of the object that could then use a base material to operate further selections and interpolations.

Fig. 1 Chairs models. Generated by Ideogram^{TM} AI

The first images were created using three different neural networks (from the top to the bottom): Ideogram^{TM} AI (Figure 1), Leonardo^{TM} AI (Figure 2), and RunwayML^{TM} (Figure 3). As can be seen, there are some common elements in the generation in terms of forms, shapes, and geometries used, leading to some consideration regarding correspondence between image embeddings and image descriptions that have been encoded. Indeed, no further specifications were made considering style, designers, materials, etc. The main objective was to challenge the neural network on the idea of a chair, while specifying only the photographic representation style. The enunciative string of code used is the following: "*a series of nine-by-nine grid compositions featuring stark, high-contrast black and white photographs of the different details of some chairs in an empty room, reminiscent of the typological photographic approach of the Bechers*".

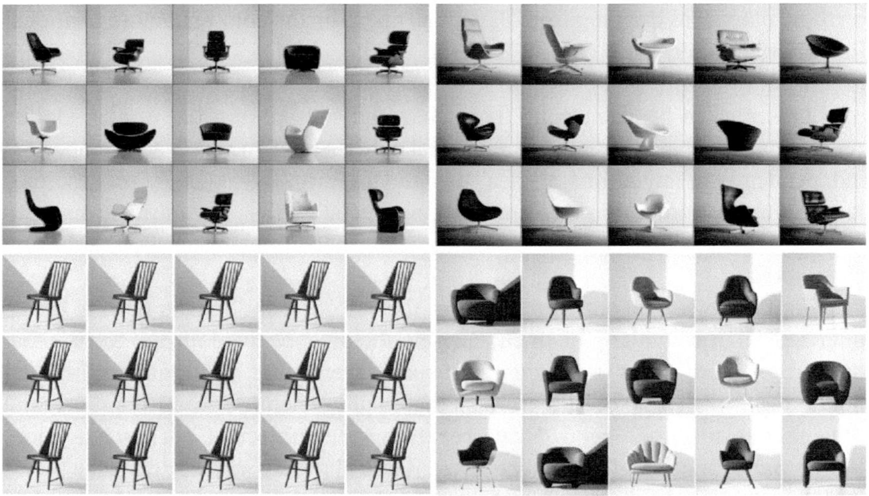

Fig. 2 Chairs models. Generated by LeonardoTM AI

Fig. 3 Chairs models. Generated by RunwayML™

As a final confirmation, another round of generation with the help of the ChatGPT4™ model was run to provide a comprehensive textual description that was then used to reroll another set of images. ChatGPT™'s suggestion was more elaborate and nuanced rather than the previous ones used and scripted together with the class:

"Generate an image series capturing the 'platonic ideal' of a chair in a neutral, typological style inspired by Bernd and Hilla Becher's photography. Each chair should be centred in the frame against a blank, neutral background, with no visible context beyond the object. Use grayscale tones with balanced, even lighting to emphasize form and structure over colour or material. Chairs should vary slightly in shape and design but share core attributes that represent the essence of 'chair-ness.' Aim for a uniform, symmetrical arrangement that allows the viewer to appreciate each chair individually while observing variations as a cohesive whole".

Nevertheless, in terms of consistency, we do not see major differences between its suggestions and our previous generations (Figure 4). The translation of the revised prompt into the image is fairly successful as it captures all the compositional elements outlined in the description. Each specified component—such as layout, colour scheme, and spatial relationships—was represented effectively, demonstrating a coherent interpretation of the prompt's instructions. This fidelity to the description means that the image successfully encapsulates the intended visual concept, offering a comprehensive rendering of the requested elements where the practice of enunciation creates interesting and coherent utterance in the investigatory phase challenging the idea of a chair.

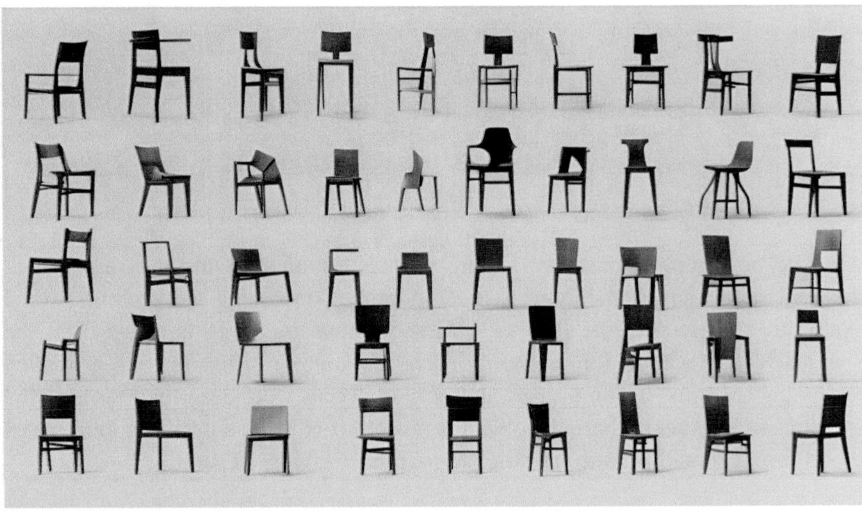

Fig. 4 Chairs models. Generated by RunwayMLTM + ChatGPT4TM

The second part of the work consisted of adding further steps and starting to define the chair, adding more information to the prompts such as material, colour, quality of the space, and ergonomics, to push the students to a final model that could be further used for modellation and 3D printing purposes. The main purpose behind this phase was to challenge the students in their generative practice on exploring the relationship between the verbal language system and the utterances that are produced through speakers' appropriations of this system.

Fig. 5 Chairs contextual renders. Generated by MidjourneyTM AI

Fig. 6 Chairs contextual renders. Generated by Dall-E^{TM} 2

The best generations were catalogued and evaluated and are reported in the generated images in Figure 5 and Figure 6. Concerning the prompts, some examples are given in the following to show how the students move from the abstract idea of a chair for taxonomy purposes to more in-formed and grounded reality generations. In some cases, not only materials were described but also the specific environment where the object is located (texting the database in terms of repertoire of semantic forms) or the photographer taking the shot.

Example Prompt 1 (Figure 5): "*photo of a beautiful ovoid shape armchair made of black fabric and with the bottom part made of triangulated golden panels, in the middle of a beautiful living room with a decorated white wall on the background, cold side light coming from a big vertical window on the side, photographed by Zachary Goulko*".

Example Prompt 2 (Figure 6): "*perspective photo of a beautiful ovoid shape armchair made of black fabric and with the bottom part made of triangulated golden panels, in the middle of a beautiful living room with*

a decorated white wall on the background, cold side light coming from a big vertical window on the side, photographed by Nick Knight".

All the images produced were analyzed and evaluated according to some specific categories reported in Table 2. Such categories touched on some important points such as the overall composition of the image, the form clarity, and the specificity/familiarity with the idea of a chair previously explored without contextual data.

Table 2 Evaluation categories for the second round of AI-generated images

Category	*Score*
Composition	0-5 score
Ergonomy	
Form Clarity	
Realism	
Form Interest	
Specificity-familiarity	
Consistency	0-5 score (with the previous generated images)

5. *Ceci n'est pas une chaise* (This is not a chair): From 2D Sketches to 3D

The field of generative AI is in continued expansion. What looked impossible just five years ago (generating images through text and semantic values) is today not just a certainty but has also entered the workflow of many studios and artists training their datasets and implementing such technologies in their everyday practice.

Taking for granted the use of it as a generator for two-dimensional raster images, new research is pushing the boundaries of AI to the generation of High-quality 3D Models almost ready for 3D printing and digital fabrication procedures. Recent advancements in AI-driven tools (Heewoo and Nichol, 2023) allow for text- and image-based manipulation of 3D models, enabling users, including beginners, to interact with and alter 3D models without needing extensive training in 3D modelling. These methods leverage prompts and image references to guide model transformations, providing a more intuitive way to reshape and refine 3D objects by adjusting textures, colours, and structures directly from the interface. Such methods also often utilize optimization techniques to enhance visual output, focusing on maximizing image quality and realism rather than technical precision in 3D construction.

The main challenge is then to blend the aesthetical appearance of the generated images and the required trade-off with functional aspects which are fundamentals for their prototyping and realization through digital fabrication procedures. Indeed, without the ability to encode physical principles and material properties in the enunciation phase, it is very challenging to obtain a proper transfer into a 3D modelling environment and reach the final purpose of digital fabrication (Oscar et. al., 2022; Tsalicoglou et al., 2024).

Following these premises, the last part of the experimentation conducted with the students was thus to investigate the chances to take the selected images according to the evaluation categories, process them again to 3D-based neural networks, and then test the possibility of their prototyping through the use of the NURBS software Rhinoceros™.

The steps were as follows: the set of best images was analyzed and a series of sketches were produced to summarize the aesthetic and functional categories through which the 3D model should be created.

Following an internal pool in the class, one of the sketches (Figure 7a) was selected as the one embedding all the categories required for a functional prototype to be produced. The drawing was then uploaded to the Meshy platform, as today one of the best online tools to transfer a 2D outcome into a 3D object with reasonable respect to components such as form, topology, and details (Figure 7b).

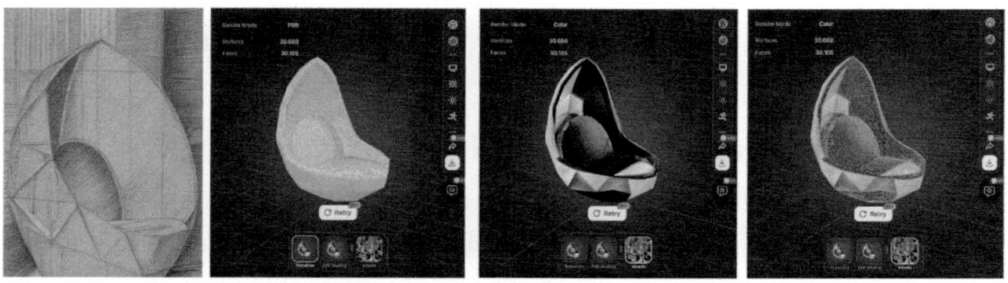

Fig. 7 (a) Selected final sketch; (b) 3D model generated by Meshy™ AI; (c) & (d) Texturing and topology verification; generated by Meshy™ AI

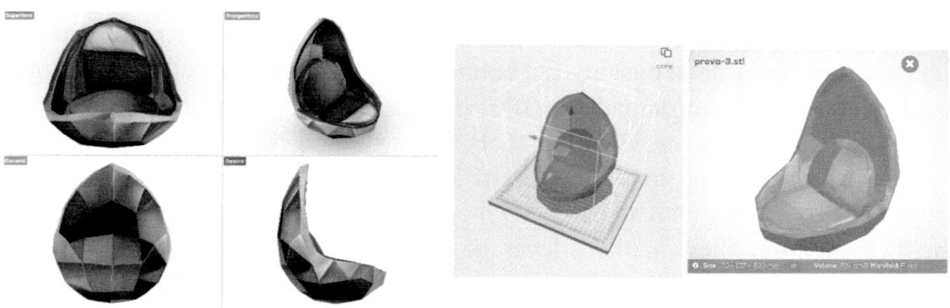

Fig. 8 (a) Rhinoceros™ viewports and the final model ready for 3D printing; (b) The final exportation for 3D printing (Printer: Prusa™ i3 MK3)

After a series of iterations and an acceptable reply from the neural networks, the texturing phase was implemented again within the software through a prompt suggesting which kind of materials, colours, and textures should be used to replicate the original idea contained in the sketch (Figures 7c and 7d). The final result (7D) was considered satisfying and then verified in terms of geometrical topology before exporting into Rhinoceros™ (Figure 8a) to avoid bottlenecks or surface errors and then verified in .stl format before being printed with a Prusa™ i3 MK3 3D Printer (Figure 8b).

Conclusions and Discussion

In this chapter, the functioning of generative AI and the possibility of its implementation into a pedagogy methodology was explored to overcome some of the actual barriers from 2D-generated images to 3D outcomes ready to be prototyped and printed. The overall idea behind this work is the possibility to empower the students in their design processes and experimentations with the new methodologies at hand that are more nuanced if filtered through the lens of the philosophy of technology (Simondon) and the semantic and semiotic component behind the functioning of LLM.

What is proposed is a new understanding that goes beyond the database functioning with the aim to debunk the black box philosophy contained in Von Neumann's work and pursue a more relational and open approach based on the theories of Wiener and his idea of variability.

The first part analyzed the concept of technicity as one of the main theoretical references to understand the possibility for technologies to not be considered as "neutral" but rather have a predominant position in a double-edged process where man shapes the form of technology itself but also later conforms to the new intellectual categories that come with it.

Furthermore, by introducing the figure of Ludwig Wittgenstein, the importance of linguistic structures when approaching AI-generated design was understood and the debate for a more in-formed use of these tools was set under the aegis of semantic communication between humans and machines.

In the second and third sections, the genealogy of AI was explored and two of the main tendencies that have characterized the debate until today were analyzed. The first one, inherited from Von Neumann, is based on a deterministic approach to the use of machines and focused on the idea of automation and top–down processes, where the inner systems are not to be revealed or understood while rather focusing on inputs and outputs for prediction tasks. The second one, based on Norbert Wiener's assumptions, where the whole problem behind the development of machines is subtly related to the relation and evolution of such systems in terms of results and datasets following the theory of nonlinearity, often connected to concepts like chaos and emergence, suggests how systems can undergo sudden, unpredictable shifts from one state to another based on variability and reconfiguration.

The fourth part deals with the similarities between semantic systems and novel neural network technologies where the enunciative part is fundamental to reach meaningful utterances that are the results of machines trained on specific existing data which are deconstructed and reprocessed according to the linguistic codes generated from the human counterpart.

Finally, the fifth section represents the original contributions to the field where the concepts explained and isolated earlier, starting from several theories reconnecting heterogeneous fields of knowledge (chaos, variability, intelligence, linguistic, enunciation, utterance), are then implemented as keywords in the development of a design exercise with a class of students in the field of product design. The selected object is a chair, one of the most common pieces of design used in pedagogy assignment, that is tested first as a platonic pure idea in its basic components and then enriched with more semantic values following Deleuze's vision of the simulacrum. One of the outcomes is presented at the end of this work and proves how such theoretical concepts can be merged and used in a path that encompasses the use of neural network and generative ideas from the conceptual phase to generation, ending up providing 3D prototype materials to be printed and verified in real life.

At the present moment, there are many methodologies and technologies to be used, each one of them enunciating differently while actualizing the content and peculiarities of the specific database they are trained with. For this reason, the design experimentation proposed here does not aim to create theories from the particular but rather serves as a moment of reflection on how holistic features can be coherently extrapolated from the burgeoning possibilities related to generative AI in the fields of design. More tests and shreds of evidence definitely need to be produced; nevertheless the implantation of Wiener's view based on variability and relational recombination can be applied in contemporary design research to enrich and enhance students' perception and usage of new digital tools.

References

Ananto, P.K.F., Hsieh, C.C. and Mahendrawathi, E.R. 2021. Competition between Online and Offline Retailer Mass Customization. *Procedia Computer Science* 197: 709–717.

Benveniste, É. 1970. L'appareil formel de l'énonciation. *Langages* 17: 12–18.

Boden, M.A. and Edmonds, E.A. 2009. What is Generative Art? *Digital Creativity* 20: 21–46.

Cao, Y., Li, S., Liu, Y., Yan, Z., Dai, Y., Yu, P. S. and Sun, L. 2023. A Comprehensive Survey of AI-generated Content (AIGC): A History of Generative AI from GAN to ChatGPT. *arXiv*, arXiv:2303.04226.

Carpo, M., 2013. *The Digital Turn in Architecture 1992–2012*. Hoboken, NJ: John Wiley & Sons.

Cramer, N. and Guiney, A. 2000. The Computer School. *Architecture Magazine* 9(89): 94–98.

Davies, M. 2000. *The Universal Computer*. New York: Norton.

del Campo, M., Carlson, A. and Manninger, S. 2019. Machine Hallucinations: A Comprehensive Interrogation of Neural Networks as Architecture Design. In *Proceedings of IASS Annual Symposia* (Vol. 2019, No. 17), 1–12. International Association for Shell and Spatial Structures (IASS).

del Campo, M. and Leach, N. eds. 2022. *Machine Hallucinations: Architecture and Artificial Intelligence*. Hoboken, NJ: John Wiley & Sons.

del Campo, M., ed. 2024. *Diffusions in Architecture: Artificial Intelligence and Image Generators*. Hoboken, NJ: John Wiley & Sons.

Heewoo, J. and Nichol, A. 2023. Shap-e: Generating Conditional 3D Implicit Functions. *arXiv* preprint arXiv:2305.02463.

Everypixel. 2024. People are Creating an Average of 34 Million Images Per Day. Statistics for 2024. Available at https://bit.ly/4hr8lJd (Last access: 5 November 2024).

Fry Richardson, L. 1922. *Weather Prediction by Numerical Process*. Cambridge, UK: Cambridge University Press p. 184.

Gould, S.J. 2022. *The Structure of Evolutionary Theory*. Massachussets: Harvard University Press.

Gourdoukis, D. 2020. Variability: Architecture and Its Fight against Chaos and Opinion. *archiDOCT* 8(1): 13–21.

Greimas, A.J. and Fontanille, J. 1991. *Sémiotique des passions. Des états de choses aux états d'âme*. Paris: Éditions du Seuil.

Hodges, A. 1983. *Alan Turing: The Enigma*. New York: Simon and Schuster.

Lévy, P. 2010. From Social Computing to Reflexive Collective Intelligence: The IEML Research Program. *International Journal for Information Sciences* 180(1): 71–94.

Liu, C., Ren, Z. and Liu, S. 2021. Using Design and Graphic Design with Color Research in AI Visual Media to Convey. *Journal of Sensors* 2021(1): 8153783.

Lorenz, E.N. 1963. Deterministic Nonperiodic Flow. *Journal of Atmospheric Sciences* 20(2): 130–141.

McCarthy, J., Minsky, M.L., Rochester, N. and Shannon, C.E. 1955. A Proposal for the Dartmouth Summer Research Project on Artificial Intelligence. Available at http://jmc.stanford.edu/articles/dartmouth/dartmouth.pdf (Last access: 5 November 2024).

McCulloch, W.S., Oettinger, A.G., Rochester, N. and Schmitt, O. 1956. The Design of Machines to Simulate the Behavior of the Human Brain. *IRE Transactions on Electronic Computers* EC-5: 240–255.

Magritte, R. 1929. La trahison des images (The Treachery of Images) [Painting]. California, Los Angeles County Museum of Art, Los Angeles, United States [https://collections.lacma.org/node/239578]

Oscar, M., Bar-On, B., Liu, R., Benaim, S. and Hanocka, R. 2022. Text2mesh: Text-driven Neural Stylization for Meshes. In *Proceedings of the IEEE/CVF Conference on Computer Vision and Pattern Recognition*, 13492–13502.

Pepperell, R. and Yazdani, M. 1995. *The Post-human Condition*. Oxford: Intellect.

Schmitt, O. 1956. The Brain as a Different Computer. IRE Transactions on Electronic Computers EC-5: 244–246.

Shannon, C.E. 1950. Programming a Computer for Playing Chess. *Philosophical Magazine* 41: 256–275.

Simondon, G. 1958. *Du mode d'existence des objets techniques*. Méot, Paris.

Smith, D.W. 2012. *Essays on Deleuze*. Edinburgh: Edinburgh University Press.

Tsalicoglou, C., Manhardt, F., Tonioni, A., Niemeyer, M. and Tombari, F. 2024. TextMesh: Generation of Realistic 3D Meshes From Text Prompts. In *International Conference on 3D Vision (3DV)*, 1554–1563.

Varela, F.J., Bourgine, P. and Dumouchel, P. 1989. Autonomie et connaissance: essai sur le vivant, Éditions du Seuil, Paris

Vogiatzaki, M. 2017. Editorial. *archiDOCT* 6(1): 6–13.

Wang, L., Liu, Z., Liu, A. and Tao, F. 2021. Artificial Intelligence in Product Lifecycle Management. *International Journal of Advanced Manufacturing Technology* 114(3–4): 771–796.

Wittgenstein, L. 1921. Tractatus Logico-Philosophicus, Annalen der Naturphilosophie, Weimar Republic.

Xi, Z., Chen, W., Guo, X., He, W., Ding, Y., Hong, B., Zhang, M., Wang, J., Jin, S. and Zhou, E. 2023. The Rise and Potential of Large Language Model based Agents: A Survey. *arXiv*, arXiv:2309.07864.

Zuoxu, W., Jihong, L. and Lianyu, Z. 2022. The Evolution, Framework and Future of Cognitive Intelligence-enabled Product Design. *Procedia CIRP* 109: 526–531.

11 | Next-generation Digital Design Strategies for the Textile Industry

İbrahim Üner[1*] and Gizem Karakan Günaydın[2]

1. Introduction

The textile and fashion industry has undergone significant transformations over the centuries, from handcraft to computer-aided design (CAD) and computer-aided manufacturing (CAM) systems (Tao et al., 2016). This evolution has brought about a revolutionary change by increasing precision, efficiency, and creativity in textile and clothing design and production processes. The basis of textile design and fashion dates back to the times when dressing was a necessity. In later periods, craftsmen created complex patterns and garments using simple tools (such as looms, needles, and threads) to produce textiles that were uniquely artistic.

Due to its reliance on the ingenuity and ability of each individual craftsman, this method proved impractical. It was necessary to manually draw patterns on paper or fabric, which took a great deal of care and concentration. Although there was a profound grasp of materials and procedures throughout this time, large-scale production was challenging due to the labor-intensive nature of the task. The textile industry became mechanized during the Industrial Revolution which took place throughout the eighteenth and nineteenth centuries. Production capacity was significantly improved by inventions like the sewing machine, mechanical loom, and spinning jenny. Future technological advancements were made feasible by these developments, which allowed for the more efficient creation of increasingly complicated designs. All these advancements notwithstanding, the design process remained essentially archaic (Singh et al., 2022).

Even with the advent of increasingly sophisticated technologies, designers were still creating prototypes and sketching patterns by hand. Although mechanization improved consistency and speed in production, it had no significant impact on the design process. The earliest computer-aided design (CAD) systems appeared in the middle of the 1800s and were initially employed in the automobile and aviation industries. These technologies started to seep into the fashion and textile sectors in the 1960s and 1970s. CAD/CAM refers to a system controlled by computers and includes design, analysis, and manufacturing activities. CAD focuses on design and product development processes, while computer-aided

[1,2] *Assistant Professor*, Department of Textile and Fashion Design, University of Pamukkale. Email: ggunaydin@pau.edu.tr[2]

* Corresponding author: iuner@pau.edu.tr

manufacturing (CAM) supports control operations and machines. CAD, part of the computer integrated manufacturing (CIM) system, is a combination of computer-aided engineering (CAE) and CAM and is simply called CAD/CAM. Early CAD systems, while primitive by today's standards, were a significant step forward for the textile industry (Collier and Collier, 1990; Dwivedi and Dwivedi, 2013). Early CAD systems allowed designers to create digital representations of patterns and garments. These systems provided tools for drawing, editing, and manipulating designs on a computer screen, reducing the need for physical prototypes. This digital approach accelerates the design process, allowing for faster iterations and more precise adjustments.

The 1980s and the 1990s were a period when advances in CAD technology accelerated and CAM systems advanced at the same pace. The integration of CAD/CAM technologies revolutionized the textile and fashion industry by increasing the relationship between design and production. CAD/CAM systems provided seamless transitions from digital design to automated production processes. Designers could directly convert their work into machine instructions for processes such as weaving, knitting, and ready-to-wear. With this smooth transition, errors were reduced, consistency was increased, and the production process was significantly accelerated (Singh et al., 2022).

Today, CAD/CAM systems used in the textile and fashion industry are highly developed and offer a wide range of design tools. Modern CAD software, which includes features such as 3D modeling, virtual prototyping, and simulation, allows designers to visualize products in a realistic environment before they are produced. These systems provide designers with unprecedented creative freedom by supporting processes such as complex pattern creation, color matching, and fabric simulation. CAM systems that have developed in this direction have also incorporated robotic and automation technologies and ensured that designs are produced with high precision, ensuring quality and consistency. Innovations such as laser cutting, 3D knitting, and automatic embroidery have created new opportunities for textile design and production.

2. Key Features of Modern Textile CAD Systems

2.1 2D–3D Modeling and Visualization for Textiles

The textile industry is known for being a labor-intensive sector, requiring a significant amount of manual effort at various stages of production. From the initial processing of raw materials to the final stages of fabric creation and garment manufacturing, much of the work relies on skilled hands and detailed craftsmanship. While automation and technology have been integrated into many processes, the need for human involvement remains high, particularly in tasks such as quality control, intricate design work, and finishing touches. This reliance on human labor highlights the industry's role as a key employer, especially in regions where textile production forms a substantial part of the economy.

Apart from being labor intensive, the textile industry is highly competitive, making it necessary for companies to expand their horizons and respond to the ever-shifting consumer behaviors. For companies that wish to perform well in such a playfield, there are several areas they need to stress on, such as innovation, quality, and efficiency. The design and sampling process, for example, is one of the vital processes that help in maintaining the competitive advantage. It is important to point out that creative and original designs differentiate a brand from its competitors, while the ability to produce and provide timely, high-quality samples assists more in tracking the market and the customers. CAD/CAM systems transform the whole system in such a way that the cycle of other activities takes less time since the design work overlaps with production. Companies are, therefore, not only able to come up with creativity in product development but are also able to develop new products that meet the extreme expectations of their customers and consumers alike (Sinha, 2020).

The advances in the use of CAD have changed the textile and fashion industry. With the advent of 2D and 3D CAD systems, conceptual designers are able to visualize their concepts in a form of designs which helps them in effective presentation of the ideas. These CAD systems have ramped up the productivity of the design cycles and processes by making it possible for the designers to alter their designs very fast without having to physically build models. This technology has equally brought about new avenues for the creativity of the design works as designs of different colors, patterns, and textures can be tried with more ease. In addition, CAD systems have led to reduction of waste in the industry by reducing the use of physical samples and prototypes. Separately, textile CAD systems are grouped into two basic types: 2D CAD and 3D CAD. The 2D CAD systems are the ones aimed at making the primary layouts and constructions of patterns that will be used, while the 3D CAD systems are oriented toward the generation of computer-based models of clothing, ornaments, and cloth items. These systems are further classified under engineering CAD systems and art-and-design ones. In engineering CAD systems, attention is paid to the technical details of surroundings such as the yarn count, the structure of fabrics, and properties of materials, whereas the art-and-design CAD systems are concerned with the creative and aesthetic details such as motifs, colorways, or styles of the designs. As an example, Dai and Hong (2024) addressed the continuity concepts of fabrics such as tensile, shear, and bending aspects, and impact of these parameters for garment simulation and draping. In contrast, in another paper, Angelova and Sofronova studied the role of CAD applications in woven textiles design, and what benefits these systems provide. It outlines that such systems improve the scope of textile production by allowing a quicker turnaround time for design processes coupled with increased accuracy of the output (Angelova and Sofronova, 2021).

The CAD/CAM systems help to see how the fabric will be draped and fitted over a three-dimensional body, which provides almost real look of the end product. This chapter reviews and analyzes a number of commercial 2D CAD systems in the textile and fashion industry. Each one has its own set of advantages and disadvantages. Some of such widely used 2D CAD systems include Adobe Illustrator®, Lectra Modaris®, Gerber Accumark, Optitex®, EAT, and Nedgraphics Texcel® (Sayem, 2023).

2.1.1 2D CAD Software for Textiles

The affordable 2D CAD/CAM system is widely used in clothing pattern construction, as well as fabric cutting and spreading, to save time, improve efficiency, and lower labor costs. By using this 2D digital platform, pattern construction enables fast, standardized parametric drawing, which enhances collaboration and communication among designers, manufacturers, and retailers, both remotely and on a global scale (Collier and Collier, 1990).

- *Adobe Illustrator®*: While not specifically designed for textiles, Illustrator is widely used in the fashion industry for creating 2D sketches, patterns, and technical drawings.
- *NedGraphics Texcelle®*: It is a popular 2D CAD software tailored for textile design, offering features like pattern creation, editing, and grading.
- *Lectra Modaris®*: Although primarily known for its 3D capabilities, Modaris also includes robust 2D pattern-making functionalities.
- *Optitex®*: Optitex offers solutions for both 2D and 3D textile design, with its 2D module focusing on pattern creation, editing, and marker making.

2.1.2 3D CAD Software for Textiles

- *Browzwear*: Browzwear offers 3D CAD solutions specifically tailored for the apparel industry, allowing designers to create virtual garments and simulate fabric behavior.
- *CLO Virtual Fashion™*: CLO is widely used for virtual prototyping in the fashion industry, enabling designers to create realistic 3D garments and visualize fabric drape and movement.
- *Marvelous Designer*™: While not exclusively for textiles, Marvelous Designer™ is popular among fashion designers for its intuitive 3D modeling tools and realistic fabric simulation capabilities.
- *Optitex 3D Suite*™: The 3D suite of Optitex™ provides tools for virtual prototyping and simulation in the textile and apparel industry, allowing designers to visualize garments in 3D and assess fit and fabric behavior.

The textile industry is characterized by the need for swift commercial transactions and the development of products tailored to customer expectations and demands. Just as the production sector adapts to the rapid changes in fashion and technology worldwide, educational institutions are also updating their curricula and programs to reflect the latest innovations. In textile design, the process varies depending on the specific features of the products, meaning design programs must evolve to meet these demands. CAD systems, in particular, need to be compatible with CAM, which has led software companies to focus on creating seamless integration between the two. In recent years, these companies have increasingly prioritized both technological advancements and customer-focused strategies, continuously updating their software to stay aligned with industry trends.

3. Weaving Design

One of the oldest arts in human history is undoubtedly the art of weaving. Weaving emerged from the need for people to protect themselves from the cold, dress according to the seasons, cover themselves, and adorn their surroundings (Başaran, 2019).

Design encompasses all products such as projects, drawings, and models, which emerge as a result of a design action and serve as guides during the realization of the main work. Designing is a comprehensive process and involves defining an object, a system, or an event according to a desired outcome. Also, design is described as a phenomenon that includes art, science, engineering, humanities, and professional sciences. On the other hand, it is categorized into five types, namely, engineering design, industrial design, process design, visual design, and product design (Gürcüm and Üner, 2016).

Just like other design categories, pen and paper are the simplest tools for pattern work in the textile design industry. For a designer to transfer the shapes envisioned in their mind onto paper, they must first know how to look and perceive what they see well. They must then be able to transfer it onto paper in line with the result they want to achieve. Every line drawn is an effort that takes the designer forward in small steps in terms of pattern, composition, and overall design. Therefore, the maturity of the lines is also the result of accumulated experiences.

4. From Craft to CAD/CAM: The Weaving Design

In textile design, although computers bring certain conveniences, shortcuts, and applications, a designer who lacks sufficient drawing skills, cannot translate their ideas onto paper or screen, and cannot correctly use composition elements such as proportion, emphasis, balance, and unity will not achieve successful results. In other words, a textile designer who knows very well how to use CAD/CAM software and what the menus and icons do must also possess drawing skills, knowledge of color, and technique developed through education and experience in their field. This is because a textile designer must master every stage from the initial story of the targeted product, the mood board, the first sketch drawings, to the final finished product. They must be able to control the entire process and even have environmental awareness (Başaran, 2022).

To design a textile product, a designer needs knowledge of color, drawing skills, creativity, and technical knowledge of the product. For example, a pattern with a full rapport applied by printing on a duvet cover cannot be produced on a dobby loom, and even jacquard weaving has limitations. Therefore, the producibility of a design is directly proportional to its success. A successful design should be aesthetically pleasing and implementable with the correct production technique (Başaran, 2022).

With the correct production methods utilized today, textile design and production have been transformed by digital technologies, from concept development to final production. Design processes have been accelerated and streamlined through the use of CAD/CAM systems. Specific to weaving design CAD systems allow designers to create patterns, simulate fabric structures, and visualize the final product before production (Studd, 2002).

Weavers draw from a rich history and continuously innovate to create unique fabric blends and textures. They combine different fibers and weaves, and success in weave design depends on translating ideas both visually and texturally within the loom's limitations. The design process for woven fabrics varies based on the designer's context and the fabric's intended use. Inspiration often starts similarly to other fabric types, but the final translation is unique due to the woven fabric's nature. Color, texture, and feel are usually incorporated into the final product, except in jacquard fabrics, where designs can be directly represented.

4.1 Traditional Weaving Design before CAD/CAM

In the past, creating a weave design could take weeks or even months. The process began with the fabric designer drawing the design on graph or point paper, which was then meticulously transferred onto a board by a painter, using the exact colors specified (Figure 1). This structure correctly conveys that the panels were first sent to the mill and that, if any changes were needed, feedback was relayed back to the studio from the mill (Sinha, 2020). Designers used their experience and yarn knowledge to visualize their designs, which they could test directly on the loom. While these processes were lengthy, the available thinking time, knowledge base, and ability to check designs on the loom ensured a high rate of success (Briggs-Goode and Townsend, 2011).

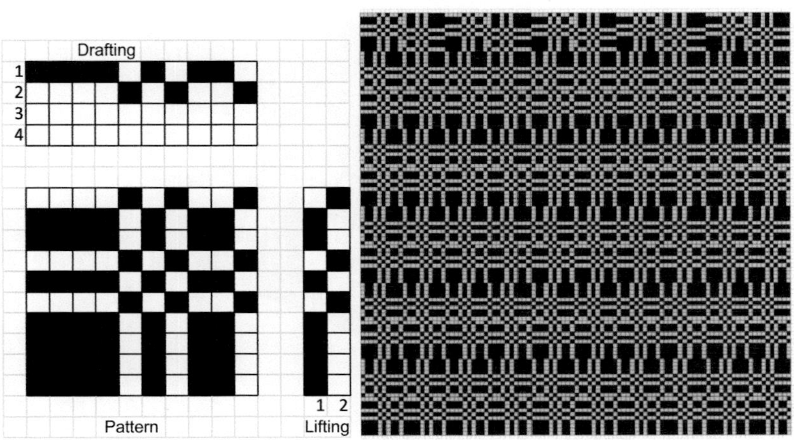

Fig. 1 Traditional weaving design process and color management (originally created by the author)

The Jacquard loom, invented by Joseph Marie Jacquard, a silk weaver from Lyon, introduced the first automated system for weaving intricate patterns into silk fabric. Jacquard's innovation centered around a series of linked punched cards, each containing part of the weave's design. These punched cards controlled the movement of warp threads by guiding hooks when lifting and lowering. This mechanized process allowed a single weaver to create intricate designs up to 24 times faster than two people working on a traditional loom (Nawab et al., 2017). What made Jacquard's loom revolutionary was not just the automation but also the concept of using punched cards to store and transmit pattern data. While earlier inventors had attempted to automate weaving, their mechanisms were inefficient and costly. Jacquard's use of punched cards was a breakthrough, simplifying the process and making intricate pattern weaving more accessible. This system inspired future technological advancements beyond the textile industry. One notable example is Charles Babbage, an English inventor from the mid-1800s. Babbage recognized that Jacquard's punched card system could be adapted for storing and processing data, not just weaving patterns. He incorporated a similar mechanism into his "Analytical Engine," a machine designed to calculate and sort information—considered a key ancestor of modern computers.

The success of the Jacquard loom also influenced the development of the dobby mechanism, a pattern-control system used in weaving plain and textured textiles. Though widely used in the British cotton and wool industries by the mid-1800s, the dobby loom's origins are less clear, with no definitive inventor or date recorded. The dobby is a shedding system, and looms with this system are designated as dobby looms, irrespective of the picking media (Figure 2). Dobby is a relatively complex shedding system, and it controls up to 32 heald shafts (Nawab et al., 2017).

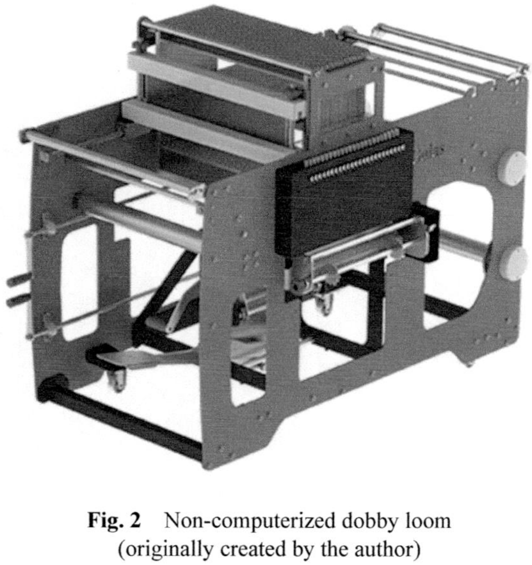

Fig. 2 Non-computerized dobby loom (originally created by the author)

Similar to Jacquard, the dobby mechanism significantly increased production speed. Designing patterns for either the Jacquard or dobby loom traditionally began by plotting the weave pattern on graph paper, which would then guide the punching of cards for the Jacquard or the pegging of lags for the dobby. This preparation process was time-consuming, especially for the dobby loom, where the pegged bars often needed to be reconfigured for each new production order. Full automation in weaving would take another century and a half to develop, but Jacquard's invention laid the groundwork for much of the progress that followed (George, 2020).

Before the advent of CAD/CAM systems, weaving design was primarily done manually and through mechanical processes. The design creation was laid on drawing by hand usually on graph paper. The motifs or patterns were then either woven directly on the loom or transferred to the loom mechanism by punch card system. And then, weaving processes were carried out to control design correction (Figure 3).

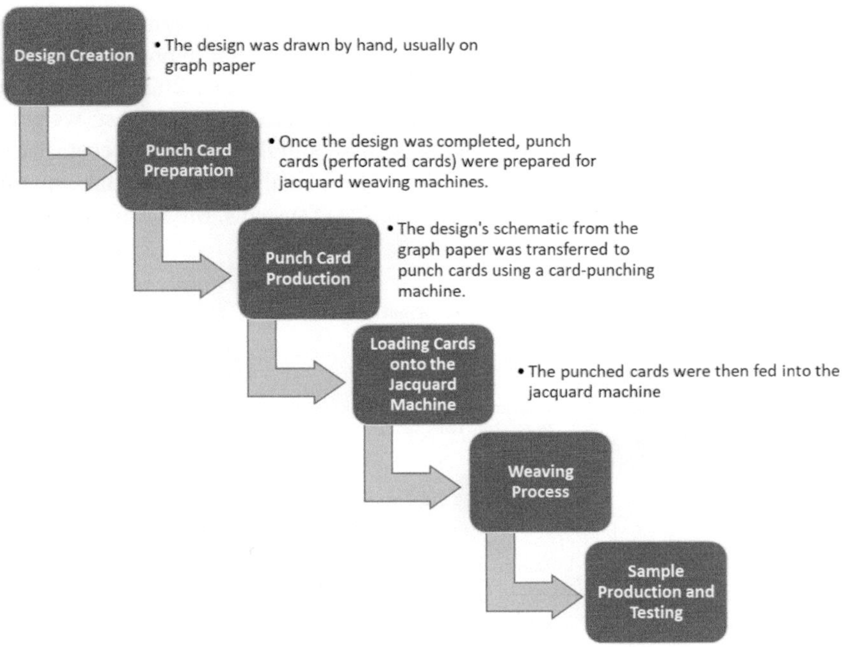

Fig. 3 Pre-CAD/CAM term Jacquard weaving design process (originally created by the author)

4.2 The Transformation of Weaving with CAD/CAM

The first CAD/CAM components for woven textiles made their debut at International Textile Machinery Association exhibition, ITMA 1983 in Milan, Italy. AVL Looms UK, a branch of AVL USA, showcased the first computer-interfaced handloom. Bonas^{TM} Machine Co. Ltd, a loom manufacturer from the UK, introduced the first electronic Jacquard loom. CAD design systems for jacquard textiles were also presented by EAT/Designscope™, Sophis Systems™, and ScotWeave™ at the 1983 ITMA. Among them, EAT™ and Sophis™ particularly set high standards in jacquard woven design systems. Today, EAT™ is considered a leading jacquard design system, widely used by manufacturers and educational institutions worldwide, known for its advanced 2D and 3D textile simulation capabilities. NedGraphics™ from the Netherlands showcased a carpet design system that featured a large monitor with a 256-color display. Since color digital printers were not yet available, the NedGraphics™ system projected its digital design onto a white carpet displayed on the wall of their booth. Several beta versions of digital design systems for textile printing, carpet design, jacquard design, and knitting were also displayed at ITMA 1983.

Today, with CAD/CAM technology, designers can directly choose the yarns, sketch the weave design on the computer, and easily send it to clients via print or email as a JPEG or other digital formats. Modifications can be made swiftly as needed (Hao and Ni, 2022). Advances in screen and printer calibration have introduced a "virtual" sampling method, though final approvals may still be required before moving to full-scale production (Sinha, 2020) (Figure 4). However, textile design phases are easier and quicker than in the past; weaving designers need to be multiskilled and understand both the supply and cost issues, the technicalities of the production such as fabric analysis, drafting the pattern, tensions, and yarn lengths required as well as CAD systems. Most aspiring textile

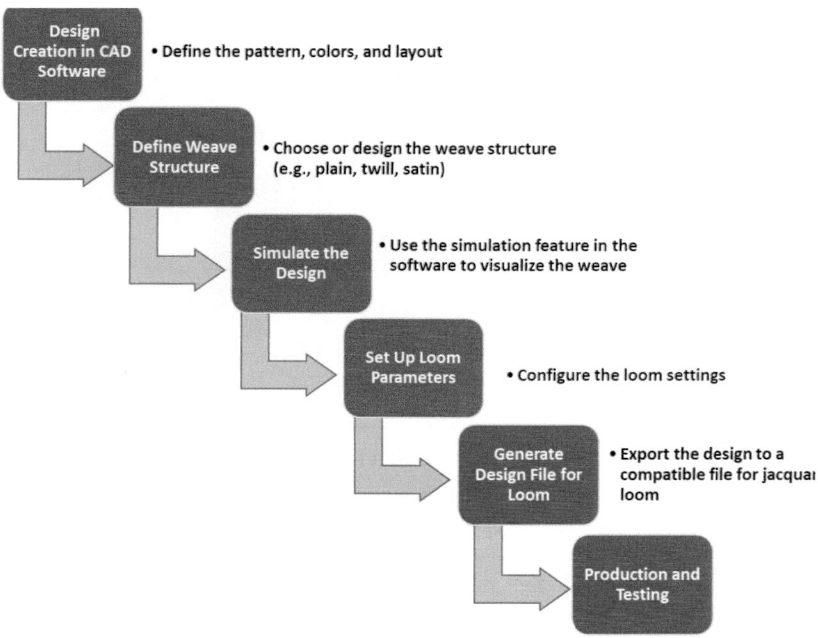

Fig. 4 Digital jacquard woven fabric design process (originally created by the author)

designers acquire their skills through universities or textile colleges. Mastery of CAD/CAM technology often relies on a solid understanding of traditional hand-weaving techniques, which typically requires a year of study before these digital tools can be effectively used to create woven designs.

While traditional weaving techniques lay a strong foundation, modern advancements in technology have revolutionized the way textile designers approach their craft. Today, designers increasingly rely on specialized software to streamline the design process, allowing for greater precision and creativity. These tools enable the visualization and manipulation of complex patterns that would be challenging to achieve by hand, merging the art of weaving with cutting-edge innovation. The commonly used weave CAD packages are: ScotWeave™, Textronics™, Arahne™, Penelope™, Booria™, NedGraphics™, and EAT™ (Table 1). Windows™ is the most widely used operating platform for for these softwares, though it is also compatible with Apple Mac™ and Linux™. Designs are typically accessible across different platforms, minimizing the risk of technology "lock-in." Generally, weaving software can handle designs for dobby, jacquard, and carpet patterns.

Table 1 Most popular weaving software and their modules (Sayem et al., 2023)

Software	*Modules*
EAT™	Dobby/Jacquard/Multilayer/3D/Knitting
NedGraphics™	Dobby/Jacquard/Textile Designer/2D or 3D geometry
ScotWeave™	Dobby/Yarn/Jacquard/2D or 3D geometry
Penelope™	Dobby/Jacquard/Terry
Point Carre™	Dobby/Jacquard/Textile Designer
TexGen™	2D/Multilayer/3D geometry
Textronics™	Dobby/Jacquard/Carpet
Arahne™	Dobby/Jacquard
DB Weave™	Dobby
Weave Point™	Dobby
WeaveMaker™	Dobby
Wise Tex™	Geometrical model of fabrics (woven, knitted, laminates, braided)
YXENDIS™	Dobby

As is seen from Table 1, design CAD systems have some modules such as dobby, Jacquard, 3D weave, and terry to ensure the support for different fabric design requirements. In addition to this, color library helps to create, save, and load the color pallets when required. The color specifications are also available, which allow the user to edit a certain color using its specifications. Also, weave library and yarn library help to design excellent weave fabric.

The common feature of the widely used design software in the market is that they enable virtual designing and support multi-colored designs. Also, ease of design change is the most prominent feature of the design systems. The traditional way of manual design of weave patterns requires time, skills, and patience. Even a small error or need for a change requires corrections, which could lead to mistakes in the looming process (Figure 5). The best possibility is to start the design from the beginning. With the use of a CAD system, every change is possible at any time. With the help of CAD technology, the design process has been significantly shortened.

CAD technology enables designers and manufacturers to make quick and efficient iterations on their designs without waiting for physical samples, thus significantly speeding up the pre-production process. At the same time, various fiber and yarn attributes, as well as weave and knit characteristics, can be tested within a selected range in the computer environment, allowing different combinations to be visualized without the need to produce trial fabrics. Designs can be viewed on high-resolution monitors, high-quality photographs can be generated, and outputs can be produced using plotters, thereby reducing costs considerably. These features make the design process more efficient and economical (Nawab et al., 2017).

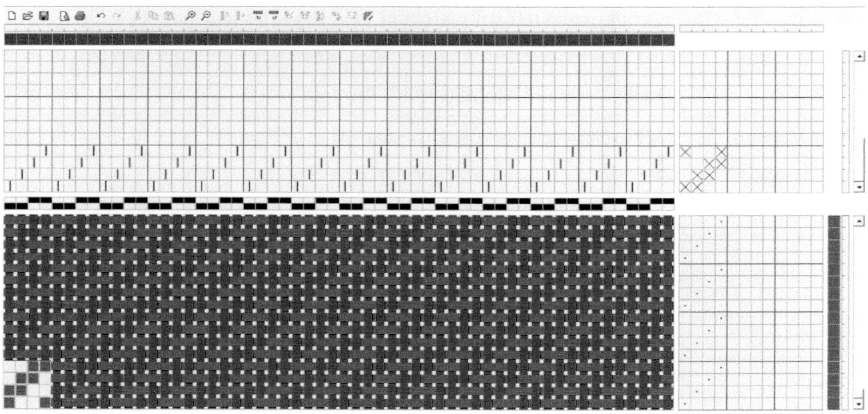

Fig. 5 DB™ weave design module—twill weave simulation (produced in DB™) (originally created by the author)

Arahne™, a company specializing in textile design software, assists fabric designers and weavers in addressing three main challenges: enhancing communication with customers via fabric simulations, reducing product launch times, and preventing production issues. ArahWeave™ is used by international weaving mills that design fabrics for renowned global brands and must consistently work within tight time and budget constraints. This comprehensive solution is also utilized in educational institutions, such as textile colleges and universities, where both designers and technical staff rely on it (Arahne, 1992). In the ArahWeave™ Weave Editor™ window, use the quadruple function to easily create quadruple weaves with either clockwise or counterclockwise rotation. This method is a traditional approach for designing complex weaves (Figure 6).

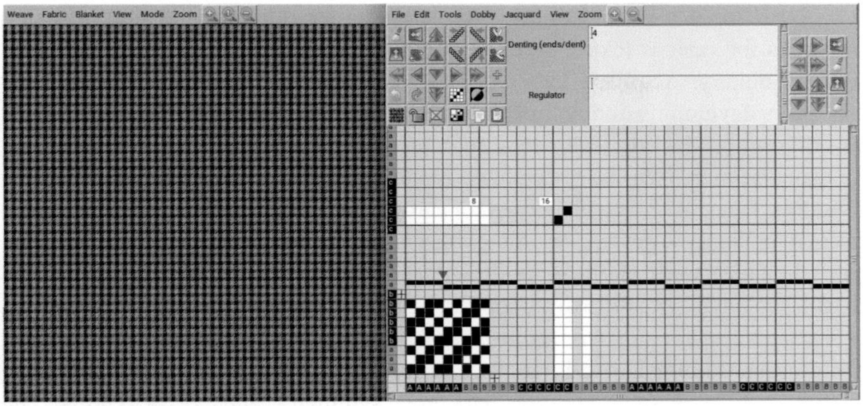

Fig. 6 Easy motif generation in ArahWeave™
(originally created by the author in ArahneWeave)

ArahWeave™ CAD program already has a large collection of weaves included in the standard installation. The entire weave database has about 41,000 weaves, including historical weaves from notable collections. The weaves were collected and digitized from known weave collection books including Serrure, Poma, Donat, Fressinet, Griswold, Gunetti, Posselt, and Pinchetti. The 1180 weaves collection comes from J. Moiret's twentieth century French book *Recueil d'Armures Fantaisies*, published in Lyon in 1908. The rotation feature of unit raport of pattern gives flexibility to designer; and color support is provided by Pantone Fashion, Home+interiors.

The EAT™ company, based in Germany, has been supporting the textile industry for over 30 years with software specifically designed for jacquard and dobby fabric production. Its key features include simplifying pattern creation with a variety of color and yarn options, and offering detailed work on weave structures. The software's simulation tool enhances the design process by accurately reflecting the fabric's real appearance, including details like shadows, light, and folds. This allows designers to visualize both the front and back of the fabric in a realistic way, aiding in decision-making and reducing errors before production (EAT, 1983).

One standout feature is EAT's 3D Mapp™, a virtual showroom tool that presents textile designs in their most lifelike form. Thanks to its high Dots per inch, DPI capability, it can simulate large environments without any loss in resolution, making it ideal for realistic presentations to clients or stakeholders. Additionally, the software is regularly updated to incorporate the latest trends in textile technology, ensuring that users always have access to the most efficient and cutting-edge tools. EAT™ also offers a specialized software called 3D Weave Composite™, which is designed for technical textiles with multi-layered structures. This tool allows designers to view detailed cross-sections of their designs and set specific parameters, making it ideal for complex projects in industries like automotive or aerospace textiles, where precision and structural integrity are key.

NedGraphics™ is a leading CAD software provider, specializing in design solutions for various textile sectors, including fashion, retail, home furnishings, and upholstery. Acquired by FOG Software Group in 2014, NedGraphics™ has been developing innovative solutions over the years. The software caters to different production techniques such as printing, knitting, carpet, dobby, and jacquard fabrics, offering specialized modules for each process. One of its key modules, Texcelle™, is widely used for carpet and jacquard fabric design. It allows users to create new designs or work with scanned hand drawings or fabric photos, providing a versatile platform for designers. Texcelle™ and its variations, such as TexFlash™ and Texcelle Pro™, support tasks like drawing for jacquard weaving, assigning yarns for single or multi-layer fabrics, and generating color variants for the designs (Figure 7). Additionally, users can arrange the pattern compositions with repeat options like straight, half-drop, symmetry, and mirror repeats. The design process is completed with the Loom Editor and Product Creator modules, which help refine and finalize the pattern. For further repeat functions and motif placement adjustments, the Design and Repeat module can also be utilized outside of Texcelle™, allowing for seamless integration of the design into the production workflow. In fabric simulations, fabric densities can be adjusted. The software's library includes accurate yarn sizes based on systems such as English cotton yarn count (Ne), denier (den), tex, metric number (Nm), worsted count, and more. For the use of colored warps and wefts, ready-made sets are available, or designers can create their own custom color schemes (Başaran, 2022).

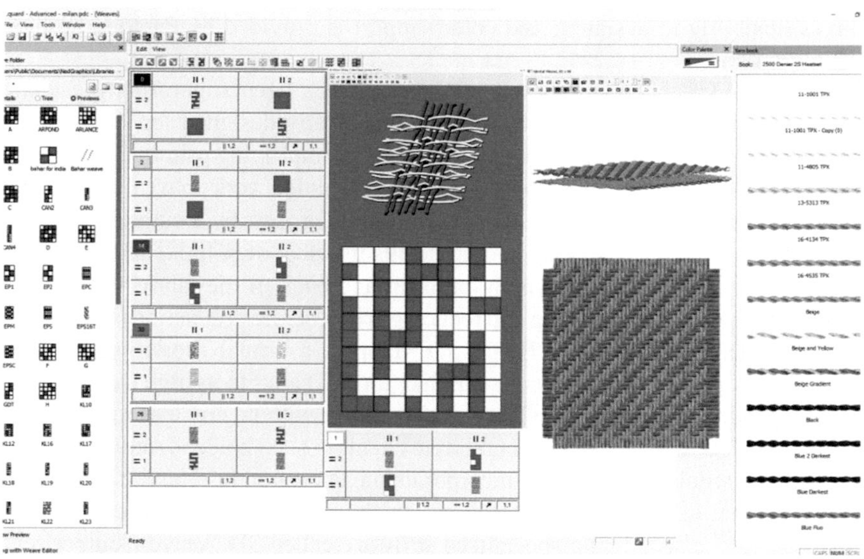

Fig. 7 NedGraphics™ design fabric from simple weaves to multi-layered structures (https://nedgraphics.com/" accessed on 21.04.2024)

In addition to visual design capabilities, ScotWeave™ has recently introduced software tailored to the specific requirements of highly technical weaving sectors, including automotive, aerospace, and medical industries. This software incorporates real-world yarn data, covering fiber types and the cross-sectional structure of weave architecture, allowing for the visualization of multi-layer fabrics, yarn interlacing, and anticipated tension calculations during weaving (Briggs-Goode and Townsend, 2011).

ScotWeave™ Technical Weaver is specifically developed to address the precise demands of technical textile manufacturers and engineers in industries such as industrial, commercial, medical, and geotextiles. This all-inclusive CAD/CAM software facilitates the design of dobby and Jacquard preforms, focusing on the critical aspects of structure and performance. Through detailed cross-sectional visualizations, it simplifies the creation of multi-layered architectures, incorporating features like stacking orders, stuffers, and varying densities (Figure 8). This allows for clear visualization of weft tows, binding warps, and stuffer warps necessary for constructing complex textile structures. The software provides tools for easily building intricate 3D structures in XYZ directions, with functionalities such as copy and paste to streamline the design process. Additionally, the weave notation is automatically generated from cross-sectional diagrams, ensuring an efficient workflow from design to production. Accurate yarn data, including yarn count, fiber type, and cross-sectional shapes, are integrated to produce detailed and realistic 3D weave schematics (ScotWeave, 1984).

ScotWeave™ also enables direct transmission of production data to electronic loom controllers like Bonas and Staubli, supporting up to 32,000 hooks for Jacquard looms via the ScotWeave™ Jacquard Looms program. Dobby looms can utilize up to 64 shafts, and the software provides support for Staubli bitmap information for multiple warps shed positions. This integration ensures precise execution of complex textile designs and efficient production workflows.

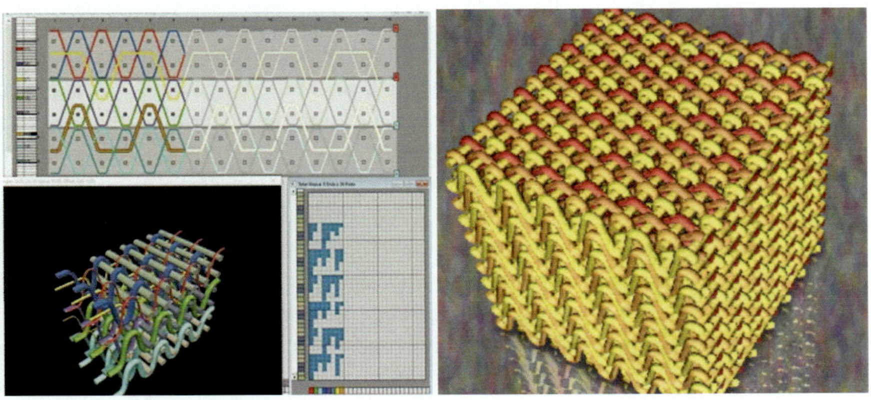

Fig. 8 ScotWeave™ technical weaving modeling (https://ScotWeave.com/" accessed on 21.04.2024)

The CAD/CAM design programs, developed through various collaborations to be comprehensive and responsive to the needs of the industry, are typically used by both industry and educational institutions. These systems are designed to be compatible with industrial weaving machines. Students who receive training in design software at educational institutions apply their experience within companies in the textile sector. In addition, there are software programs developed for sample weaving looms to bring sample production and individual design projects to life during the educational process. Ala and Celik (2024) present the overview of a CAD/CAM software called DNC.GARM.DOBBY WEAVING DESIGN that was developed using Java SE, for the purpose of designing weaves and also generating machine readable files for the semi-automated sampling looms, manufactured by the company Gulas Weaving Machines. The developed weaving design CAD/ CAM system does not have the capability to display design simulations unlike others.

In the field of woven fabric design, both aesthetic and texture considerations are enhanced through a combination of advanced software and technical modeling techniques. Software applications such as Adobe Illustrator™ and EAT™ are widely used for visualizing and creating intricate textile patterns and designs. These tools enable designers to efficiently experiment with various color schemes, textures, and patterns. Alongside these aesthetic tools, technical modeling plays a crucial role in analyzing the structural integrity and performance of woven fabrics. Techniques such as weave simulation and analysis involve assessing parameters like yarn density, weave structure, and fabric drape. These analyses provide insights into the fabric's mechanical properties, including its tensile strength, abrasion resistance, and dimensional stability. Additionally, tools such as MATLAB™ and ANSYS™ are employed for more in-depth computational analysis, allowing designers to predict the fabric's behavior under different conditions. This combination of creative software and technical analysis ensures that woven fabrics are not only visually appealing but also functionally robust and suitable for their intended applications. Fabric performance largely depends on the geometric structure of the yarns within it. Understanding how textiles respond to different types of loads—mechanical, thermal, or fluid—is crucial for designers. Geometric models of woven fabrics enable the estimation of certain structural and physical properties, such as areal mass and porosity. These models are especially valuable as they provide accurate geometric data on textile assemblies, supporting finite element analysis (FEA) for performance simulations.

TexGen™ is open-source software developed by the Polymer Composites Group™ at the University of Nottingham. Designed as a modeling pre-processor for textiles, TexGen™ supports a wide range of applications, including solid mechanics, fluid dynamics, and thermodynamics (Lin et al., 2011).

In their study, Stig and Hallström propose a finite element-based approach for the spatial modeling of 3D-woven textiles, utilizing TexGen™ software to define the initial geometry of yarn paths. This method accurately represents the internal strand geometry, ensuring that overlaps are avoided during the simulation (Figure 9). The approach is validated through a comparison with Computed Tomography Scan, CT scans of a carbon fiber weave, showing excellent correlation between the simulated and real geometries. Their robust method can be applied to predict mechanical properties and permeability of composite materials with 3D reinforcement (Stig and Hallström, 2012).

Fig. 9 2D and 3D fabric geometry modeling by TexGen™ (Orginally created by the author in TexGen)

5. Development in Knitting Design

Automatic knitting machines have been innovated since 1967, when Shima Seiki introduced the world's first fully automated collar knitting machine. Today, these machines are widely used in fabric manufacturing. Knitting is known for its high flexibility and ability to create various fabric structures by interlocking loops of yarn. Over the past 20 years, automated flat and circular knitting machines have become increasingly common in the knitwear and fabric industries, producing fabrics that are both soft and elastic. During the knitting process, yarn loops are interlocked using needles to create knit structures. In weft knitting, loops form horizontally across the fabric, while in warp knitting, they form vertically along its length. A basic knit loop consists of a head (H) and side leg (L), supported by a foot (F) and a sinker loop (S) at the base. The foot interlocks with the previous loop's head, and the yarn forms a full loop from one loop's foot into the next. The sinker loop, located at the bottom, connects with neighboring needle loops (Spencer, 2001). In a basic knit unit, three or more interlocked loops form in the same column (or wale) and join together in both the horizontal rows (course) and vertical columns (wale) to create the knit structure. The plain structure, or single jersey (SJ), is the most common and basic stitch type in knitting. It is a single-faced structure, where only one type of loop is visible on either side of the fabric (Ray, 2012). The curvature of the sinker loop base and top loops can be adjusted to achieve horizontal stretch, but the frictional forces between yarns can sometimes exceed the stretch strength at the binding areas. Yarn can then shift from the vertical side loops into the top and bottom loops. Plain knitted structures can stretch approximately 15%–20% in width and 5%–10% in length without significant yarn extension (Cooke, 2011).

The integration of CAD/CAM technologies into the textile industry, especially for knitting structures, has gained significant attention due to their transformative impact on design and production. CAD/CAM enables the creation of intricate knitted forms that are not only visually appealing but also functionally advanced, allowing for the customization of material properties and structural configurations (Sinke et al., 2023).

One of the key advantages of CAD/CAM in knitting is its ability to minimize production waste through precise digital manufacturing techniques. By controlling stitch distribution numerically, CAD/CAM technology allows for the creation of customized three-dimensional surfaces tailored to specific needs (Sinke et al., 2023). Additionally, the inherent flexibility of knitted materials gives designers the opportunity to explore innovative forms and functions that traditional woven textiles cannot accommodate, expanding the creative possibilities within textile design. Research emphasizes the role of CAD/CAM technologies in improving the efficiency and effectiveness of textile design. CAD/CAM systems streamline the design process while increasing the competitiveness of textile products by enabling rapid prototyping and customization in response to market demands (Liliana et al., 2019). This adaptability is crucial in an industry shaped by rapidly changing trends and consumer preferences. Moreover, integrating CAD/CAM in textile design allows designers to experiment with programmable materials that can dynamically respond to environmental stimuli, such as humidity, pushing the boundaries of what knitted structures can achieve.

5.1 CAD/CAM Integrated Whole-garment/Seamless Knitting Machines

In knitwear production, seamless garments are often referred to as "whole garments." These garments can be produced using both circular and flat knitting machines, with minimal or no cutting and sewing required. This type of production involves the use of multiple feeders and eliminates the need for seams, resulting in garments that fit the body well and are stretchable. Whole-garment machines are capable of creating intricate designs, transforming traditional 2D fabrics into 3D knitwear, including shaped and pleated fabrics (Seiki, 2004). Full-fashion knitting is similar to integral knitting, as all design elements are created directly on the machine. However, in full-fashion knitting, additional features like pockets, trims, and buttonholes can be incorporated during the knitting process. Both full-fashion and integral knitting reduce the amount of cutting and sewing needed to produce a whole garment (Anderson, 2008).

One of the earliest shaping techniques in knitting is known as integral knitting, or short row knitting. This method achieves shaping by holding stitches at the fabric edges using short needle strokes, a technique called flechage. Unlike traditional sewing, pre-shaped panels are assembled using a linking process with chain stitches, where each stitch holds the loops of adjoining panels (Power, 2008). In flatbed knitting machines, the needles are arranged in an inverted "V" shape, positioned opposite each other on two different beds (Spencer, 2001). Similarly, in machines designed for knitting whole garments, the needles are arranged in an "X" shape with four beds. Each set of knitting needles faces the opposite set, allowing for easy loop transfers between them. The two sets of needles work sequentially to transfer loops from one bed to the other. This arrangement also permits the transfer of loops between adjacent needles (Spencer, 2001). There are four main types of loop transfer techniques: rib loop transfer, needle-to-adjacent needle transfer, lace stitch transfer, and sinker loop transfer (Onal and Yıldırım, 2012). However, all these techniques can reduce the structural stability of the knitted fabric (Ionesi et al., 2010). High-fashion garments produced using fully fashioned techniques often feature pre-shaped armholes created on flatbed machines, but the shoulder shaping is typically refined by cutting and joining with an overlock seam. Some whole-garment knitting machine examples are now discussed.

SHIMA SEIKI™

Researchers are concentrating on advancing design capabilities to increase productivity and enhance fabric quality. The WHOLEGARMENT™ knitting machines by Shima Seiki™ offer gauges of 5–18 needles per inch and widths of 50–80 inches. To enable shaping and to improve both quality and efficiency, Shima Seiki has reengineered the traditional latch needle (Millington, 1996; Anderson, 2008). Their Intelligent Digital Stitch Control Device (IDSCS), seamlessly built into the knitting machine, ensures uniform loop formation.

The IDSCS calculates yarn consumption for each stitch, programs the desired loop length, and reduces yarn stress. It also allows for color changes during the knitting process, accommodating multi-colored designs. The use of air jet splicing to eliminate knots significantly enhances fabric quality, particularly in garments with intricate intarsia and stripe patterns. Users can also access a database to select from various collars, cuffs, plackets, and other garment components (Millington, 1996).

The ultra-fine gauge flat knitting by Shima Seiki™, introduced in 2011, enables the production of lightweight flat knitted fabrics. These lightweight fabrics are typically used for garments like T-shirts and lingerie. The Shima Seiki™ model SWG 154 S21 demonstrated a significant advancement in fully fashioned knitwear, with the ability to produce 18–26 gauge single jersey and 18–20 gauge rib fabrics (Pasek and Starosielski, 2021). While flat knitting traditionally uses plied yarns, the machine's modified needle and cam arrangement allows it to handle single yarns made from delicate materials as well (Pasek and Starosielski, 2021)

SANTONI™

Lonati™ owns the brands Sangiacomo™ and Santoni™, both of which are capable of producing shaped tubes. Santoni™ machines are more widely used in the market compared to Sangiacomo™ machines. Unlike Shima Seiki™ and Stoll™ machines, which have a flat V-bed configuration, Santoni™ and Sangiacomo™ machines feature a circular knitting bed. Although sewing is necessary to complete garments, Santoni™ and Sangiacomo™ machines are notably fast, with shirts taking just 2–4 minutes to produce, while seamless shirts on full garment knitting machines can take hours. Shape creation is facilitated through CAD by using combinations of floating, tucking, and knit stitches. Santoni^{TM} has introduced in 2004 two new machines: one with advanced design capabilities and another that produces shapes via reciprocating motion (Spencer, 1989; Penciuc et al., 2010). The SM4TL2 machine can create all shapes through reciprocating motion, moving in opposite directions like a sock machine. Santoni^{TM} machines can knit up to 16 colors per course, with most fabrics being subsequently piece-dyed (Santoni, 2004).

STOLL^{TM}

Stoll^{TM} produces flatbed knitting machines capable of creating seamless whole garments. These machines are available in various gauges and can knit up to 32 colours, with widths ranging from 72 inches to 84 inches. They can handle different knitting constructions and produce both coarse and fine areas within a garment. Stoll^{TM} machines are ideal for product development due to their design flexibility and versatility, and they require fewer yarn packages. The CAD system provided by Stoll^{TM} is known for its simplicity (Schlotterer, 1998). During garment development, two computer screens are used: one displays technical information while the other shows the fabric design. The yarn tensioning system is highly reliable and adaptable, even accommodating challenging yarns such as strong ones with many knots (Schlotterer, 1998). The 84-inch model offers a range of gauges and is equipped with intarsia carriers to support more complex design possibilities (Anderson, 2008).

Conclusion

The continuous effort to introduce technological advancements in the fabric manufacturing industry is not driven solely by the pursuit of high production and profit. Instead, it also aims to establish a reputation as a Green Manufacturing (GM) entity. The objective of this GM approach is to promote the sustainability of both industry and the environment. This may be achieved by conserving materials, reducing costs and energy consumption, and preventing pollution and environmental harm through innovations and improvements at every stage of the production process. Utilizing CAD/CAM technologies in textile production stages may have a big contribution toward reducing costs and energy consumption in a sustainable manner.

This study has been undertaken in order to emphasize some digital design strategies in the textile and fashion sectors considering CAD technology. CAD is extensively used in the textile industry across various departments such as spinning, weaving, finishing, and testing. It plays a key role in developing, editing, and viewing 2D and 3D images from different angles. In the finishing process, CAD/CAM makes it easy to experiment with different finishes, while in the weaving department it enables the creation of realistic designs that can be presented to clients, giving them a clear preview of the final product. Overall, these systems allow for the rapid production of many products in a short time, delivering high-quality results with minimal human error. One of the key benefits of CAD in most textile departments is the reduction in labour costs.

The advantages of utilizing CAD/CAM have grown significantly over time in the textile industry. Simulating knitwear or woven fabric using CAD/CAM is more efficient and provides a realistic approach to developing new designs. This reduces the need for multiple physical trials on machines, saving both time and money. Additionally, CAD minimizes errors, allowing the design to be easily transferred between systems. Without CAD/CAM, poor design quality and low production rates are likely. Therefore, to produce an optimal product, manufacturers must be encouraged to use CAD/CAM and they must also be familiar with the CAD systems of every machine they use.

References

Ala, D.M. and Celik, N. 2024. A Computer Software Developed for Designing Woven Patterns and Generating Machine Readable Files for Sampling Looms. *The Journal of the Textile Institute* 115(9): 1423–1431.

Anderson, K. 2008. *Seamless Technology*. Tech Exchange.

Angelova, R.A. and Sofronova, D. 2021. Application of CAD/CAM Systems in the Design of Woven Textiles. *2021 IEEE 7th International Conference on Computing, Engineering and Design (ICCED)*, 1–6.

Arahne. 1992. Available at https://www.arahne.si/tr/about/companyandfounders/ (Last access: 10.11.2024).

Başaran, F.N. 2019. *Simple Weaving Techniques*. Nobel Academic Publishing, Karınca Agency.

Başaran, F.N. 2022. CAD systems used in weaving design. *Textile and Enginnering,* 29(127): 168–184.

Briggs-Goode, A. and Townsend, K. 2011. *Textile Design Principles, Advances and Applications* (The Textile Institute). Woodhead Publishing Limited.

Collier, B.J. and Collier, J.R. 1990. CAD/CAM in the Textile and Apparel Industry. *Clothing and Textiles Research Journal* 8(3): 7–13.

Cooke, B. 2011. The Physical Properties of Weft Knitted Structures. In K. Au and Textile Institute Publishing (eds), *Advances in Knitting Technology*, 37–49. Cambridge: Woodhead Publishing.

Dai, X. and Hong, Y. 2024. Fabric Mechanical Parameters for 3D Cloth Simulation in Apparel CAD: A Systematic Review. *Computer-Aided Design* 167: 103638.

Dwivedi, A. and Dwivedi, A. 2013. Role of Computer and Automation in Design and Manufacturing for Mechanical and Textile Industries: CAD/CAM. *International Journal of Innovative Technology and Exploring Engineering* 3(3).

EAT. 1983. Available at https://www.designscopecompany.com/ (Access date: 10.11.2024).

George, P.F. 2020. The Digital Dawn: Introducing Computer-aided Design to Woven Textiles 1966–2000 [Master of Arts]. S.U.N.Y. Fashion Institute of Technology.

Gürcüm, B.H. and Üner, İ. 2016. Application of Conceptual Design in Textile Design: A Sample Fabric Collection. *Art Journal of SDÜ Art-E* 9(17): 26–52.

Hao, M. and Ni, T. 2022. Application of CAD Technology in Textile Art Design. *Computer-aided Design and Applications* 19(S8): 11–22.

Ionesi, D., Ciobanu, R., Vircan, A., Blaga, M. and Budulan, C. 2010. Three-dimensional Knitted Fabric with Technical Destination. *Buletinul Institutului Politehnic Din Iaşi* LVI(LX): 3.

Lin, H., Brown, L.P. and Long, A.C. 2011. Modelling and Simulating Textile Structures Using Texgen. *Advanced Materials Research* 331: 44–47.

Liliana, I., Mutlu, M.M., Efendioglu, N.O., Simona, T., Garcia, P.D. and Soler, M. (2019). Computer aided design of knitted and woven fabrics and virtual garment simulation. Industria Textila, 70(6).

Millington, J. 1996. Knitting: A High Technology Industry. *Textile Outlook International*, 98–112.

Nawab, Y., Hamdani, T. and Shaker, K., eds. 2017. *Structural Textile Design: Interlacing and Interlooping*. CRC Press.

Onal, L. and Yıldırım, M. 2012. Comfort Properties of Functional Three-dimensional Knitted Spacer Fabrics for Home-textile Applications. *Textile Research Journal* 82(17): 1751–1764.

Pasek, A. and Starosielski, N. 2021. Fiber, Fingers, and Code: Manufacturing Software and Seamlessness in the Garment Industry. *Television & New Media* 22(5): 465–481.

Penciuc, M., Blaga, M. and Ciobanu, R. 2010. Principle of Creating 3D Effects on Knitted Fabrics Developed on Electronic Flat Knitting Machines. *Buletinul Institutului Politechnic Din Iasi* LVI (LX): 4.

Power, J. 2008. Developments in Apparel Knitting Technology. In C. Fairhurst (ed.), *Advances in Apparel Production*, 178–196.Cambridge, UK: Woodhead Publishing

Ray, S.C. 2012. *Fundamentals and Advances in Knitting Technology*, 1st ed. New Delhi: Woodhead Publishing.

Santoni. 2004. Know-How. Santoni: Gruppo Lonati.

Sayem, A.S.M., Chakraborty, S., Hoque, S.A., Saha, K., Mica, M.T. and Ahsan, M. (2023). Defining Digital Fashion and Tracking the Developments in Relevant Technologies. *In Digital Fashion Innovations* (pp. 3–14). CRC Press.

Sayem, A.S.M. 2023. *Digital Fashion Innovations: Advances in Design, Simulation, and Industry*, 1st ed. CRC Press.

Schlotterer, H. 1998. Stoll-pattern Innovations from Pitti Imagine Filati. *Knitting Technology* 20(3): 97–99.

ScotWeave. 1984. Available at https://www.scotweave.com/ (Last access: 10.11.2024).

Seiki, S. 2004. *Creating Shaping Guide. SDS-ONE Knit Paint (Shaping HelpE)*. Version A38. https://www.shimaseiki.com/product/design/system/, (Last accessed: 20.08.2024).

Singh, I., Singh, K. and Sethi, N. 2022. CAD/CAM—Historical Aspects, Processing and Various Systems: A Literature Review. *IAR Journal of Medical Sciences* 3(1): 1–6.

Sinha, P. 2020. CAD/CAM in the Woven Textiles Industry. In K.L. Gandhi (ed.), *Woven Textiles*, 273–289. Elsevier.

Sinke, Y., Ramsgaard Thomsen, M., Tamke, M. (2024). Integrated Design Models for Materially Differentiated Knitted Textile Membranes as the Means to Sustainable Material Culture Within Membrane Architecture. In: Thomsen, M.R., Ratti, C., Tamke, M. (eds) Design for Rethinking Resources. UIA 2023. Sustainable Development Goals Series. Springer, Cham. https://doi.org/10.1007/978-3-031-36554-6_23

Spencer, D. 1989. *Knitting Technology*. Cambridge: Woodhead Publishing Limited.

Spencer, D.J. 2001. A comprehensive handbook and practical guide. *Knitting Technology,* 3rd Edition, Woodhead Publishing Ltd.

Stig, F. and Hallström, S. 2012. Spatial Modelling of 3D-woven Textiles. *Composite Structures* 94(5): 1495–1502.

Studd, R. 2002. The Textile Design Process. *The Design Journal* 5(1): 35–49.

Tao, Y., Lu, N., Zhang, C., Wang, G., Yao, C. and Ying, F. 2016. Compuwoven: A Computer-aided Fabrication Approach to Hand-woven Craft. In *Proceedings of the 2016 CHI Conference Extended Abstracts on Human Factors in Computing Systems*, 2328–2333.

TexGen. 2007. Available at https://texgen.sourceforge.io/index.php/Main_Page (Access date: 10.11.2024)

12 Digital Tools to Improve Aesthetic and Functionality of Fashion Products: Footwear

Tatjana Spahiu[1*], Henrique Almeida[2], Liliia Chertenko[3] and Ermira Shehi[4]

1. Introduction

Product design is a creative task that implies new ideas or improvements of existing ones. Creating products that are functional or fulfill consumers' requirements is a process that involves following various steps to meet these requirements. The whole process of product design includes all the steps followed from market research to the final product. All the steps from 2D sketches, 3D computer-aided design (CAD) modeling, simulation, etc., are needed to create a prototype/product to fulfill customer needs. These are continuously changing and linked with various applications. Highly important applications are addressed for specific users. Obviously, consumers' requirements have evolved over time, and the products offered try to accomplish not only the functional part, but to also include the aesthetic part. As a result, more and more consumers are involved in the product design process.

Footwear design and production have changed over time, thanks to technological advances at every stage of production. As a mainly manpower-based industry and, of course, considering the costs related to product development, today the integration of computer-aided design/manufacturing (CAD/CAM) systems in footwear production has brought many advantages, i.e. 3D technologies, part of product design in footwear industry as modeling, simulation, and visualization. These digital tools improve the whole process with regard to related costs, time, waste reduction, etc.

The present research introduces product design methodologies in the footwear industry. As one of the main processes of product development, it plays an important role by becoming a decisive factor in producing footwear with adequate comfort and sustainable production.

1 *Associate Professor*, Polytechnic University of Tirana.
2 *Associate Professor*, School of Technology and Management, Polytechnic Institute of Leiria.
3 *Associate Professor*, Kyiv National University of Technologies and Design.
4 *Professor*, Polytechnic University of Tirana
Email: henrique.almeida@ipleiria.pt[2]; lily-che@ukr.net[3]; eshehi@fim.edu.al[4]
* Corresponding author: tspahiu@fim.edu.al

A rapidly growing awareness of sustainability issues and impact of the footwear industry is receiving attention from companies operating in this sector and dealing with these issues. This chapter discusses the attempts to evaluate sustainability as an emerging need in the footwear industry by bringing it about through various applications. Obviously, they are linked with technological advancement, occurring continuously, and are known under the term Industry 4.0.

The most important properties of footwear such as creativity, functionality, sustainability, and customization are provided by a complex set of characteristics as described by the following statements: creativity plays an essential role in the product design process. It must meet requirements as an original design and follow the latest trends. Regarding the functionality of footwear products, they must fulfill the requirements of quality, regarding fit and comfort for consumers' feet. These last two requirements can be achieved by customization, which is a growing trend, including in footwear products. Customization can include aesthetic, fit, and comfort for individualized shapes. Moreover, customization can enhance sustainability of footwear production, by reducing the number of footwear products and the use of recycled materials. All these properties can be improved using digital technologies such as 2D/3D modeling and additive manufacturing (Figure 1).

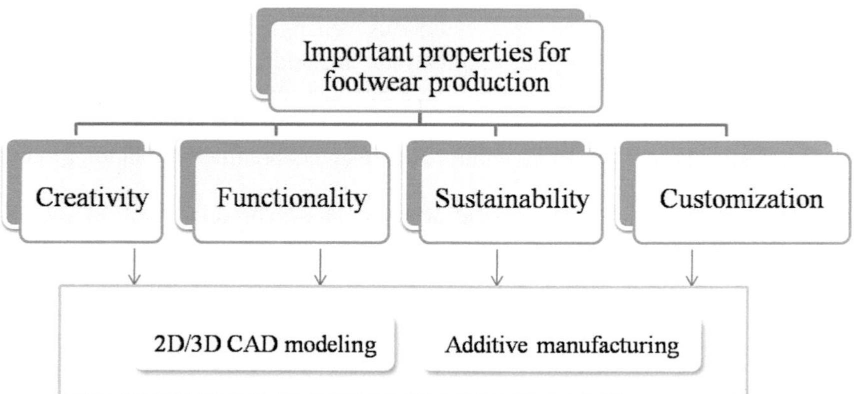

Fig. 1 Important properties for footwear production

2. Footwear Products and Role of Design in Comfort Perceived

Footwear products are considered the most important wearable items, especially related to functionality and comfort that they should offer. Comfort is a term related to subjective assessment from consumers on how they perceive it. It is linked with materials used and product shape or design. Producing footwear products with the right fit is a complicated process due to the complex geometry of the foot. Footwear products, in contrast with other items, are modeled in the same way by traditional methods and advanced 3D modeling software over the last shape, which is even called as the heart of shoe production. It is said that should be the translated form of the shoe, but not exactly. In other words, the last shape is responsible for the fit and comfort of the produced shoes.

Traditionally, lasts shapes are created by experienced people working in this sector and having good knowledge of human feet. But due to the introduction of digitalization, the process of last designing is much shorter, more precise, and includes virtual simulation. It is of high importance to have a preliminary evaluation of the products prior to their production. It allows the pattern maker, also known as technical designer, to add extra features that include various changes to achieve the right fit required by the consumers.

Wearing shoes without the right fit impacts our health. Evidence found by various researchers identifying the role of shoes and their impact on our health demonstrates problems that are evident in various parts of the body. The nature of the foot and its linkage with other parts of the body confirm this evidence.

Consumers have become more aware about the role of shoes regarding foot comfort and whole-body health by changing or using products that are more comfortable such as sport shoes.

3. Factors Affecting Footwear Comfort

Various consumers experience different problems with footwear products during their lifetime. There are different factors that indicate footwear comfort. These factors are related to the last shapes and materials used for their production. But, on the other hand, comfort is defined as perceived sensation from the user. It can be affected by shoe microclimate, such as temperature and humidity (West et al., 2019). Studies conducted to evaluate comfort for shoes indicate various characteristics such as suitable heel height (Goonetilleke and Karwowski, 2016), good fit in the forefoot region, and attractiveness as well (Au et al., 2005).

Shoe inserts are another part of the shoe responsible for comfort. Their texture can change gait pattern and can be important to determine movement strategies (Nurse et al., 2005). Comfortable shoes are the ones with the correct shape of the last. It is mandatory to create a last with the correct shape based on the anthropometric data of consumers' feet. A shoe last is a complex shape due to its free form geometry. To model a new shoe last, an existing shoe last can be used as a reference and its virtual copy can be created based on reverse engineering method. Later, using digital tools, the shape of the virtual last is merged and compared with the shape of the consumer's foot. The parameters of shoe lasts are generated by features based on the results of 3D scanning, and considering the practical experience, the most important footwear products are fashionable footwear, everyday comfort footwear, orthopedic footwear, and special footwear for increased comfort. Figure 2 depicts the shoe last design according to two different ways. The first one is based on the adapted shape of shoe last that is designed according to the main foot parameters. Meanwhile, the second case creates individual shapes of shoe lasts that are designed according to the shape and all measurements of the foot (Figure 2).

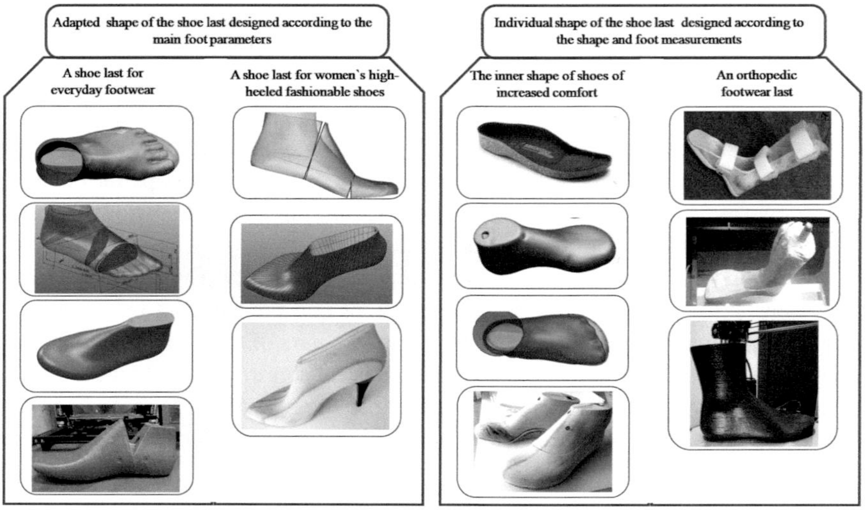

Fig. 2 Shoe last design according to foot parameters and according to shape and foot measurements

Analyzing foot pressure distribution is crucial for data collection to improve human performance. These data provide important information about static and dynamic weight distribution of our body under different activities performed during our daily life. The technology used helps to register and analyze these data reported in real time not only for athletes but also for people with various foot problems. Figure 3 depicts pressure maps of the feet taken with T-Plate for two different participants.

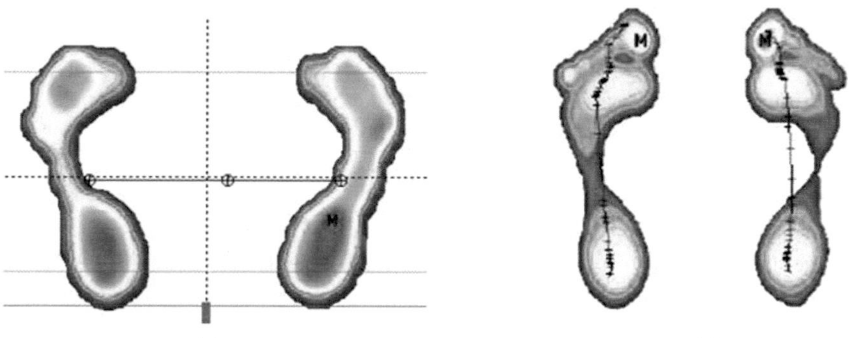

Fig. 3 Static analysis of the feet (a) participant 1 and (b) participant 2

Figure 4 depicts the dynamic analysis for both feet for the second participant. This measure is valuable in terms of forces, which act in various parts of the foot.

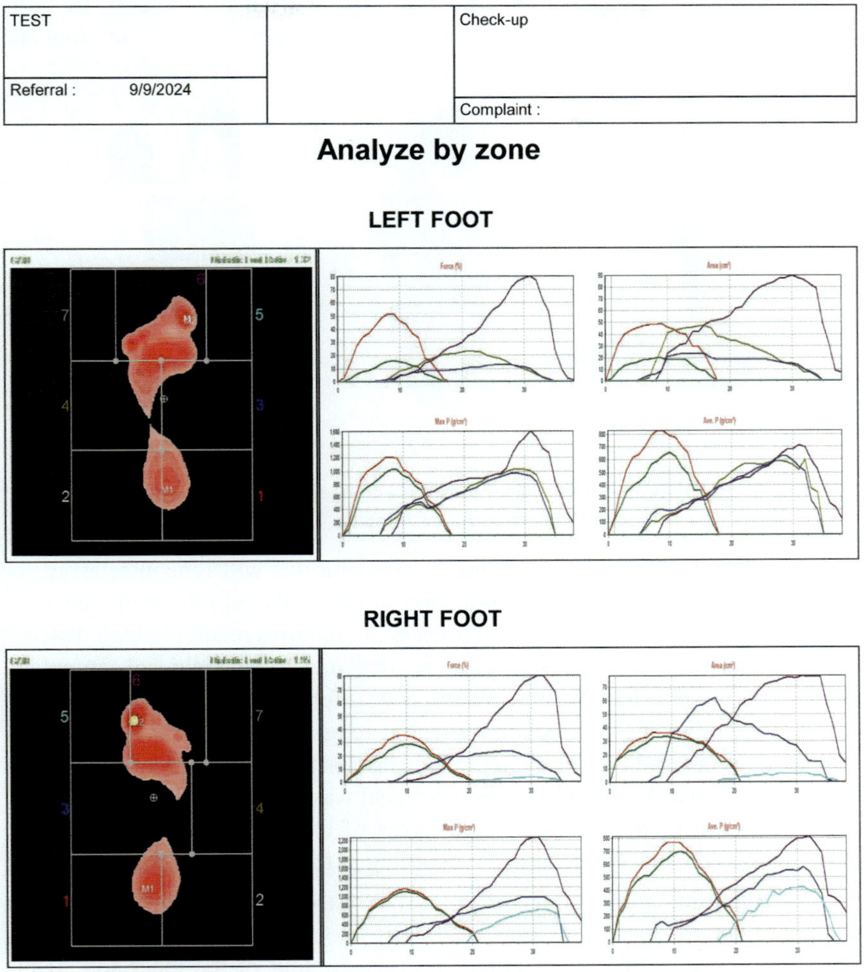

Fig. 4 Feet dynamic analysis of the second participant

During the gait process, the higher the pressure points that the feet apply on the walking surface, the higher the discomfort. To overcome the discomfort, the insoles should be designed considering two design schemes: (a) the pressure map showing the force distribution during static pose of the participant and when walking bare foot, and (b) based on specific forces on the foot (the force sensors were used at nine different points of the foot area) (Almeida et al., 2019). During this process, the participant wore casual shoes (Figure 5). From the first test conducted on the static analyses, the resulting pressure distribution map is presented in Figure 6. In the design process, the insole was divided into nine parts (Figure 5b). This division was important to evaluate for each single part the mechanical characteristics according to the pressure distribution by changing material used for production or manufacturing parameters. The resulting insole produced was later tested by the participant to evaluate their comfort. During the evaluation process, from the study, the participant exerted less force during the gait analysis of the left foot and the force values were reduced from pick values.

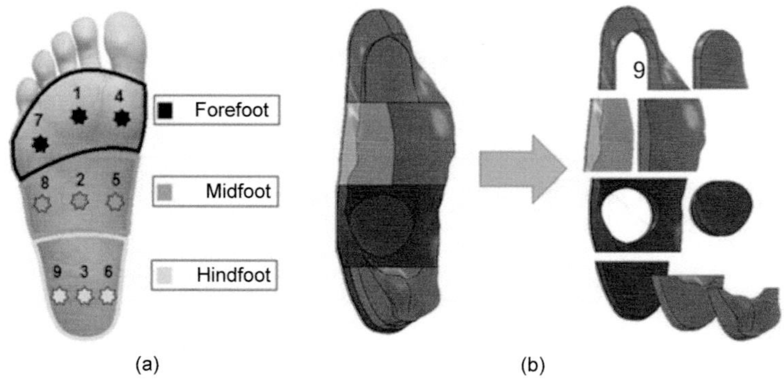

Fig. 5 (a) Division of foot plantar sensor positions outlines and (b) Insole design (Henrique Amorim Almeida, 2019)

The classical methods of making orthoses/insoles will remain relevant for a long time. The layer-by-layer lamination of materials with different densities, followed by thermal and mechanical processing, requires relatively high material costs (a variety of materials and tools are needed) and proper technical equipment. 3D printing of custom orthopedic insoles may offer advantages, such as reduced time costs, automated production, potential long-term cost reduction through equipment usage, and the ability to produce insoles that better match the contours of the foot. By using a single material, our goal was to achieve different physical and mechanical properties in different areas of the product. The insole shape modeling was done in Rhinoceros™ according to the following algorithm:

- designing the upper surface of the insole based on the 3D scan of the foot imprint on the foam;
- fitting the constructed 3D surface into the contour of the inner sole of the footwear;
- aligning the resulting model with the foot pressure map to determine areas with high pressure;
- creating counters that limit areas with high pressure;
- constructing surfaces to slice the 3D insole model into individual areas;
- save the individual areas of the insole;
- convert and import them into Cura™ slicing software; and
- selecting 3D printing parameters according to each individual area.

Figure 7 presents pictures during the process for insole modeling. The first step includes creating a digital copy of the bottom part of the foot. The second step deals with the elaboration of the 3D model of the bottom part of the foot and

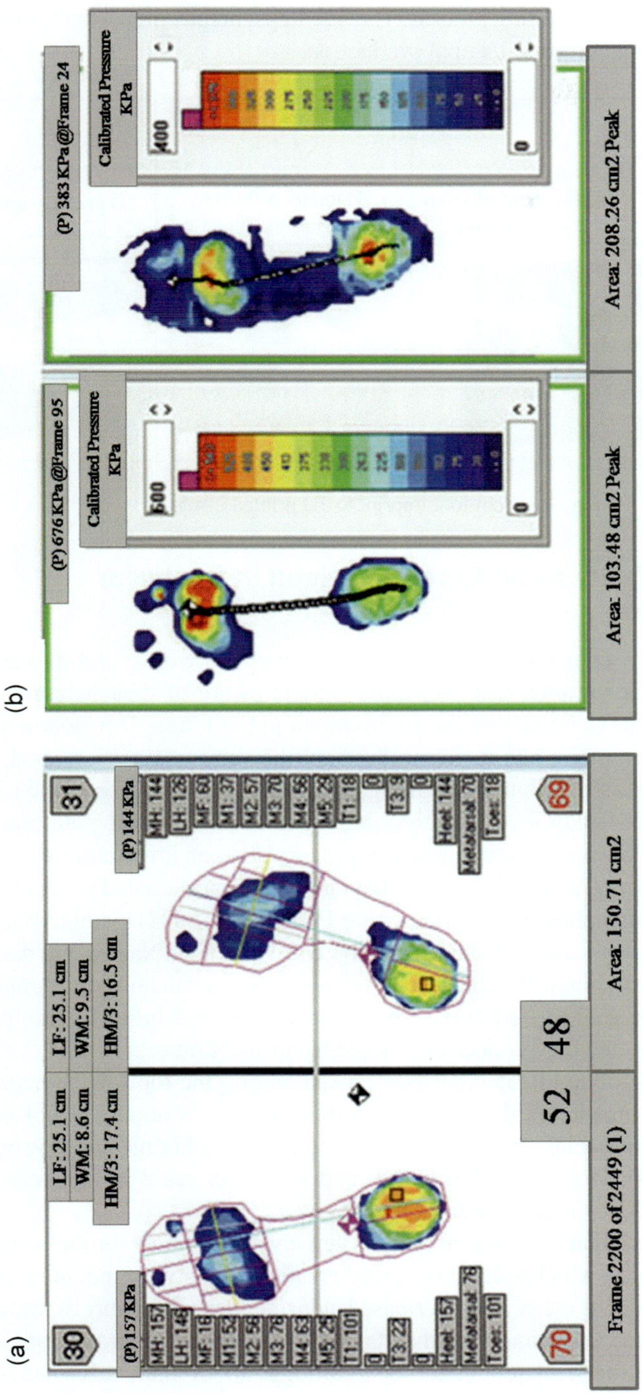

Fig. 6 (a) Plantar pressure distribution mapping of the foot during static state and (b) Plantar pressure distribution mapping for the left foot during walking analysis without insole (left) and with the produced insole (right) (Henrique Amorim Almeida, 2019).

preparing it for the 3D printing process. The last steps include preparing the model for 3D printing machine and its final production.

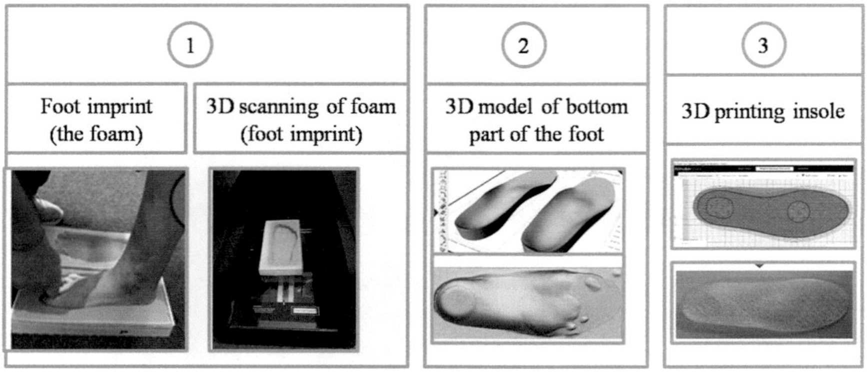

Fig. 7 From foot imprint to 3D printed insole

4. Sustainability as an Emerging Need in Footwear Production

Sustainability is a common goal for people all over the world, and its various definitions have a common task—to make people aware of their actions so as not to have a negative impact on the environment in the future. Developments in technology have brought about overproduction; the various materials used, and the means of product distribution, retailing, and usage—in other words, the whole product life cycle—play a role in contributing to environmental pollution. To maintain sustainable production, the design process plays an important role, as the core of the circular economy. It is evaluated that 80% of sustainable production is defined at the design phase (Ellen MacArthur Foundation, 2022) by underlying as the most important process in product development. The United Nations has defined various goals that should be considered for securing sustainable development. Among them, the goal related to industries, innovation, and infrastructure plays a key role in sustainable production, including in the footwear sector (United Nations, 2021). As one of the industries ranked among the top for the negative impact on environment, textile production is responsible for about 20% of water pollution. This is especially related to the dyeing process and finishing (European Parliament, 2024). Meanwhile, business is expected to increase and grow to 147 million tons in 2030 (Geneva Environment Network, 2024).

The increased focus on sustainability is perceived as a boost for the adoption of new technologies, which are sustainability-enablers in footwear manufacturing. Footwear production is mainly a power-based industry due to the nature of footwear products. Of course, over time, the advantages offered by the integration of CAD/CAM systems in footwear production have brought about many advantages. At first, they were mainly focused on 2D patterns and have evolved with time by including 3D technologies for the process of product development in the footwear industry. Modeling, simulation, visualization, and so on, offer digital tools that improve the whole process regarding costs, time, waste reduction, etc. There is a growing number of well-known companies operating in footwear production that are trying to push further the integration of sustainable processes and materials in the footwear industry. In this scenario, product design becomes a decisive factor to produce footwear with adequate comfort and sustainable production. Understanding consumer behavior on the role of sustainability in buying decisions has been the focus on several researches. Consumers exhibit a positive attitude toward sustainability and eco-fashion but these factors are less important than other factors—i.e. price, value, size, quality, style, convenience of purchase, materials, and many others—that prevail while making their buying decisions (Mandarić et al., 2022). Based on other studies, this can be attributed to lack of information, education, and transparency among consumers, where companies can play a crucial role in educating consumers to implement sustainability in their buying choices (Pereira et al., 2021). The correlation between educational level and knowledge of sustainable fashion shows a high purchase intention for these products (Han et al., 2024). These highlight the need to enhance the efforts toward consumers' education for the high impact of their purchases on sustainability (Papamichael et al., 2024) and the need for strategies to support environmental consciousness (Lopes et al., 2024), where marketing sustainable fashion can contribute to consumer awareness and reduce the negative impact of their buying choices (Ray and Nayak, 2023).

The findings may differ for the footwear industry with sustainability being one of the main factors taken into consideration or ranked as the least important factor (Polese et al., 2019) or not considered at all by the new generation (Bernardes et al., 2018). Among the three pillars of sustainability, studies have found that social sustainability has resulted in the most important impact on the supply chain of the footwear industry (Munny et al., 2019).

In the footwear industry, leather processing through chrome tanning releases waste that has a negative impact on the environment. Among the various ways to achieve sustainable production in the footwear industry, one of the steps is developing a new tanning technology and using the waste generated in cosmetics, agriculture, etc. (Deselnicu, 2014). Solutions can be found by focusing on design principles and circular economy to come out with reduced waste and negative impact at a high level (Marques et al., 2017).

Changes in footwear production are also related to the materials used that contain of materials like plastics, rubber, leather, and textiles, which in most cases makes it impossible to recycle them. Running shoes, for example, contribute to waste generated due to their use for a short time.

To have a more sustainable production of shoes, plantar pressure maps were used together with topological optimization schemes to develop soles with the best material design according to the feet's pressure, while maintaining biomechanical performance and aesthetics. The design process considered the feet's plantar pressure maps (Figure 8a) which were then converted into numerical simulation data for the topological optimization scheme (Figure 8b). After undergoing the topological optimizations, an optimized biomechanical sole was obtained with the best material design (Figures 8c and 8d) (Spahiu et al., 2021).

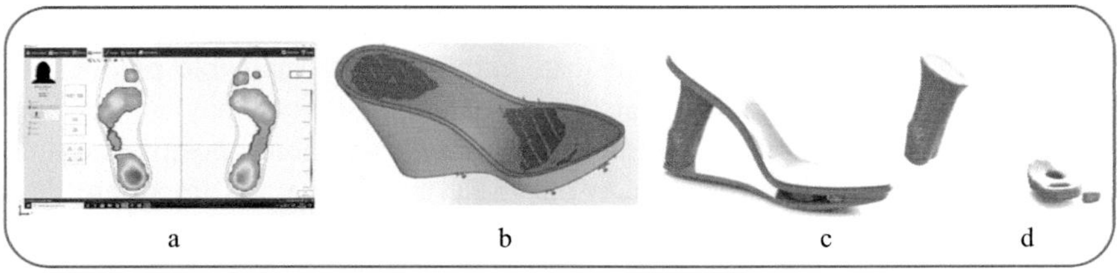

Fig. 8 (a) Plantar pressure maps; (b) boundary conditions for the topological optimization and illustration of the optimized designs; namely, (c) the design region and (d) the optimized sole (Spahiu et al., 2021)

Conclusions

Designing process is a crucial step in product development by including new ideas or improving existing ones. Creating products that are functional or fulfill consumer requirements is a process that includes various steps followed by 2D sketches, 3D CAD modeling, simulation, etc. These are continuously changing and linked with various computer-based applications. Consumers requirements have evolved over time including the aesthetic part too, which includes consumers in the process of product designing. Highly important applications are addressed for specific users.

The role of digital tools is indispensable in product development due to their numerous advantages. The manufacturing part through additive technologies as 3D printing present an efficient technology to transform 3D models into a physical product by reducing time, costs and realizing 3D structures that can not be produced using traditional technologies.

References

Almeida, H.A., Costa, A.F., Ramos, C., Torres, C., Minondo, M., Bártolo, P.J., Nunes, A., Kemmoku, D. and Lopes da Silva, J.V. 2019. Additive Manufacturing Systems for Medical Applications: Case Studies. In E. Pei, M. Monzón and A. Bernard (eds), *Additive Manufacturing—Developments in Training and Education*. Springer. 187–209.

Bernardes, J.P., Ferreira, F., Marques, A.D. and Nogueira, M. 2018. Generation Y's Sustainability Attitude-Behaviour Gap. In A.C. Broega, J. Cunha, H. Carvalho, M. Blanco, G. García-Badell and D. Goméz-Chacón (eds), *Reverse Design*, 623-628. London: Taylor & Francis Group.

Deselnicu, V., Crudu, M., Zăinescu, G., Albu, M.G., Deselnicu, D.C., Guþă, S.A., Ioannidis, I., Gurău, D., Alexandrescu, L., Constantinescu, R.R., Chirilă, C., Macovescu, G. and Bostaca, G. 2014. Innovative Materials and Technologies for Sustainable Production in Leather and Footwear Sector. *Leather and Footwear Journal* 14: 147–158.

Ellen MacArthur Foundation. 2022. An Introduction to Circular Design. Available at https://www.ellenmacarthurfoundation.org/news/an-introduction-to-circular-design (accessed on 10 October 2024).

European Parliament. 2020. The Impact of Textile Production and Waste on the Environment (infographics) https://www.eumonitor.eu/9353000/1/j9vvik7m1c3gyxp/vlf0bgvqy6vx?ctx=vhsjgh0wpcp9 (Accessed on 2025).

Geneva Environment Network. 2024. *Environmental Sustainability in the Fashion Industry*. Available at https://www.genevaenvironmentnetwork.org/resources/updates/sustainable-fashion/ (Accessed on 10 October 2024).

Goonetilleke, R.S. and Karwowski W. 2016. *Advances in Intelligent Systems and Computing*. Florida, USA: Springer.

Han, J., Woodside, A.G. and Ko, E. 2024. Does Consumer Knowledge about Sustainable-Fashion Impact Intention-to-Buy? *Asia Pacific Journal of Marketing and Logistics* 36(10): 2390–2410.

Lopes, J.M., Gomes, S. and Trancoso, T. 2024. Navigating the Green Maze: Insights for Businesses on Consumer Decision-making and the Mediating Role of Their Environmental Concerns. *Sustainability Accounting, Management and Policy Journal* 15(4): 861–883.

Mandarić, D., Hunjet, A. and Vuković, D. 2022. The Impact of Fashion Brand Sustainability on Consumer Purchasing Decisions. *Journal of Risk and Financial Management* 15(4): 176.

Marques, A., Guedes, G. and Ferreira, F. 2017. Leather Wastes in the Portuguese Footwear Industry: New Framework According Design Principles and Circular Economy. *Procedia Engineering* 200: 303–308.

Munny, A.A., Ali, S.M., Kabir, G., Moktadir, M.A., Rahman, T. and Mahtab, Z. 2019. Enablers of Social Sustainability in the Supply Chain: An Example of Footwear Industry from an Emerging Economy. *Sustainable Production and Consumption* 20: 230–242.

Nurse, M.A., Hulliger, M., Wakeling, J.W., Nigg, B.N. and Stefanyshyn, D.J. 2005. Changing the Texture of Footwear Can Alter Gait Patterns. *Journal of Electromyography and Kinesiology* 15(5): 496–506.

Papamichael, I., Voukkali, I., Economou, F., Ciro Liscio, M., Sospiro, P., Naddeo, V. and Zorpas, A.A. 2024. Investigation of Customer Behavior Regarding Circular Fashion. *Sustainable Chemistry and Pharmacy* 41: 101675.

Pereira, L., Carvalho, R., Dias, A., Costa, R. and António, N. 2021. How Does Sustainability Affect Consumer Choices in the Fashion Industry? *Resources* 10(4): 38.

Polese, F., Ciasullo, M.V., Troisi, O. and Maione, G. 2019. Sustainability in Footwear Industry: A Big Data Analysis. *Sinergie: Italian Journal of Management* 37(1): 149–170.

Ray, S. and Nayak, L. 2023. Marketing Sustainable Fashion: Trends and Future Directions. *Sustainability* 15(7): 6202.

Spahiu, T., Almeida, H., Ascenco, R.M.T., Vitorino, L. and Marto, A. 2021. Optimization of Shoe Sole Design According to Individual Feet Pressure Maps. *Computers in Industry* 125: 103375.

United Nations. 2021. *The Sustainable Development Goals Report 2021*. Available at https://unstats.un.org/sdgs/report/2021/The-Sustainable-Development-Goals-Report-2021.pdf (Accessed on 10 October 2024).

West, A.M., Schönfisch, D., Picard, A., Tarrier, J., Hodder, S. and Havenith, G. 2019. Shoe Microclimate: An Objective Characterisation and Subjective Evaluation. *Applied Ergonomics* 78: 1-12.

Yim Lee Au EY. and Goonetilleke, R.S. 2007. A Qualitative Study on the Comfort and Fit of Ladies' Dress Shoes. *Applied Ergonomics* 38(6): 687-696.

13 Use of Computational Design Technology to Automate Sew Pattern Design for Dress Garment

Prodromos Minaoglou[1], Gheorghe Oancea[2], Manoj Gupta[3] and Panagiotis Kyratsis[4*]

1. Introduction

Apparel design procedure is a method for designing and fabricating clothes that has been used for many years. The design, aesthetics, materials, and manufacturing techniques have been changed and replaced many times over the years. New trends appear and clothing is characterized by novelty and industrialization. In recent years, new technologies have been created with an aim to improve the design and fabrication, as well as the user experience. Tools such as virtual design and computer simulation of garments come to assist in the design process and as a result improve the final garment (Volino et al., 2005). Virtual reality is the continuation of technological developments with which the approach to design can be significantly improved. All this comes in conjunction with the digital 3D computer-aided design (CAD)–based design. Some of the features provided by a CAD system include 2D/3D digital modeling, 3D garments design and modification, 2D pattern creation, etc. (Liu et al., 2010).

Computational design is an integration of programming in the field of 3D CAD-based design. The designer, instead of using traditional design tools, uses commands of a programming language, which is adapted to the standards of design pieces of software using libraries. In most cases, a visual programming language (VPL) with nodes is used. In the VPL, several connections are made between the commands/nodes in order to automate the design process and offer additional functionality with respect to downstream applications, i.e. drawings, bill-of-materials (BOM), and render facilities (Manavis et al., 2023; Minaoglou et al., 2024a). One of the main advantages of computational design is the parametric nature of the result. The algorithm created is based on some initial parameters; if

[1] *Product Design Engineer*, Department of Product and Systems Design Engineering, University of Western Macedonia.

[2] *Professor*, Department of Manufacturing Engineering, Transilvania University of Brasov.

[3] *Associate Professor*, Department of Electrical Engineering, Guru Ghasidas Vishwavidyalaya.

[4] *Professor*, Department of Product and Systems Design Engineering, University of Western Macedonia.

Email: p.minaoglou@uowm.gr[1]; gh.oancea@unitbv.ro[2]; manojgupta35@yahoo.co.in[3]

* Corresponding author: pkyratsis@uowm.gr

these parameters are changed, the final result changes simultaneously. Examples of computational product design can be found in a variety of industries and especially in the clothing sector. The designer can automate the process of designing a garment from the initial body shape up to the calculation of the 2D patterns needed and their equivalent 3D cloth geometry (Minaoglou et al., 2024b). Automation based on specific design parameters is a feature that can significantly lower the cost and time to market, and at the same time improve the user experience. The end user selects the design parameters and creates the final product, thus feeling the satisfaction of contributing toward the design of the customized product received at the end (Manavis et al., 2024; Manavis et al., 2022). Finally, in recent years, computational design has been used in another popular fashion related industry, footwear design. A great deal of research development has been presented with an aim to handle the unusual geometries needed and their complexity (Firtikiadis et al., 2023; Minaoglou et al., 2023).

2. State of the Art

An application that takes sewing patterns and turns them into 3D clothing models was presented by Berthouzoz et al. The 2D geometries were automatically recognized together with all the appropriate fabrication features by using machine learning and integer programming. The user initially provided the sewing pattern in vector format. The app then realized how the 2D geometries fit together with an aim to create the complete garment. In the event that the app could not complete automatically the whole procedure, the user was given the option to manually make the appropriate selections and feed the correct pieces of information. The research results proved that 68% of the sewing patterns tested were correctly analyzed, while the rest needed some external assistance from the user (Berthouzoz et al., 2013).

When designing clothing patterns, different approaches can be found. Handmade/traditional design and CAD-based design were studied with an aim to report the advantages and the disadvantages of different types of templates created. More specifically, the design by hand, while offering very good quality, requires more time. On the other hand, using the CAD-based approach results in great flexibility and minimum amount of time, especially when changes are involved. At the same time, it requires expertise in the CAD system and purchasing the equipment used for fabrication. Another aspect studied was to investigate how the material selection was affecting the design and fabrication process (Jeong et al., 2015).

A digital clothing design process was proposed in order to finally design a 3D garment geometry, while starting with the 2D sketching of the garment. To start with, the user draws the basic elements such as contours and seam areas. With the use of an appropriate application the 2D outline of the sew pattern is converted into a 3D surface, i.e. a three-dimensional garment. Using the effect of gravity digitally, realistic folds and creases are calculated and applied to the 3D garment geometry.

Finally, the three-dimensional garment is separated again into sections in order to create the final sew pattern. With this method, a simple two-dimensional draft is transformed into a final and tested sew pattern. The main advantage of the proposed technology over the traditional design and testing is that by using this application the designer can design a garment with less technical difficulty, while the result is fully realistic (Decaudin et al., 2006).

In mass production cases, designers use traditional sew pattern designs, which are based on a "standard figure" body shape. As expected, in many cases the "standard figure" used cannot cover the needs of all customer body forms and dimensions. A study with an aim to improve the application of the clothes via the adjustment of the sew pattern was implemented. In this context, parabolic curves were developed using applied mathematics in the sew pattern design, in order to connect the armpit and the hand, an area where most problems of poor fit are found. Finally, the project concluded with the extremely important role of using CAD when improving the clothes' fitting for alternative body forms and dimensions. Also, reference is made to the need for personalized design. When applying new technologies, the fitting properties of the newly proposed clothes can be significantly improved (Fayoomy and Tahan, 2014). Programming industrial knitting machines is becoming more and more popular due to increased digitalization and reduction in cost. 3D knitted garments can be created based on 2D clothing designs. Special applications undertake to design the knitting itself, while at the same time, the control of the industrial machine is secured. The main requirement is the 2D design to incorporate the seam points and the pairs that should be joined. Based on the 2D design defined, the application automatically segments with simple areas so that they can be knitted at a later stage by the machine. It is important to mention that the designer can control some of the knitting options such as knitting direction. Finally, the modern needs for custom design can be incorporated via this application (Kaspar et al., 2021). With similar methodology another solution to determine and to thoroughly evaluate body dimensioning was presented. In this case, the garment achieves an increased fit on the body under study. The method used in this particular study was based on a triangular pattern development and origami techniques for the measurement and construction of the garment. With this method, sew patterns were designed with a more accurate fit to the measured body. The results, after a series of tests, showed that the fit of the newly designed clothes was equally accurate for all cases, i.e. men and women. Finally, the measurement process could be carried out with 3D scanning technologies instead of the traditional manual body measurements. In the case of scanning, the time was significantly reduced and the process was greatly improved (Scott and Muhammad Sayem, 2018).

Enes and Kipöz carried out a study on the minimization of cutting and sewing waste in the garment production line. Garment waste is the amount that is left over when sewing patterns are cut. Half the cost of a garment is made up of fabric, so reducing excess fabric is extremely important both ecologically and economically. In this study, different fabric cutting plans with asymmetric and symmetric design approaches were investigated. The results showed that the use of fabrics that do not require oriented cutting significantly reduce cutting waste (Enes and Kipöz, 2020). McQuillan examined various methods in order to reduce textile waste. While collaborating with two major fashion companies, experimental tests were carried out. From the results of the experimental tests, it appeared that reducing textile waste can be implemented with difficulty in a large size company. The study concluded that the reduction of textile waste can be achieved by increasing the collaboration of designers with all other professionals involved (McQuillan, 2019). Nursari and Djamal looked at how to reduce textile waste through the concept of Zero Waste Fashion Design (ZWFD). Environmental pollution due to clothing waste has been a major issue in recent years. The main conclusion of the research was that the designer has an important role both in the design of the garment and in the choice of materials; thus proper choice of fabrics and correct cutting techniques selected can significantly reduce the problem of textile waste (Nursari and Djamal, 2019). Ramkalaon and Sayem presented a model for integrating Zero Waste Pattern Cutting (ZWPC) into the mass production garment industry. The goal was to change the design of sew patterns so that fabric waste could be significantly reduced. The model was tested with an aim to apply ZWPC to two specific clothing styles that consisted of multiple sizes. The results showed that 98% of the fabric was used to create clothes. Traditional industrial technique rates were around 85% fabric usage. Therefore, the model of incorporating ZWPC into the mass production of garments was something that could lead to a significant reduction in textile waste (Ramkalaon and Sayem, 2021).

Apeagyei explains the importance of incorporating new digital technologies, such as 3D scanning, in the field of clothing design and anthropometry. 3D scanning is a technology that leads to digitization of real 3D morphology. This research addressed how this technology can be used and integrated into the apparel design process. It also commented on how the use of computational design could provide customized design for the manufacture of custom clothing. The 3D scanning is a technology that can break down the traditional one-size-fits-all design and shift the design trends toward customization (Apeagyei, 2010). Leong et al. developed a method for automating personalized body measurements for use in the apparel industry. The method was based on the use of a digital scanning piece of equipment. It is important to mention that the users were scanned in a standing position in order to measure their main geometrical characteristics. Creating an algorithm based on mathematical definitions resulted in automating the whole process. After testing the method, it emerged that this specific process could substitute most manual measurements within the fashion industry. The key advantages of this method over traditional manual measurement were increased accuracy and high efficiency (Leong et al., 2013). Yan and Kuzmichev designed an application which aims to improve the sew patterns intended for the creation of men's shirts, regarding the improved fit on the body. The first step of the study was the enrichment of the body measurements. Using 3D scanning technology, a database of 156 male bodies was created. Based on the exact dimensions from the database, the application designed a sew pattern for a shirt. Also, the application had the ability to change the style, depending on the aesthetics desired by the customer, i.e. body fit, slim fit, regular fit, comfort fit. At the same time, a 3D visualization was created, in which the customer could see digital details of the garment (Yan and Kuzmichev, 2020). Vitali and Rizzi developed another application that enabled the tailor to collect all the measurements needed from a human body. More specifically, the study used a set of open-source libraries, a 3D scanning device, a virtual reality device, and a hand movement tracking device. Through these tools, an application was created that digitized the human body. The main objective of the application was the representation of the workplace and the accurate measurement of body dimensions. Using virtual tape measurements, the dimensions were depicted. After testing the application, it appeared that the measurement results were very accurate, and the maximum error encountered was less than 5 mm. In future studies, various features like sensor gloves and improved digital databases were expected to be added (Vitali and Rizzi, 2017).

Spahiu et al. presented the creation of a garment made with the use of a 3D printer. 3D printing is a process of manufacturing geometries by depositing mainly plastic materials. The mechanical properties of the materials used vary depending on the designer selection, i.e. hard, rigid, flexible. The use of 3D printing in the clothing industry can greatly revolutionize the design and manufacture of clothing. In this study, in addition to manufacturing garments with a 3D printer, a survey was conducted on 100 people regarding the knowledge they had about 3D printing in the field of clothing. The results showed that those questioned had increased knowledge of 3D printing as a process and that they were willing to try a garment made with a 3D printer in the future (Spahiu et al., 2020). Valtas and Sun proposed a new idea for integrating parts, manufactured with 3D printing, on clothes. The aim of this specific research was to investigate the cost and time required for the use of printed parts on clothing. After creating prototypes of prints and integrating them into real clothes, the conclusion was that this particular technique could be applied to their production as it would offer multiple benefits (Valtas and Sun, 2016). Papachristou et al. used CO_2 laser cutting/engraving technology for cutting the sew pattern of a garment. Laser cutting has many advantages in the field of garment manufacturing. First of all, it is a technique that can cut the fabric with great precision in a very short time. Also, cutting complex shapes and fabrics with stretch can be completed easily. Finally, the modern design of garments based on CAD systems can facilitate automation that results in improved performance (Papachristou et al., 2023). Many researchers have published works on using laser technology to cut the corresponding sew patterns and provide additional advantages in both costing and time needed (Trejo et al., 2016; Stoilkova, 2021).

Nayak et al. researched the role and future challenges of mass customization in the apparel design and manufacturing industry. Mass customization enables users to have garments that perfectly fit their needs. New technologies such as 3D body scanning and design through computer-aided design and manufacturing (CAD/CAM) systems can alter the clothing design in the coming years. The accuracy and speed offered by the new technologies are their most important characteristics. The main conclusion of the study was that despite the difficulties due to complexity, the trend for personalized clothing will increase in the following years. One of the main reasons is the rapid development of technology. In fact, many large companies have already invested in mass customization (Nayak et al., 2015). Wolff et al. presented an interactive application that designed custom clothing based on customized 3D body scan data. The standard clothing sizes, because of the high rates of misfits, were the main problem that this app aimed to address. The key feature of the garment was the design of a 3D garment on the scanned digital body. The 3D scanned model was placed in various positions in order to account for all possible scenarios of the 3D garment not fitting properly. At all stages of the process, the designer could intervene and edit the garment. At the end of the process, the application exported the sew pattern of the garment. The application was evaluated with real life cases on various users. Users reported that the garments were very comfortable on their bodies, without affecting them in various postures (Wolff et al., 2023). Kakinuma et al. created a system in order to automate the design of clothing, aiming at specific elderly users. The system was named Customized Garment Design Supporting System (CGDSS). The prospective customer could order the custom garment by defining some of the basic dimensions of his body. Based on the body dimensions, the application selects the appropriate setup among eight body types. These body types were created based on the measurements taken from a sample of 52 elderly people. After selecting the body type, the digital body was deformed based on the user's dimensions. At the end of the process, the sew pattern of the custom garment was automatically created. After testing and evaluating the system, the results proved that the garments were successfully worn by the end users (Kakinuma et al., 2008). Wang et al. developed an application for designing clothes that adapted to the aesthetic needs of the end user. The basic steps in designing the garment were: designing the surfaces, designing the patterns/shapes/textures, and mapping the patterns/shapes/textures onto the garment surfaces. In this application, customers chose the patterns through an interactive system while, at the same time, they could draw on the garment. With this study, the application was presented with which customers could participate in one of the design steps and create their custom garment (Wang et al., 2011).

Kim and Kyu Park presented a process for creating flat clothing designs. The process was based on the development of an algorithm that automated the sewing pattern design of garments, while resulting in a variety of sizes and silhouettes. In this way, the time and cost of producing the garment were significantly reduced. Initially, clothing was categorized into two zones. The first zone was called the fit zone and was based on 3D scan data that was parameterized for the best fit. The second zone was called the fashion zone and was based on characteristics that affect the appearance and aesthetics of the garment. The result of this specific process was the design of the corresponding sew pattern. The shape of the sew pattern and the connection areas were automatically adjusted according to the physical properties of the fabric (Kim and Kyu Park, 2007). Datta and Seal designed an application which, based on a reference image of a garment, created a digital version of the garment as well as the sew pattern. The process consisted of four main steps. The first step was to import an image that contains a garment worn by a user. At this point, the sew pattern was automatically created and exported. Then the type of garment was recognized through a database, together with the primary body size. Finally, a 3D visualization facility was established with the texture of the garment. The results of the study showed a great deal of success and accuracy in the use of the application, mainly in simple types of clothing (Datta and Seal, 2018). Lim et al. presented a model that reconstructed 3D garments in digital format using an image of the garment. The model consisted of two main stages. The first stage involved the insertion of an image that presented the garment for reconstruction. The garment must be worn on the user, who should be standing in an upright position. In the second stage, the calculations of the sew patterns were carried out and then a three-dimensional simulation of the garment was presented. The model was trained based on different clothing categories. The results showed great reliability in the connection between garment and sew patterns. The only limitations that existed were in the ability to predict garments that were outside of the training data limits (Lim et al., 2024). Liu et al. developed an algorithm to automate the sewing patterns for jeans. The time-consuming process of traditional design was one of the reasons this study was established. The algorithm received the dimensions of the user's body and produced the final sew patterns automatically. The designer defined three basic input elements and three 3D parameters. In the case that the automatic result of the sew patterns was not satisfactory, the algorithm allowed the designer to manually edit the pattern produced and improve it. The results were tested successfully and the sew pattern was created automatically and in a short time (Liu et al., 2019). Pietroni et al. presented a method that could calculate the shape of the sew pattern through an existing 3D garment model. The process consisted of three stages. In the first stage, a surface was drawn on the 3D garment. In the second stage, the surface was cut into smaller pieces. In the last stage, the smaller pieces were placed on the plane and created a sew pattern. Data related to fabric properties and process constraints were calculated for initial surface slicing. For example, the algorithm calculated seam symmetry, fabric alignment, and strain gauge for fabric flattening. At the same time, it took into account the non-predetermined behavior of the fabric. The algorithm was tested, and the final sew pattern was cut and sewn with real fabric. The final garment was successfully used, and no significant problems were observed (Pietroni et al., 2022).

This chapter describes the development of an advanced algorithm that automates clothing design. First, specific dimensions of the user in the form of numbers are introduced in the algorithm. At the same time, it is possible to import a body model through a 3D scan facility with an aim to automate the measurements. Then the algorithm based on the dimensions received automates the design of the sew pattern. The appropriate code to guide a CO_2 laser cutting/engraving machine is produced, thus fully automating the complete process. Finally, a three-dimensional visualization of the garment is presented based on the designed sew pattern. At this point, the user can see the final result on a digital body and then decide about its final fabrication.

3. Materials and Methods

3.1 Workflow, Input, Output and Data Processing

The design and construction of a garment is the main objective of the proposed research study. It aims to model a custom garment design according to the dimensional characteristics of each user. The fabrication of the final garment can be completed using a CO_2 laser cutting/engraving machine and assembling the associated parts together. The design process consists of four main phases (Figure 1—process design development):

- The first phase is related to measuring the dimensions of the user's body. The measurement areas were the circumference of the bust, waist, pelvis, and various other body proportions.
- The second phase is aimed at developing an algorithm for designing the appropriate digital sew pattern. The algorithm was implemented using a VPL (Grasshopper^{TM}) and, more specifically, using nodes of code. The final step of the algorithm was the automatic representation of a 3D garment based on the 2D sew pattern.
- The third phase dealt with the automation of the algorithm. At this point, a union was made between the first two steps. In essence, the algorithm reads the user's dimensional data and adjusts the design of the garment based on them. Any change in the input data was affecting the design in real time, together with the shape of the sew pattern and the corresponding 3D projection of the garment. The goal was for the whole process to be completed automatically without any intervention of the designer. The third step was completed with the final shape of the sew pattern and its visualized 3D version.
- In the last phase, the algorithm was further developed with an aim to allow the automation of the garment manufacturing process. More specifically, the algorithm is based on the sew pattern created paths and output G-code to be sent to the CO_2 laser cutting/engraving machine. Various details were introduced into the design to make it easier to sew.

The idea behind the algorithm was divided into three basic stages: (a) data input, (b) data processing, and (c) data output. First, data was entered according to the category of the application. In the case study to be presented, the data entered was related to the dimensions of the user's body (numerical input). Second, based on the input provided, a set of 2D and 3D models were built together with additional features needed for the final product fabrication. Third, the 2D vector sew pattern design and the associated G-code were delivered. Figure 1 (definition and processing of data) depicts the three stages described.

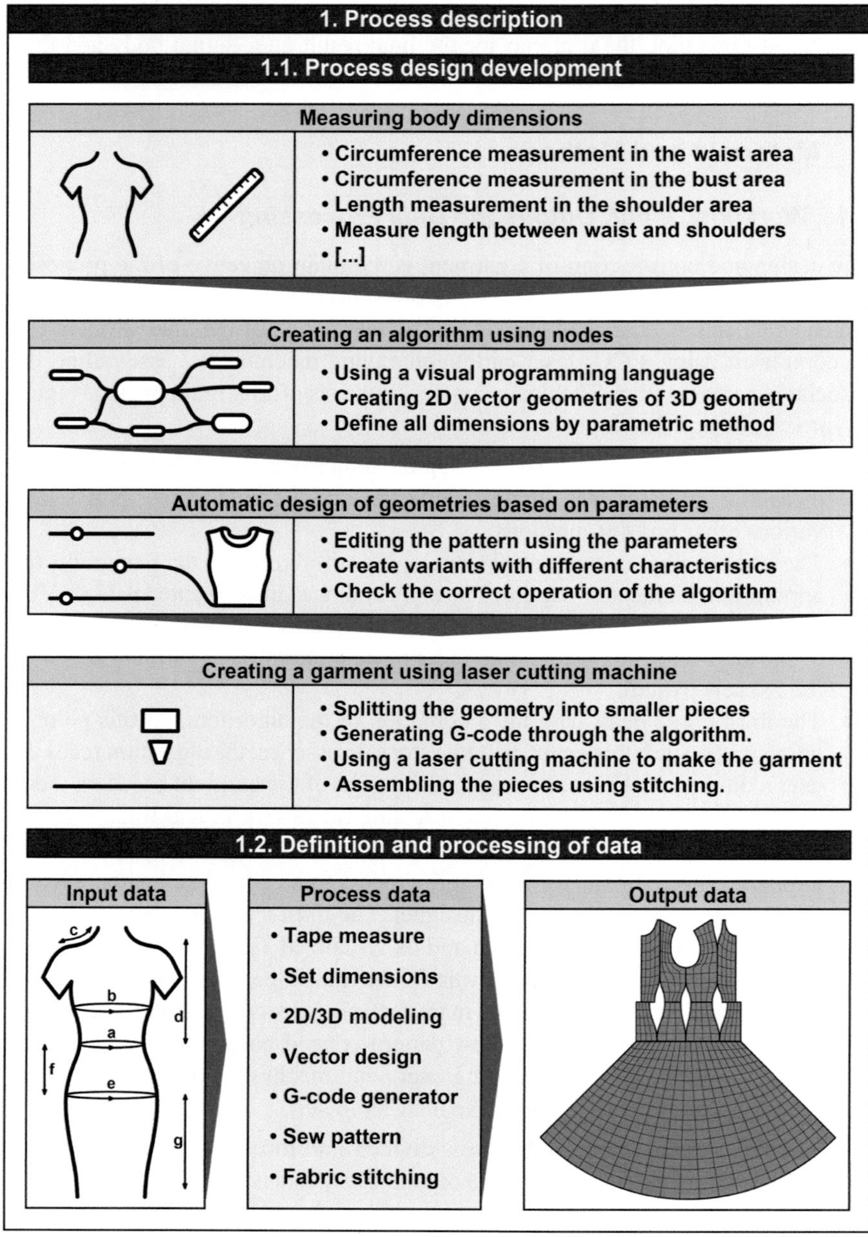

Fig. 1 An overview of the design and fabrication procedure followed

3.2 Algorithm Development and Parameterization

A VPL was used to create the application. More specifically, the development of the algorithm was implemented in Grasshopper3D^{TM}, which is an extension of the

Rhinoceros3D™, a general purposes CAD system. The main part of the algorithm was developed with VPL included, while several parts of it were developed using written code in Python™ textual programming language. The VPL was based on nodes, which were linked together. Each node had input, processing, and output of various data. The most common use of Grasshopper3D™ was to automate 3D and 2D CAD-based processes.

The algorithm developed together with its basic result is presented in Figure 2. When analyzing the result of the algorithm, the sew pattern designed was half of the complete needed for the fabrication of the complete product. This was true because of the mirroring feature required. In the result image, a number of dimensional variations were depicted, thus proving the versatility of the application created and the variety of alternatives produced based on the input parameters.

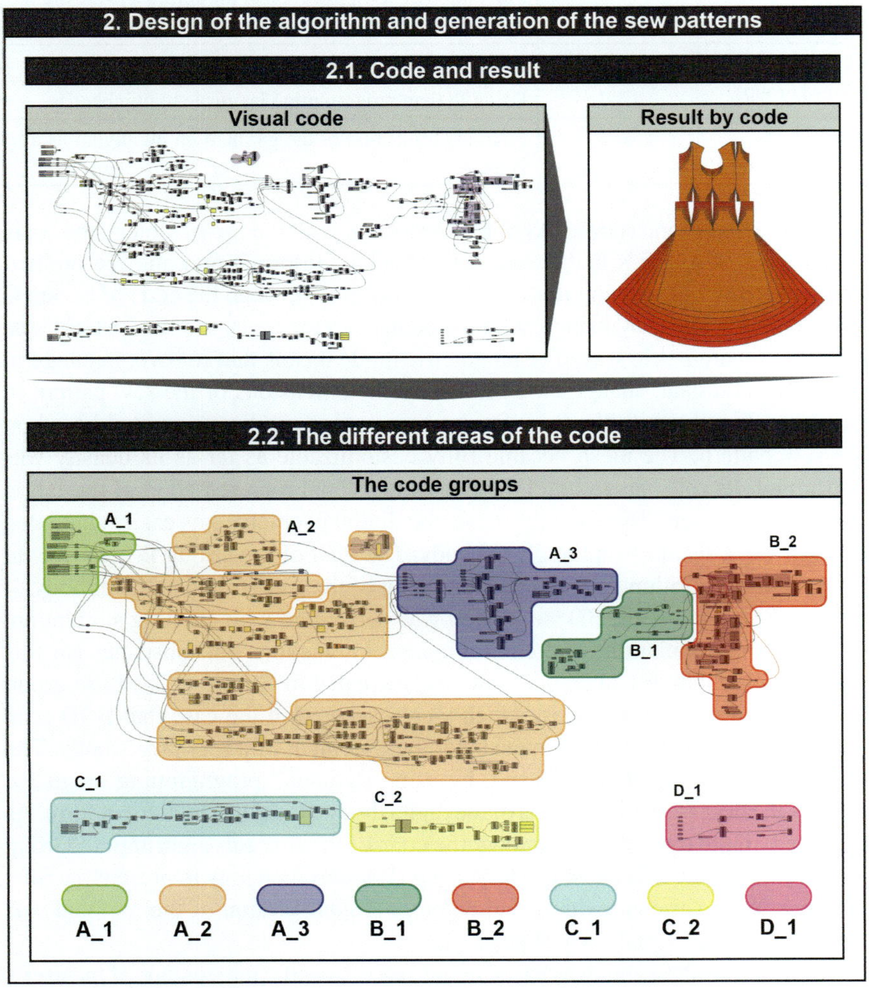

Fig. 2 The algorithm in the Grasshopper3D™ environment

To better understand its structure, the algorithm was divided into four groups and eight sub-algorithms. The groups and sub-algorithms are presented in detail in Table 1.

Table 1 Description of the steps of the algorithm

Group	*Sub-algorithm*	*Description*
Group A (first group)	A_1	Imported anthropometric data
	A_2	Design of the sew pattern according to the entered data
	A_3	Mesh generation with parametric density settings
Group_B (second group)	B_1	Definition of the seam areas and orientation of the sew pattern
	B_2	Simulation of 3D application of the garment to the digital user's body
Group_C (third group)	C_1	Collection of dimensional data from a 3D body scan
	C_2	Using Python programming language to number data
Group_D (fourth group)	D_1	Final 3D projection of the garment on the digital user's body

In the first group (Group A) of the algorithm, with the use of sliders, the basic dimensions of a user's body were defined. (This information could be omitted and replaced by processing a scanned 3D body model, when the code of Group C was applied.) Then, by using two-dimensional vector tools, the sew pattern was drawn in a parametric way. The parametric mode means that for every change in the dimension data, changes were made to the dimensions of the sew pattern in real time. Finally, with the use of a meshing tool, a mesh was created based on the sew pattern. The mesh became fully customizable as far as its density was concerned. Its role was to build the two-dimensional representation of the fabric used in the garment. By using the mesh, the digital geometry was easily simulated with physical features on the digital body. The goal of the Group B code was to evaluate the algorithm through the simulation of the garment on a digital body of similar dimensions. The 3D simulation was performed with clothing simulation. The clothing simulation used environmental interactions and obstacles (in this case, the digital body), and curved the mesh created in Group A. The third group (Group C) was a set of nodes, which were executed in the case that a 3D scan was performed instead of a dimensional measurement of a real user's body. The algorithm reads the 3D model and, by using Python™ programming language, outputs the necessary dimensions of the digital body model in a numerical form. As expected, in the case of 3D scanning, the code of the first sub-algorithm in Group A was not executed. The last group (Group D) focused on how to present the final garment. With various textures and colors, a digital version of the garment was presented on the digital body (Figure 2).

The whole algorithm was based on the use of specific dimensions of the user's body to create the sew pattern. The selection of these dimensions was done in such a way that the algorithm received all the necessary pieces of information with an aim to function properly. The dimensions chosen to be used are presented in Table 2. There were five measurements related to practical reasons, while the other five measurements were based on aesthetics. The practical dimensions were: 1. bust size, 2. waist size, 3. hip size, 7. cervical to waist, and 8. waist to hips; the aesthetic ones were: 4. front neck size, 5. back neck size, 6. shoulder length, 9. clothing height, and 10. skirt size. Practical dimensions defined the sizes in the critical areas of the garment. These dimensions should closely follow the real ones. Instead, the aesthetic dimensions were defined according to the user's preferences, i.e. the user can increase or decrease the overall height of the garment.

Table 2 Description of measurement areas

Number	*Measurement*	*Reason*	*Description*
1.	Bust Size	Practical	Half the circumference of the bust
2.	Waist Size	Practical	Half the circumference of the waist
3.	Hip Size	Practical	Half the circumference of the hip
4.	Front Neck Size	Aesthetic	The height of the front neck through an angle
5.	Back Neck Size	Aesthetic	The height of the back neck
6.	Shoulder Length	Aesthetic	Shoulder length
7.	Cervical to Waist	Practical	The distance between cervical and waist
8.	Waist to Hips	Practical	The distance between waist and hips
9.	Clothing Height	Aesthetic	The overall height of the garment
10.	Skirt Size	Aesthetic	The circumference of the skirt

Using a series of sliders tools, the 10 values of the dimensions were defined. Based on these values, the sew pattern shown in Figure 3 was automatically designed. The sew pattern consists of nine different sections for each side of the garment. The final garment incorporated the assembly of 18 separate pieces of textile. Figure 3 depicts the relation between the measurements and the sew pattern designed. Based on this figure, it was easily understood how the designed features depend on each other.

3.3 Data Collection

As mentioned in the introduction, the algorithm of the specific application was executed with two different ways of collecting data. The first way was the most traditional and was carried out with a tape-measure and the body of the user. With the tape-measure, five practical dimensions were measured and, at the end, the customer chose the features preferred due to increased aesthetics. At this point, the dimension values were manually introduced in the application. The continuation was fully automated and through the algorithm, the final sew pattern was produced. The second way used input received via 3D scanning technology. Such a tool can

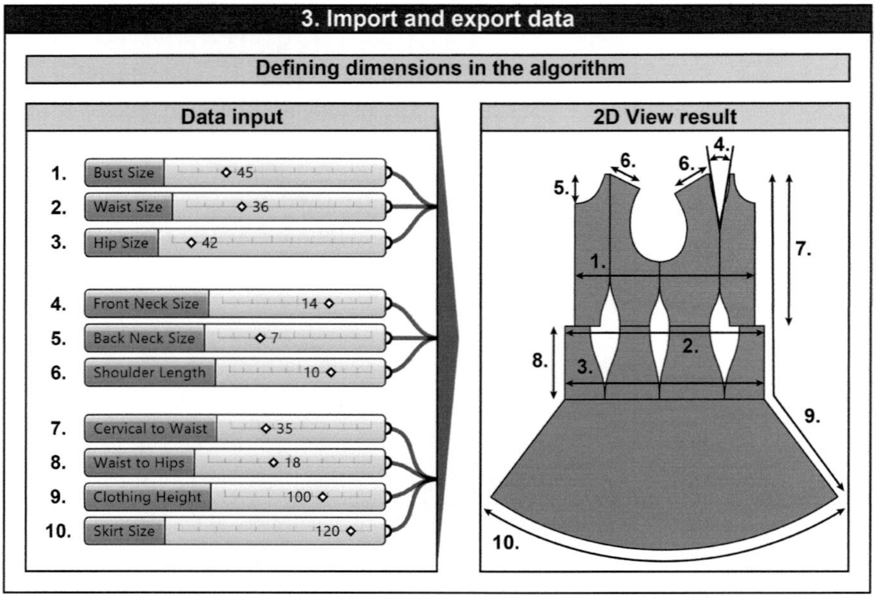

Fig. 3 The settings that designs the pattern

digitize all the morphological characteristics of a user body and create the user's digital 3D model. In this case, the algorithm undertakes a series of procedures to automatically collect the dimensions needed (Figure 4—the two methods for data collection).

Visual programming was not only limited to the use of nodes but also allowed the integration of textual code within the Grasshopper3D™ node. Each node was divided into three areas: (a) the data to be used in the written code, (b) the appropriate code was written in textual format, and (c) the results from running the code were ready to be presented as an output set. The whole process was like creating and using a function in a textual programming environment. The Python™ programming language was one of the languages supported in Grasshopper3D™ and it was the most popular in this case.

One difficulty when using the VPL within Grasshopper3D™ was creating a conditional iteration structure (i.e. re-execute a sequence of nodes for a certain number of times). The opposite was the case when textual programming was used. With an aim to measure the circumference in different parts of the digital body, it was necessary to make a flat cut in the corresponding parts. The result of the intersection created a closed polyline that had recesses. The traditional girth measurement was performed with the measuring tape, which enclosed the area following only the protrusions of the body and not the recesses. The goal of the textual code was to eliminate the indentations and create a new polyline, which would enclose each area without indentations. The Python™ code was based on segmenting the polyline by two contiguous lines. In each check, the angle between the two lines was calculated. If the angle was greater than 180°, then the corner point was deleted. On the other hand, if the angle was less than 180°, then the next point remained, and the check continued for all the subsequent lines. The whole procedure was performed repeatedly until there were no more angles greater than 180°. In Figure 4 (data collection process from 3D scanning) depicts a graphical representation of the process and the final result.

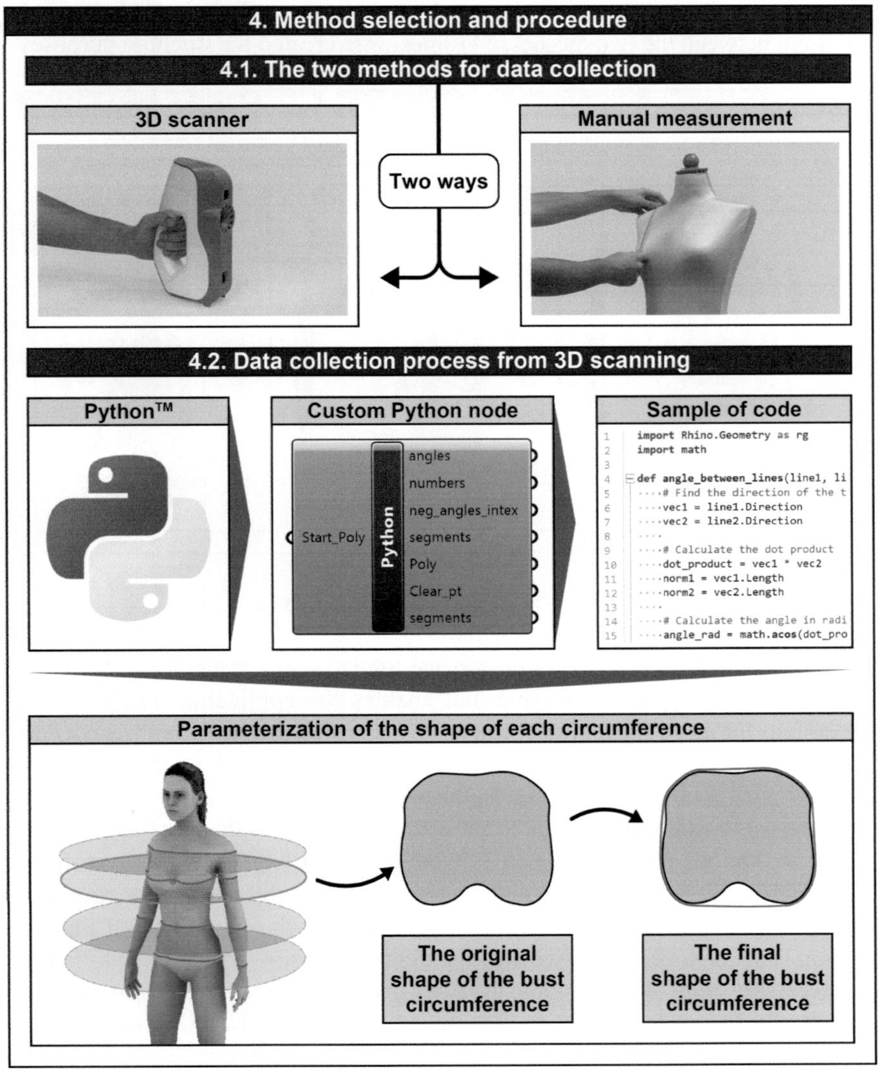

Fig. 4 The complete process of collecting measurements through 3D scanning

The final sew pattern was converted into a meshed network in order to be simulated on the digital body. The conversion of the closed curves into a mesh was carried out. The conversion process was fully parametric, so the number of surfaces were defined, i.e. the density of the mesh. The surface density used consisted of 33,120 vertices and 31,592 faces. The generated mesh was based only on quad faces instead of triangular and polygonal ones. Quad faces had many advantages over the rest. Some of the advantages were their subdivision surface capability, ease of UV wrapping and unwrapping, less polygonal errors, and smoother surface and shading. Figure 5 depicts the mesh placed on the digital user's body. In addition, between the two meshes, 347 lines were created for stitching purposes.

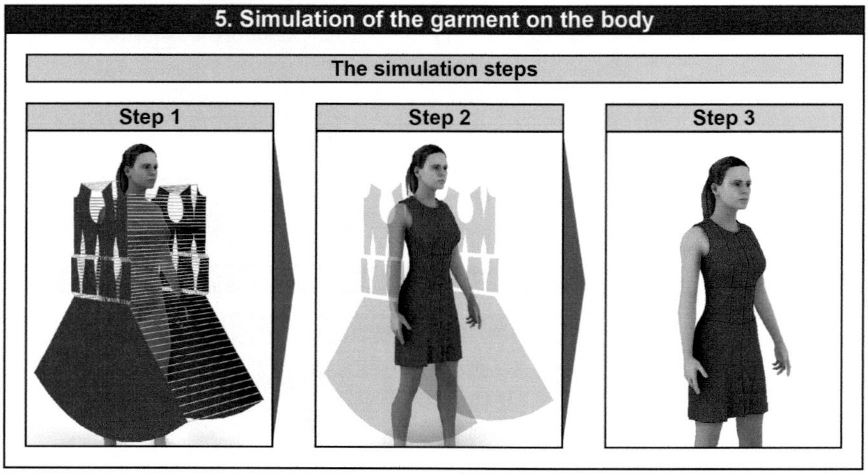

Fig. 5 The steps of the fabric simulation

The Kangaroo Physics™ extension was used to simulate the fabric. Kangaroo Physics™ is a simulation engine that enables the application of forces and constraints formulas to geometric structures. In essence, physical laws were defined and applied to the digital components. It was a useful tool for experimentation and anticipating possible future issues. In this case, when using Kangaroo Physics™, the code was written with an aim to simulate the mesh as following the properties of fabric. Various physical properties were defined, i.e. gravity, fabric stiffness, fabric elasticity, seams/threads, and obstructions. The obstacles in the Kangaroo Physics™ simulation symbolize the limits at which the fabric should limit its motion. The main obstacle used was the digital 3D body. With an aim to reduce the necessary computational time, the body geometry density was reduced. More specifically, there was a reduction in the number of surfaces by 78.4% compared to the original. As a result, the duration of the simulation was significantly reduced. Figure 5 depicts the sequentially developed three stages of the simulation. Finally, the areas where inner stitching was needed appear in black.

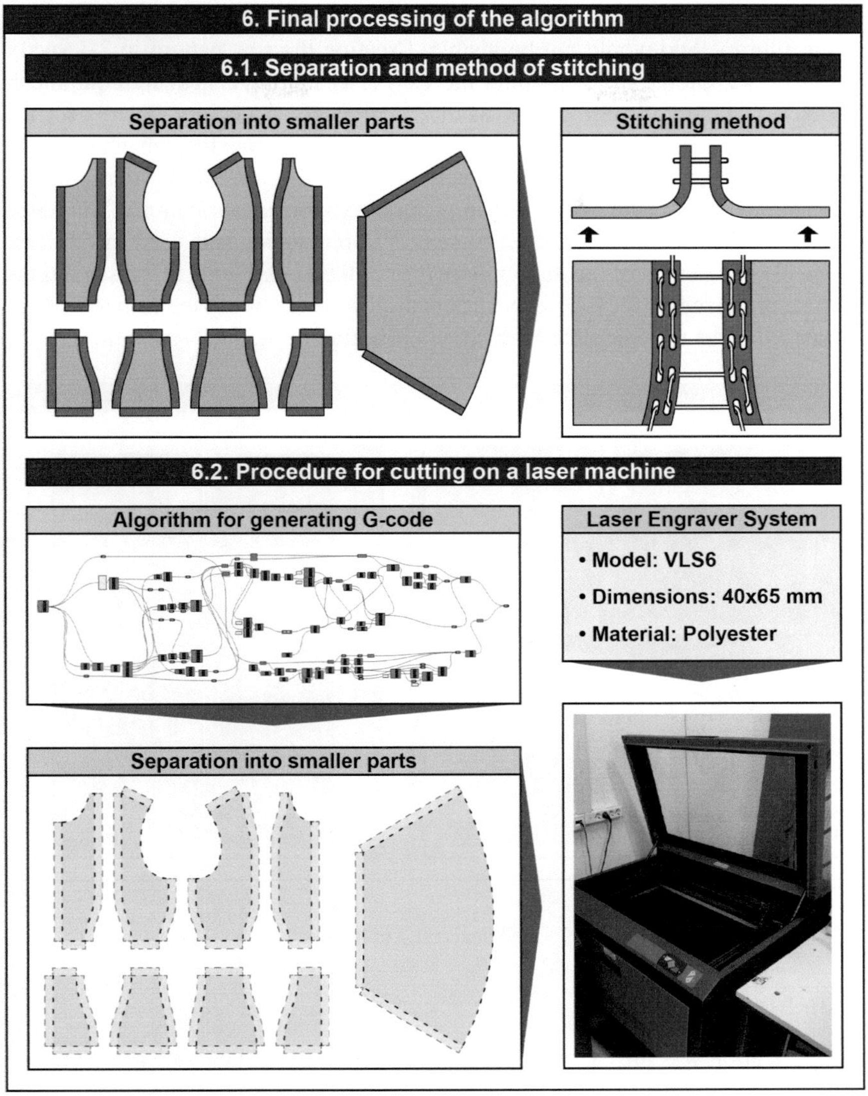

Fig. 6 The final stage of application and the garment construction

The final stage of the application was managing the final processing of the sew pattern. First, all the 2D geometries were separated, and the seam areas were defined. The edge of each piece that should be connected to another adjacent one was considered a seam and more fabric for stitching should become available. Figure 6 (separation and method of stitching) presents the 2D geometries and the stitching areas using different coloring. The management of the 2D geometries and the addition of the joint areas were performed automatically by the algorithm.

The goal was to connect all the 2D geometries with an internal seam so that the connection method would not be visible. Creating the sew pattern in 2D vector format was appropriate for guiding the CO_2 laser cutting/engraving equipment. Taking advantage of these pieces of information, the necessary G-code for the fabric cut was applied (Figure 6—procedure for cutting on a laser machine).

The project comes to an end with the actual garment fabrication. The completion of the specific study was implemented with the fabrication of the actual garment. The CO_2 Laser Engraver System VLS6 machine was used to fabricate all the 2D geometries of the burgundy polyester fabric used and, at the same time, the heat produced by the CO_2 laser system sealed the edges of the synthetic fabric. Figure 7 depicts the complete fabrication procedure of the final garment.

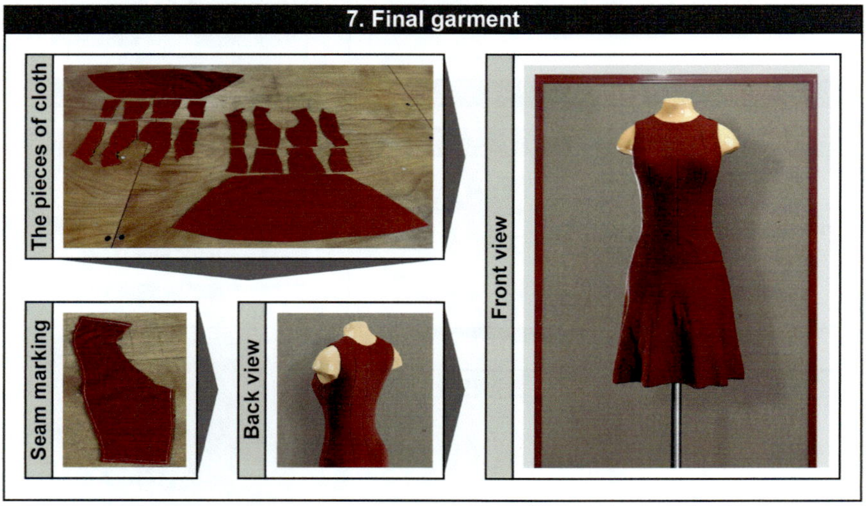

Fig. 7 The final result of the garment

Conclusions

An application was presented that was able to design and fabricate a customized garment. The garment was based on the geometrical characteristics of each user, thus offering automatic and customized design. The input could follow two approaches. One was based on manual measurements of specific dimensions related to the user's body and the second was using a scanned 3D digital model with automatic dimension recognition capability.

The application was built according to the computational product design principles, which means that not only was the product automatically designed and fabricated, but at the same time a series of design alternatives became available. In addition, the end user participated in the implementation process, which resulted in increasing satisfaction. At the end of the design process, the product was (a) completed using high-end visualization technologies; (b) built via CO_2 laser cutting/engraving facility; and (c) presented successfully as a final customized product.

Several additional features and extra functionality in the future are expected to increase the capacity of the proposed application for a variety of materials available, animation of a digital 3D body in a runway, and alternative clothing designs and sizes.

References

Apeagyei, P.R. 2010. Application of 3D Body Scanning Technology to Human Measurement for Clothing Fit. *International Journal of Digital Content Technology and Its Applications* 4(7): 58–68.

Berthouzoz, F., Garg, A., Kaufman, D.M., Grinspun, E. and Agrawala, M. 2013. Parsing Sewing Patterns into 3D Garments. *ACM Transactions on Graphics* 32(4): 1–12.

Datta, D.B. and Seal, P. 2018. Various Approaches in Pattern Making for Garment Sector. *Journal of Textile Engineering & Fashion Technology* 4(1): 29–34.

Decaudin, P., Julius, D., Wither, J., Boissieux, L., Sheffer, A. and Cani, M.P. 2006. Virtual Garments: A Fully Geometric Approach for Clothing Design. *Computer Graphics Forum* 25(3): 625–634.

Enes, E. and Kipöz, Ş. 2020. The Role of Fabric Usage for Minimization of Cut-and-sew Waste within the Apparel Production Line: Case of a Summer Dress. *Journal of Cleaner Production* 248: 119221.

Fayoomy, A. and Tahan, A. 2014. Basic Garment Pattern Design and the Development for the Standard Figure. *World Journal of Engineering* 11(2): 171–180.

Firtikiadis, L., Minaoglou, P., Efkolidis, N. and Kyratsis, P. 2023. Slipper-sole Computational Design: A Customer-based Approach. In Proceedings of VI International Scientific Conference in Contemporary Trends Innovations in the Textile Industry (CT&ITI 2023), Serbia.

Jeong, M.H., Han, D.H. and Ko, H.S. 2015. Garment Capture from a Photograph. *Computer Animation and Virtual Worlds* 26(3–4): 291–300.

Kakinuma, Y., Tsutsumi, E., Kondo, K., Mitani, J., Miyahara, S., Suzuki, H. and Iida, Y. 2008. Customized Garment Design Supporting System for Aged People Using Digital Dress Form Model. In Proceedings of the 13th International Conference on Geometry and Graphics, 1–9.

Kaspar, A., Wu, K., Luo, Y., Makatura, L. and Matusik, W. 2021. Knit Sketching: From Cut & Sew Patterns to Machine-knit Garments. *ACM Transactions on Graphics* 40(4): 1–15.

Kim, S. and Kyu Park, C. 2007. Basic Garment Pattern Generation Using Geometric Modeling Method. *International Journal of Clothing Science and Technology* 19(1): 7–17.

Leong, I.F., Fang, J.J., and Tsai, M.J. 2013. A Feature-based Anthropometry for Garment Industry. *International Journal of Clothing Science and Technology* 25(1): 6–23.

Lim, S., Kim, S. and Lee, S.H. 2024. SPnet: Estimating Garment Sewing Patterns from a Single Image of a Posed User. In Proceedings of Eurographics (Short Papers). The Eurographics Association.

Liu, K., Zhu, C., Tao, X., Bruniaux, P. and Zeng, X. 2019. Parametric Design of Garment Pattern Based on Body Dimensions. *International Journal of Industrial Ergonomics* 72: 212–221.

Liu, Y.J., Zhang, D.L. and Yuen, M.M.F. 2010. A Survey on CAD Methods in 3D Garment Design. *Computers in Industry* 61(6): 576–593.

Manavis, A., Efkolidis, N., Kakoulis, K. and Kyratsis, P. 2022. The "Do-It-Yourself (DIY)" Brand Design Strategy through Computational Design Tools. In International Symposium on Graphic Engineering, 687–696.

Manavis, A., Kakoulis, K., and Kyratsis, P. 2023. A Brief Review of Computational Product Design: A Brand Identity Approach. *Machines* 11(2): 232.

Manavis, A., Minaoglou, P., Efkolidis, N. and Kyratsis, P. 2024. Digital Customization for Product Design and Manufacturing: A Case Study within the Furniture Industry. *Electronics* 13(13): 2483.

McQuillan, H. 2019. Hybrid Zero Waste Design Practices. Zero Waste Pattern Cutting for Composite Garment Weaving and Its Implications. *The Design Journal* 22(sup1): 803–819.

Minaoglou, P., Efkolidis, N., Manavis, A. and Kyratsis, P. 2024a. A Review on Wearable Product Design and Applications. *Machines* 12(1): 62.

Minaoglou, P., Efkolidis, N., Manavis, A. and Kyratsis, P. 2024b. Customizing the Garment Design Using Computational Design and 3D Printing Technologies: A Case Study. In 7th International Scientific Conference Contemporary Trends and Innovations in Textile Industry—CT&ITI 2024 (No. Plenary Lectures, 24–31).

Minaoglou, P., Kakoulis, K., Manavis, A. and Kyratsis, P. 2023. Computational Wearables Design: Shoe Sole Modeling and Prototyping. *International Journal of Modern Manufacturing Technologies* 15(2): 143–151.

Nayak, R., Padhye, R., Wang, L., Chatterjee, K. and Gupta, S. 2015. The Role of Mass Customisation in the Apparel Industry. *International Journal of Fashion Design, Technology and Education* 8(2): 162–172.

Nursari, F. and Djamal, F.H. 2019. *Implementing Zero Waste Fashion in Apparel Design*. In 6th Bandung Creative Movement 2019, 98–104. Telkom University.

Papachristou, E., Kalaitzi, D. and Pissas, V. 2023. A Methodological Framework for the Integration of 3D Virtual Prototyping into the Design Development of Laser-cut Garments. *Journal of Engineered Fibers and Fabrics* 18: 15589250231194621.

Pietroni, N., Dumery, C., Falque, R., Liu, M., Vidal-Calleja, T.A. and Sorkine-Hornung, O. 2022. Computational Pattern Making from 3D Garment Models. *ACM Transactions on Graphics* 41(4): 1–14.

Ramkalaon, S. and Sayem, A.S.M. 2021. Zero-Waste Pattern Cutting (ZWPC) to Tackle over Sixty Billion Square Metres of Fabric Wastage During Mass Production of Apparel. *The Journal of the Textile Institute* 112(5): 809–819.

Scott, E. and Muhammad Sayem, A. 2018. Landmarking and Measuring for Critical Body Shape Analysis Targeting Garment Fit. In Proceedings of 3DBODY.TECH, 222–235.

Spahiu, T., Canaj, E. and Shehi, E. 2020. 3D Printing for Clothing Production. *Journal of Engineered Fibers and Fabrics* 15: 1558925020948216.

Stoilkova, D. 2021. Cutting as an Innovative Approach to Surface and Textile Design. In *IOP Conference Series: Materials Science and Engineering*, 1188(1), p. 012006. IOP Publishing.

Trejo, N.K., Reyes, C.G., Sanchez, V., Zhang, D. and Frey, M.W. 2016. Developing Composite Nanofibre Fabrics Using Electrospinning, Ultrasonic Sewing, and Laser Cutting Technologies. *International Journal of Fashion Design, Technology and Education* 9(3): 192–200.

Valtas, A. and Sun, D., 2016. 3D Printing for Garments Production: An Exploratory Study. *Journal of Fashion Technology and Textile Engineering* 4(3): 1000139.

Vitali, A. and Rizzi, C. 2017. A Virtual Environment to Emulate Tailor's Work. *Computer-Aided Design and Applications* 14(5): 671–679.

Volino, P., Cordier, F. and Magnenat-Thalmann, N. 2005. From Early Virtual Garment Simulation to Interactive Fashion Design. *Computer-aided Design* 37(6): 593–608.

Wang, J., Lu, G., Chen, L., Geng, Y., and Deng, W. 2011. Customer Participating 3D Garment Design for Mass Personalization. *Textile Research Journal* 81(2): 187–204.

Wolff, K., Herholz, P., Ziegler, V., Link, F., Brügel, N. and Sorkine-Hornung, O. 2023. Designing Personalized Garments with Body Movement. *Computer Graphics Forum* 42(1): 180–194.

Yan, J. and Kuzmichev, V.E. 2020. A Virtual E-bespoke Men's Shirt Based on New Body Measurements and Method of Pattern Drafting. *Textile Research Journal* 90(19–20): 2223–2244.

Index

For Product Safety Concerns and Information please contact our EU representative GPSR@taylorandfrancis.com Taylor & Francis Verlag GmbH, Kaufingerstraße 24, 80331 München, Germany